Tropical Forestry

Series Editor
Michael Köhl, Hamburg, Germany

For further volumes:
http://www.springer.com/series/5439

Jürgen Pretzsch • Dietrich Darr • Holm Uibrig •
Eckhard Auch
Editors

Forests and Rural Development

With 41 Figures and 19 Tables

Springer

Editors
Jürgen Pretzsch
Holm Uibrig
Eckhard Auch
TU Dresden
Institute of International
Forestry and Forest Products
Tharandt
Germany

Dietrich Darr
Faculty of Life Sciences
Hochschule Rhein-Waal
Kleve
Germany

Series Editor
Michael Köhl
Department of Wood Science
University of Hamburg
Hamburg, Germany

ISSN 1614-9785
ISBN 978-3-642-41403-9 ISBN 978-3-642-41404-6 (eBook)
DOI 10.1007/978-3-642-41404-6
Springer Heidelberg New York Dordrecht London

Library of Congress Control Number: 2013957698

Cover photo: ready2shoot/Fotolia.com

Copyedited by: David Butler Manning

Printed on acid-free paper

Springer is part of Springer Science+Business Media (www.springer.com)

Foreword

Two major approaches to forests were observed in the past. A utilitarian one, focused on the wood resources that could be obtained from the forests, and another one oriented to the environmental values related to the existence of forests. Both had their related constituencies and have shown limited capacity to learn from each other.

Recently, a growing interest is being observed toward a third approach based on the social and cultural dimension of forests. Many loose ends that could be hardly integrated and assimilated from the previous approaches might, from a social lens, be easier to be addressed. In the political debate, the final argument to preserve forests is much more social than environmental or economical. Political controversy is solved by social categories that are ranked by values.

The social and cultural approach to forests is a vast universe that includes issues like decent and health labor, cultural and spiritual values, traditional forest knowledge, anthropology, geographical history, indigenous people, community management of natural resources, and rural development. The growth of this approach is progressively incorporating new professions, sciences, and interested people to the forest dialogue and community, which enrich it while helping to overcome conflicts of the past. Overseeing the impressive coevolution of many ecosystems is a clear indication of the limitations of the previous approaches.

One of the most consolidated ones is rural development. In developing countries, it could be rephrased as rural poverty alleviation by forests. Rural development is related to the contribution of forests to the livelihoods of the forest-related rural communities.

Forests cover 31 % of the terrestrial part of the planet. But actually analyzing the global forest map, one can easily observe that what prevail globally are either poor forest areas or dense forest areas with some transitional mixed areas. If we analyze it in even more detail, forest-dominant areas are those disadvantaged by nature – except deserts – due to harsh climate, morphology (mountains, regularly flooded areas), or poor acidic soils. They are characterized by very low population densities. This makes those populations extraordinarily reliable on the forest resources, regardless of the degree of development of the country. The rural development perspective in those areas is by far much more crucial than the agricultural one prevailing normally in less disadvantaged and much better communicated areas accessible to public services.

In recent times, the need to move the focus from sectors (agriculture, forestry) to landscape approaches is gaining momentum. Watershed management, biodiversity preservation, enhancing landscape values, preventing disasters, or optimizing production in added value and employment terms can be done only from a landscape perspective. In that sense, a growing attention can be observed regarding land use planning, but from a different perspective than the previous technocratic approaches. If we want to answer the growing food, energy, and stable climate demands of an estimated nine billion human beings around 2050 and ensure decent living conditions for them, in no way can it be done by segregating functions or by a suboptimal use of land.

The obverse in social terms is rural development that focuses on and tries to optimize the use of existing endogenous resources in order to provide decent livelihoods to the rural population living in those areas. Rural development endeavors to ensure those populations similar living conditions as their urban peers, and this needs to be complemented with public policies related to infrastructure, education, and health.

Rural development in forest-dominant areas needs to focus strongly on small and medium enterprises (SMEs) and entrepreneurship. Wood, bioenergy, nonwood forest products, and tertiary services can only be fully valorized if a vital network of SMEs exists. Public policies need to overcome the main bottlenecks and to support their development. Firewood and nonwood forest products involve even in developed countries a considerable amount of informal labor that needs to be incorporated into the formal economy. SMEs are also crucial for the social architecture and democracy as, by increasing competition, they prevent corruption and ensure a broad middle class.

Major attention has been given to the vital environmental services which forests have been providing to the rest of the citizenry. Fragmentary answers have been given with limited results as shown by the deforestation figures. The downstream focus has failed since it forgot to keep the equity balance between what was requested to the forests including their dependent communities and the correspondent upstream flows that ensure the needed return in order to keep the process sustainable and equitable. Actually, the raising gender issue has not been much different, as far as the approach was directed to the services women provided to families and societies, so the issue was not solved. It was only when the equity issue was brought into the debate that progressively the balance was found. Payment for environmental services is, while being still in its infancy, a revolutionary tool that allows for the first time to address the distributional failures of previous approaches that, despite focusing on low-cost wood or environmental services supply, had this failed equity issue in common.

May I congratulate the excellent team of authors of this book for the quality of their respective chapters and the opportunity of this book and thanks particularly to its coordinator, Professor Jürgen Pretzsch from the University of Dresden.

This book provides very valuable information for students taking advanced courses and for practitioners in the area of forests and rural development. Surely, this book will cover an existing gap and contribute to bring the issue of forests and rural development forward.

Eduardo Rojas-Briales
Assistant Director-General
Rome
Forestry Department
Food and Agriculture Organization of the United Nations

Abbreviations

A/R	Afforestation/Reforestation
AFCC	Agriculture and Forestry Towards Food Security
AFDB	African Development Bank
AG soil	Agricultural Soil
AGB	Above Ground Biomass
AM	Adaptive Management
AMC	Adaptive Management Cycle
ASEAN	Association of Southeast Asian Nations
BC	Before Christ
bdt	Tonnes Dry Bulk
BMU	Federal Ministry for the Environment, Nature Conservation and Nuclear Safety
C&I	Criteria and Indicator
CAPRI	Collective Action and Property Rights
CBC	Caribbean Biological Corridor
CBFM	Community-Based Forest Management
CBNRM	Community-Based Natural Resource Management
CDM	Clean Development Mechanism
cf.	Confer or "Compare with"
CIFOR	Center for International Forestry Research
COMIFAC	Central African Forest Commission
CPF	Collaborative Partnership on Forests
CRTM	Consejo Regional Tsimane' Moseten
DAC	OECD Development Assistance Committee
DDI	Domestic Direct Investment
e.g.	Exempli Gratia or "For Example"
ECA	European Commission on Agriculture
ES	Environmental Services
EU	European Union
FAO	Food and Agriculture Organization of the United Nations
FDI	Foreign Direct Investment

FLEG	Forest Law Enforcement and Governance
FLEGT	Forest Law Enforcement, Governance and Trade
FSC	Forest Stewardship Council
FMU	Forest Management Unit
FUG	Forest User Group
GEF	Global Environmental Facility
GFEP	Global Forest Expert Panel
GIS	Geographic Information Systems
GIZ	Gesellschaft für Internationale Zusammenarbeit
GPS	Global Positioning Systems
ha	Hectare
HEM	Human Ecological Model
i.e.	Id est or "That Is"
IADB	Inter-American Development Bank
Ibid.	Ibidem or "In the Same Place"
IBRD	International Bank for Reconstruction and Development
ICRAF	World Agroforestry Centre
ICSID	International Centre for Settlement of Investment Disputes
IDA	International Development Association
IFAD	International Fund for Agricultural Development
IFC	International Finance Corporation
IFF	Intergovernmental Forum on Forests
IFM	Improved Forest Management
IKEA	Trade Mark
ILO	International Labour Organization
IPCC	Intergovernmental Panel on Climate Change
IPF	Intergovernmental Panel on Forests
ITTO	International Tropical Timber Organization
IUFRO	International Union of Forest Research Organizations
LIS	Land Information Systems
LUT	Land Utilization Type
M	Million
MAR	Monitoring, Assessment and Reporting
MDG	Millennium Development Goals
ME	Middle East
MIGA	Multilateral Investment Guarantee Agency
NFP	National Forest Programs
NGO	Non-government Organization
NPP	Net Primary Production
NR	Not Relevant
NS	Not Suitable
NTFP	Non-timber Forest Products
ODA	Official Development Assistance
OECD	Organisation for Economic Co-operation and Development

OTCA	Amazon Cooperation Treaty Organization
PES	Payments for Environmental Services
PGIES	Project de Gestion Intégré des Ecosystems Senegalaises
PROFOR	Program on Forests
PRSPs	Poverty Reduction Strategy Papers
PWS	Payments for Watershed Services
REDD	Reducing Emissions from Deforestation and Forest Degradation
REDD+	Reducing Emissions from Deforestation and Forest Degradation, Including the Role of Conservation, Sustainable Management of Forests and Enhancement of Forest Carbon Stocks
RIL	Reduced-Impact Logging
SAM	Social Accounting Matrix
SFM	Sustainable Forest Management
SLM	Sustainable Land Management
SME	Small and Medium Enterprises
SOM	Soil Organic Matter
SRC	Short Rotation Coppice, Short Rotation Crops
SSA	Sub-Saharan Africa
sub-VFM	Forest User Group
tCO_2e	Metric Tonne Carbon Dioxide Equivalent
TFAP	Tropical Forestry Action Plan
TIMO	Timberland Investment Management Organization
T-REITS	Timberland Real Estate Investment Trusts
UNCBD	United Nations Convention on Biological Diversity
UNCED	United Nations Conference on Environment and Development
UNCTAD	United Nations Conference on Trade and Development
UNEP	United Nations Environment Programme
UNFCCC	United Nations Framework Convention on Climate Change
UNFF	United Nations Forum on Forests
US	United States
USD	United States Dollar
USDA	United States Agriculture Department
VFM	Village Forest Management
WGI	Worldwide Governance Indicators
WRI	World Resources Institute

Contents

Contributors

Asmamaw Alemu Abtew Institute of International Forestry and Forest Products, Technische Universität Dresden, Tharandt, Germany

Klaus Ackermann Deutsche Gesellschaft für Internationale Zusammenarbeit (GIZ) GmbH, Bonn, Germany

Eckhard Auch Institute of International Forestry and Forest Products, Technische Universität Dresden, Tharandt, Germany

Benjamin Blom Job Creation and Enterprise Development Department, Alumnus of Yale University's School of Forestry and Environmental Studies, International Labour Organization, Genève, Switzerland

Patrick Bottazzi Centre for Development and Environment (CDE), University of Bern, Bern, Switzerland

Benjamin Cashore Environmental Governance and Political Science, Yale University's School of Forestry and Environmental Studies, New Haven, CT, USA

Dietrich Darr Hochschule Rhein-Waal, Kleve, Germany

Barbara Darr Institute of International Forestry and Forest Products, Technische Universität Dresden, Tharandt, Germany

Ingo Gentes European Union, Program for Public Administration and Regional Integration (PAAPIR), Honduras

Daniela Göhler Deutsche Gesellschaft für Internationale Zusammenarbeit (GIZ) GmbH, Eschborn, Germany

Simone Helmle Institute of Social Sciences in Agriculture, Universität Hohenheim, Stuttgart, Germany

Andre Hilbrich Leibniz Institute of Ecological and Regional Development, Dresden, Germany

Volker Hoffmann Institute of Social Sciences in Agriculture, Universität Hohenheim, Stuttgart, Germany

Gérard Hutter Leibniz Institute of Ecological and Regional Development, Dresden, Germany

François Jost Institute of International Forestry and Forest Products, Technische Universität Dresden, Tharandt, Germany

Gerald Kapp Climate Policy & Finance/World Bank Institute – Climate Change, Washington, DC, USA

André Lindner Institute of International Forestry and Forest Products, Technische Universität Dresden, Tharandt, Germany

David Butler Manning Institute of International Forestry and Forest Products, Technische Universität Dresden, Tharandt, Germany

Mauro Masiero Department of Land, Environment, Agriculture and Forestry, University of Padova, Legnaro (Padova), Italy

Davide Pettenella Department of Land, Environment, Agriculture and Forestry, University of Padova, Legnaro (Padova), Italy

Peter Poschen Job Creation and Enterprise Development Department, International Labour Organization, Genève, Switzerland

Jürgen Pretzsch Institute of International Forestry and Forest Products, Technische Universität Dresden, Tharandt, Germany

Stephan Rist Centre for Development and Environment (CDE), University of Bern, Bern, Switzerland

Merten Sievers Job Creation and Enterprise Development Department, International Labour Organization, Genève, Switzerland

Holm Uibrig Institute of International Forestry and Forest Products, Technische Universität Dresden, Tharandt, Germany

Marolyn Vidaurre Institute of International Forestry and Forest Products, Technische Universität Dresden, Tharandt, Germany

Chapter 1
Introduction

Jürgen Pretzsch, Dietrich Darr, André Lindner, Holm Uibrig, and Eckhard Auch

Keywords Rural development • Forestry • Co-evolution • Adaptive management

The conditions pertaining to both rural development and forestry have changed fundamentally over recent decades. The challenges associated with globalization offer up new options but also threats to rural development, especially in developing countries. Occupying about one third of the global land cover, forests are not only home to a significant population of indigenous people and forest farmers, but also provide important services and goods necessary for overall socio-economic development. Simultaneously, forestry depends much on the prevailing framework conditions and policies for rural development. The compilation of various perspectives on current challenges of and innovative solutions for rural development and forestry in one book is a worthwhile undertaking, because to understand the complexity of this relationship a holistic view is required, a view which may then allow us to learn from past experiences and derive scenarios for future development.

This book is embedded in a constructivist approach referring to various development strategies, their implementation and evaluation. It is assumed that the formulation of future development paths will need to be based upon several prerequisites, namely the availability of both traditional and modern technical knowledge, the qualified implementation of land use management processes as well as innovative institution building (Walker et al. 2002). The constructivist philosophy

J. Pretzsch (✉) • A. Lindner • H. Uibrig • E. Auch
Institute of International Forestry and Forest Products, Technische Universität Dresden, Postfach 1117, Tharandt 01735, Germany
e-mail: juergen.pretzsch@forst.tu-dresden.de; andre.lindner1@mailbox.tu-dresden.de; holm.uibrig@forst.tu-dresden.de; eckhard.auch@tu-dresden.de

D. Darr
Faculty of Life Sciences, Hochschule Rhein-Waal, Marie-Curie-Street 1, Kleve 47533, Germany
e-mail: dietrich.darr@hochschule-rhein-waal.de

J. Pretzsch et al. (eds.), *Forests and Rural Development*, Tropical Forestry 9,
DOI 10.1007/978-3-642-41404-6_1,

behind the book implies the rule that all steps in the factual implementation of these processes must be discussed with the relevant stakeholders. It is for this reason that technological prescriptions are deemed in part to be counterproductive. The argumentation adheres to the concept of adaptive management (Olsson et al. 2004; Williams 2011). In line with the rapidly altering global environmental and socio-economic situation, the contents of the book are a plea for more innovative model and theory building.

The socio-ecological co-evolution model is applied as a theoretical fundament and framework for understanding rural development and tropical forestry, and their links and relations to overall socio-economic development. This model has advanced considerably over the last two decades. It allows us to differentiate between the human and the ecological subsystems and at the same time to outline the dynamic interactions between these two systems (Norgaard 1994; Berkes et al. 1998). This co-evolution approach is of special relevance for tropical forestry. Conventional systems thinking favors discipline-oriented research, but this covers the reality only partially. The interpretation of social-natural relations is settled outside the traditionally anchored concepts of sustainable forest management with their static and linear character. In this way an improved management of the ever increasing complexity of forestry and rural development can be achieved. It is held that the socio-ecological systems are subject to ongoing change and that the continuous need for adjustment requires a flexible and adaptive management concept (Rammel et al. 2007). This concept must be rooted in institutional rules; rules that support the strategic development of targets, their implementation and the monitoring of ecological, social and economic standards. Priority is given to the resilience of the systems and their capacity to maintain flexibility with respect to future options.

When tackling the complex subject of forestry and rural development, an orientation towards model building is essential. Models facilitate the transfer of experience, even into other contexts. In recent decades, the number of scientifically investigated case studies of tropical forestry subjects has increased. However, the majority of the findings lack integration within overall general constructs; this even though the main objective of case study research is the generation of new models and theories.

The co-evolution model is used in a simplified way to structure this book (Fig. 1.1). The content follows four model components, which are tackled with different levels of intensity: the natural system, the social system, management as the interface of both, and the formal and informal regulations that provide the institutional framework. Ecosystem functioning and resilience are merely outlined briefly in an overview. Likewise, the social system with its social, cultural and economic characteristics is described only briefly. A particular emphasis has been placed on the relationship between both elements. The forest management systems are shaped by the prevailing physical geography, the objectives and capabilities of the owners of the land and other resources, societal frame conditions, historical land use patterns at the landscape level and current market opportunities for products and services. The framework of formal and informal regulations is an important driver of future forest-based rural development. Obvious mistakes of the past provide a

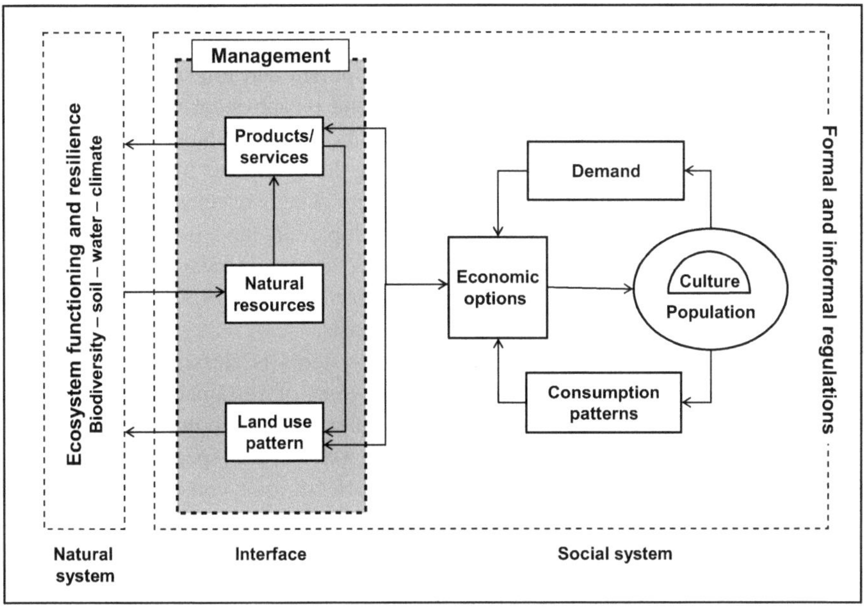

Fig. 1.1 The co-evolution model as a global explanatory/analytical model (Source: Derived from Hurni and Messerli (1981), Norgaard (1994), Berkes and Folke (1998))

reason to discuss the right institutional setting from the current perspective, one of the core aspects of this book. The increasing number of global challenges gives rise to a discussion of market, state or community driven development.

In this book 11 specific contributions deal with subjects pertaining to tropical forestry and rural development as mosaic stones, without claiming to cover the complexity of each respective field in its entirety. For example, topics like the demographic situation in rural areas, worldwide outward-migration, land grabbing, education and corruption have emerged as well-recognized concerns of forest-based rural development in the literature and are not covered in separate contributions in the current book. In a final chapter the conclusions are integrated into the co-evolution model.

The overview of paradigms regarding the situation of forests in rural development presented in Chap. 2 is rooted in a historical analysis of tropical forestry and development policies. The paradigms are underlined in the discussion of development models, theories and the review of historical facts. Beginning with traditional forest use systems, a six stage scheme is derived and the structural and functional parameters of the development paradigms and their change are outlined. The resultant genesis of tropical forestry development models facilitates a discussion of possible steps towards future development (Chap. 13).

Building upon a brief description of the major features of the environmental and social systems prevailing in rural tropical and subtropical regions, Chap. 3 highlights challenges in terms of their contribution to forest-based socio-economic development. With this the ground is prepared for the presentation of various innovative solutions in the subsequent chapters. The current state and general development of factors relevant for rural development and forest management on a global scale are documented. Starting with the bio-physical characteristics of tropical climate, soils, forest types and functions, their trends and evolution are summarized (Nagendra et al. 2004). Challenging developments include soil degradation, climate change and the loss of biodiversity. The overview of socio-cultural conditions focuses on human population development, land use change with a special emphasis on the implications for natural forests, infrastructure expansion and forest land tenure in the tropics and subtropics. Challenges with relevance for rural development are evaluated on a regional basis.

The manner of the utilization of forest ecosystems is classified according to different management systems with varying degrees of sustainability and equity. Agroforestry systems have a high potential to cater for both ecological and human system benefits. The combinations of forests or woody plant species with agricultural crops or pasture produce high synergy effects as a consequence of their complementary elements. Agroforestry offers improved and innovative options with respect to challenges arising from increased pressure on land and as a means to overcome constraints in rural development, especially in terms of the domestic supply of goods and services. Chapter 4 provides an overview of agroforestry practices and develops upon recent innovations.

Forest management systems are practiced by socio-economic units. Their management objectives, resources and capacity determine the management outcomes and impacts for both the social system and the ecosystem. In Chap. 5 the genesis of relevant organizational models is characterized, and the great variety of organizational models is structured according to production factors and legal concepts. Selected models with relevance for rural development are outlined, analyzed, discussed and complemented by case study examples.

Products and commodities produced in forest management systems are traded, processed and marketed in value chains from local up to international levels. The values created along the links in these chains represent significant proportions of rural incomes and livelihoods. Chapter 6 addresses the creation of rural jobs in forest-based value chains, as one component of sustainable forestry and of rural development strategies. Attention is given to the creation of green jobs and to the necessary market demand, which drives the creation of value chains. Innovations are focused upon market development as well as on technological, economic and social aspects as means to upgrade the chain.

The intimate association between culture and development is shown in Chap. 7, underlining the extraordinarily high relevance of cultural aspects in rural development. Culture is an important component of rural development and corresponding, innovative development tools are presented. Several concepts pertaining to access culture are introduced. Case studies demonstrate how forest management and rural development are shaped by different cultures.

Chapter 8 deals with innovative extension approaches. The paradigm change towards common technology innovation, testing and implementation is of particular relevance. New approaches are necessary to cope with the threats posed by climate change and the challenges associated with globalization. In addition to the traditional orientation on production technologies, a special emphasis is placed on social organization, including the creation of social capital.

The relevance for rural development of the widely promoted ‘payments for environmental services’ (PES) instrument is analyzed in Chap. 9. The concept is introduced, its evolution and intellectual roots briefly highlighted. Approaches to quantify the value of an environmental service and the effectiveness, efficiency and equity of PES schemes are discussed critically, including the trade-offs between these three objectives.

Major funding mechanisms for forest-based rural development projects are reviewed in Chap. 10. Long term capital fixation and relatively low rates of return served to limit investments in forest-related rural development activities in the past. In this chapter the authors assess recent trends and the current state of public and private forest sector investment, presenting examples of innovative funding schemes that take into account risks associated with forest investments and the non-marketable character of certain forest products and services.

The crucial role played by ‘land use planning’ in sustainable forestry and rural development is addressed in Chap. 11. Rarely are forests and forest land incorporated in the cross-sector concepts of integrated land use planning. With ongoing forest degradation and forest conversion to other land uses, the remaining forests represent important assets in rural land use planning. Planning approaches have shifted from ‘conventional technocratic’ to ‘participatory’. Today innovative land use planning integrates general development strategies and spatial planning.

Formal and informal institutions create the framework for the human and social system, and the interface to the ecosystem. In contrast to natural laws, which structure ecosystems, regulations for the human and social system are modifiable and of special relevance for rural development. The related field of forest governance and policy is reviewed in Chap. 12. Several conceptual pathways for the promotion of ‘good forest governance’ are identified.

In the final chapter, Chap. 13, the specific contributions are placed in the human-ecological co-evolution model presented at the outset. Findings from the individual chapters are synthesized in the form of vision-like, guiding principles. Two polarizing principles have dominated since the last century. Both the state based regulatory approach and ‘the market will fix it’ approach of the globalized liberal world have failed to achieve the dual objective of sustaining forests while simultaneously providing for sustainable (rural) development. A synthesis of the findings made during the quest to find possible solutions to the many challenging questions suggests a ‘third way’, a combination of market drive and societal steering. Some elements and aspects of this third way are identified and mapped over the course of this book.

References

Berkes F, Folke C (1998) Linking social and ecological systems for resilience and sustainability. In: Berkes F, Colding J, Folke C (eds) Linking social and ecological systems. Management practices and social mechanisms for building resilience. Cambridge University Press, Cambridge, pp 1–25

Berkes F, Colding J, Folke C (eds) (1998) Linking social and ecological systems. Management practices and social mechanisms for building resilience. Cambridge University Press, Cambridge

Hurni H, Messerli B (1981) Mountain research for conservation and development in Simen, Ethiopia. Mt Res Dev 1(1):49–54

Nagendra H, Munroe DK, Southworth J (2004) From pattern to process: landscape fragmentation and the analysis of land use/land cover change. Agric Ecosyst Environ 101(2–3):111–115. doi:10.1016/j.agee.2003.09.003

Norgaard RB (1994) Development betrayed: the end of progress and a coevolutionary revision of the future. Routledge, London/New York

Olsson P, Folke C, Berkes F (2004) Adaptive comanagement for building resilience in social-ecological systems. Environ Manage 34(1):75–90

Rammel C, Stagl S, Wilfing H (2007) Managing complex adaptive systems – a co-evolutionary perspective on natural resource management. Ecol Econ 63:9–21

Walker B, Carpenter S, Anderies J, Abel N, Cumming G, Janssen M, Lebel L, Norberg J, Peterson GD, Pritchard R (2002) Resilience management in social-ecological systems: a working hypothesis for a participatory approach. Conserv Ecol 6(1):14–30

Williams BK (2011) Adaptive management of natural ressources – framework and issues. J Environ Manage 92:1346–1353

Chapter 2
Paradigms of Tropical Forestry in Rural Development

Jürgen Pretzsch

Abstract The integration of tropical forestry in rural and general socio-economic development is reviewed from a historical perspective. Development is synthesized in six paradigms, whereby practice is integrated in theories and ongoing discourses. The paradigms are the basis for an outlook on future tropical forestry development models. The one-dimensional, static thinking prevalent in market-driven tropical forestry is complemented by more complex models; models in which the regulating and constituting functions of the state are revitalized and civil society and communities are involved in a balanced way. Future forestry development models have to be diversified, flexible and able to adapt to the increasing external influences exerted by the human ecological system.

Keywords Conceptions • Rural development • Paradigms • Historical stages • Development theories • History of tropical land use • Pre-colonial land use management • Colonial forestry • Capital formation • Growth orientation • Green economy • Knowledge systems • Globalization • Social capital

2.1 Introduction

Conceptions of the role of forestry in rural development have changed fundamentally over the last centuries. They oscillate around altering physical environments, increasing human impact, alternative models of political economy and 'development trends', which are often induced by short term politics. Although forestry is deemed to be the origin of the sustainability principle (Carlowitz 1713), there is a lack of dispute and conceptualization of the integration of forestry in socio-economic development, especially in tropical countries.

J. Pretzsch (✉)
Institute of International Forestry and Forest Products, Technische Universität Dresden, 1117, Tharandt 01735, Germany
e-mail: juergen.pretzsch@forst.tu-dresden.de

J. Pretzsch et al. (eds.), *Forests and Rural Development*, Tropical Forestry 9,
DOI 10.1007/978-3-642-41404-6_2, © Springer-Verlag Berlin Heidelberg 2014

In this chapter the changing conceptions of trees and forests and their interrelations between general socio-economic development models are outlined. The analysis focuses on tropical and subtropical countries. The historical and interpretative approach permits a classification in 'paradigms'. Six consecutive and partly overlapping paradigms are differentiated. They represent a genesis of models for tropical forestry, rooted in contemporary development theories. The categorization as paradigms results in a high level of generalization, which limits their explanatory value for specific cases. Physical conditions, forest management strategies, historical development, traditions and policies in countries of the tropics and subtropics are rather site-specific and thus may deviate from the six paradigms derived. Furthermore, the large spectrum of development theories available, partly overlapping in time, may lead to fuzziness. For this reason the contribution focuses primarily on a discursive review of conceptions, paradigms and theories on a meta-level. In spite of these methodological limitations, this synthesis is believed to be worthwhile and necessary because it facilitates new insights into the dynamic interaction between forestry, rural development and national development strategies. The paradigm approach serves as an interpretative framework for this book. Important lessons can be learnt about future forestry development strategies. In the current situation of dramatic and ongoing destructive change to rural landscapes, a lack of theories, models and instruments that successfully reach the rural population is evident (Pretzsch 2005; Douglas and Simula 2010; Escobar 2012). Rural development theories need to include the highly complex environment with a low predictability of events, like climate change-based weather fluctuations. A profound discussion of governance strategies is necessary, addressing state intervention, endogenous experiences with the self-organization of communities and market orientation, as these are the foundation for the further development of tropical forest management strategies (Uphoff 1993; Cooper and Packard 2005).

2.2 Definitions and Conceptual Framework

Trees and forests represent an important physical component in the global landscape. In the framework of the Global Forest Assessment Project of the FAO, forests are defined as land use with a tree crown cover of more than 10 %, a minimum height of 5 m and a minimum area of 0.5 ha (FAO 2001, p. 363). Figures for the current global forest cover differ. Recent FAO estimates come to about four billion hectares (FAO 2010, p. 228). In their study of frontier forests in 1997, the World Resources Institute estimated global forest cover to be roughly 3.2 billion hectares (Bryant et al. 1997). The forest area prior to human intervention was calculated by Bryant et al. (1997) to be about 6.2 billion hectares, a bit less than half of the land surface area. The estimated present forest cover accounts for roughly 30 % of the terrestrial surface area. Although in recent years the forest area has increased in countries such as China and in the European region, globally there has been a net decline of roughly five million hectares per year in the last decade (FAO 2010, p. 233). Due to eco-physical conditions, and the level of human intervention, the existing forests are unevenly distributed. In the tropical and subtropical regions the forest cover differs greatly, with 23 % in Africa, 19 % in Asia and 49 % in South America (FAO 2010, p. 225 et seqq.). The indicator 'forest cover' is not very precise, because the level of disturbance of the forests differs significantly and is poorly reflected in the available statistics. With the reduction of forest areas the existence of individual trees, groups of trees and agroforestry systems increases in importance for ecosystem services and for supporting livelihoods. In the transition theory the forest cover follows a Kuznets curve and the initial focus of forest use on market and survival-oriented product functions is, with increasing welfare in society, complemented by many social and ecological functions such as recreation and biodiversity conservation (Rudel et al. 2005). The conceptualization of the relationship society–forests has resulted in drawn out quarrels over definitions. Traditionally the term 'forest functions' was used to describe both direct and indirect values. The differentiation between eco-centric values related to the functioning of the ecosystems and anthropocentric, human demand-based values led to deviating concepts (Hanewinkel 2011). Viewed from an ecosystem perspective, De Groot (1992, p. 8) differentiated between regulation, carrier, production and information functions. The forest valuation, as viewed from the side of the demand by society, has led to the creation of the concept of ecosystem goods and services. The recent discussion of payments for environmental services (PES) has delivered a boom in publications, in which four types of environmental services are differentiated: provisioning, regulating, cultural and supporting services (Millenium Ecosystem Assessment 2005; Bizikova 2011). The definition provides a first insight into the importance of forests and tress as global land cover, for their ecosystem functions and as a means to satisfy human needs and their changes over time (Chap. 9).

Increasing global population and ecological disturbances have led to a high vulnerability of ecosystems and of social systems. Especially in tropical countries, forests and trees contribute to reducing this vulnerability and support coping strategies. Examples from an ecological point of view are the stabilization of the fragile tropical soils, support for the climatic cycles under changing land use, mitigation of climate change and carbon storage, increased water retention through reforestation programs and biodiversity conservation. Coping strategies include the contribution of forests to poverty alleviation and to livelihood development (Wunder 2001). People in and around forests use forest products directly as food, firewood and medicine, they transform them on a micro-level to sawn wood for use in construction, furniture, tools for agriculture or have local kilns for charcoal production (FAO 1998).

Differing claims to forest products and services lead to controversial discussions and a certain complexity with respect to integration in a general land use pattern. The parallel, and competing, development of agricultural and forestry institutions has hindered the formulation of a comprehensive landscape and livelihood-oriented rural development strategy. This is due in part to the specific characteristics of forestry, namely long planning cycles with a resultant inflexibility in short term decision making, the dual characteristics of forests as product and site of production and the bulky nature of timber, which makes it difficult and expensive to transport. In the spectrum of forest uses, timber for long had the highest priority, because it was always the main cash provider. Furthermore, western forestry has always followed a specific ethos; traditionally it has been mainly state and administration-oriented, whereas agriculture was preferably linked to private farms. The integrative view of landscapes is challenging but necessary to tackle the future demands of rural areas and especially in terms of developing climate change adaptation and mitigation strategies. It broadens the conventional view of forestry. New challenges have given rise to agroforestry systems, short rotation energy plantations (Chap. 4) and the maintenance of individual trees in agricultural fields. From the point of view of the methods employed, policy and practical management, this integration is as yet not very advanced and further steps are necessary.

The specific role of forests and trees depends greatly on the eco-zones in general, and on the particular land use patterns, which have brought about an alteration of the original vegetation. There are marked differences between humid tropical lowland zones, savanna forest zones and dry land. In each of the different eco-zones, the dynamic development of land use systems follows different drivers. These are embedded in very specific sets of influencing factors, characterized by ecological conditions such as climate, soil and vegetation, as well as socio-economic parameters like markets, infrastructure, knowledge, extension services and technology. Tropical forestry is extremely site-specific and the corresponding knowledge is locally embedded. This is also applicable to the cultural dimension of forestry, which has been investigated more profoundly in recent times (Ritter and Dauksta 2011). In general, forests play an important role both as an important form of vegetation cover of the planet earth and as a land use system. It is assumed that to this very day policy makers responsible for rural and national development under-estimate both values.

The term '**rural**' is generally used as an antonym to 'urban', to refer to 'things of the countryside' (Wiggins and Proctor 2001, p. 427). Although any generalization may be misleading (Hoggart 1990), rural areas are often characterized by a relatively low population density (Moseley 2003, p. 1 et seqq.). The three following characteristics of rural areas were outlined by Galston and Baehler (1995, p. 3) as a contrast to urban areas: "(1) a dense relation to nature, (2) good social networks and (3) a well determined understanding of and integration into historical processes." Wiggins and Proctor (2011) differentiated between peri-urban zones, the (middle) countryside and remote rural areas with a further separation between good and poor natural resources in the last two categories. Each of the resulting five categories is characterized by specific activities and development options. The character of rural areas changes with increasing urbanization. Similar to the intensity rings in the von Thünen model, peri-urban areas are strongly connected to urban activities. The intensity of commercial forest use increases with the distance from the centers (Thünen 1866). According to Wiggins and Proctor (2011, p. 433), forest production in peri-urban areas is focused mainly on recreation and 'deconcentrated' wood based industries such as sawmills and furniture factories. In the middle countryside and in remote rural areas forest production is restricted to primary activities with limited added value and employment. In tropical and subtropical regions further characteristics of rural areas often include weak infrastructure and low market access. Strong traditions and a lack of capital and employment limit the development options available to young people. The resulting outward migration leads to a 'brain drain' as often it is the dynamic and well qualified young inhabitants who leave these rural areas.

This book deals with tropical and subtropical countries, mostly termed 'third world countries' or 'developing countries'. The denomination 'third world country' dates back to the Bandung Conference, where the countries of the south were placed in the context of the capitalist Western countries, denoted as 'first world' countries, and the countries of the socialist bloc, termed 'second world countries' (Rist 2002, p. 80 et seqq.). Representatives of the Southern countries never agreed with this denomination and since the end of the 'cold war' it has become obsolete. The emerging term 'Countries of the Global South' might come to substitute other denominations in the future, because it includes 'emerging countries' as well as the 'least developed countries', and it is neutral in relation to the development status of a country. Here the old fashioned term '**developing countries**' will be used because it is still common in most literature and the main contents of this chapter pertain to development models. At all stages there was a strong interaction between development in practice and models. Development models have always been much driven by Western thinking and at the same time they were used as driving force for 'planned change' in developing countries.

The meaning of '**development**' has changed fundamentally over time (Nederween Pieterse 2001, p. 5; Rist 2002). The meaning and indicators of development were strongly criticized in the 1970s because there was a primary focus on capital formation and economic growth (Seers 1979; Rist 2002). Over time a number of social and ecological indicators have been integrated and today the livelihood index permits a more complete assessment of development. In this publication development is perceived as a multi-facetted process which integrates economic as well as cultural, social and ecological elements. In the "postdevelopment" discourse a pluriverse, endogenously embedded development path is observed in some places (Escobar 1988; 2012).

The discussion of the so called **development theories** was initiated in the 1960s after the Second World War and with the beginning of the cold war. Due to their complex and often normative character they frequently lacked the character of 'theories'. In many cases they failed to or only just stood up to 'scientific verification'. It is preferable, therefore, to term these 'explanation models' (Nederween Pieterse 2001, p. 150; Rist 2002). It must be taken into account that most of the so-called development models for third world countries were deeply rooted in experiences from the so called developed countries, often following a purely Western philosophy. Thus, they did not fully reflect the reality in the tropics and respective traditions (Rist 1997, 2002, p. 69). Beginning with these development models, their application to tropical forestry development is reviewed and the respective interactions documented.

This leads to the formulation of **paradigms**, the meaning of which is similar to that of the German 'Leitbild'. Kuhn (1962, p. 175) defined paradigms as "the entire constellation of beliefs, values, techniques, and so on shared by members of a given community". Or "it denotes one sort of element in that constellation, the concrete puzzle-solutions which, employed as models or examples, can replace explicit rules as a basis for the solution of the remaining puzzles of normal science." Hunt (1989, p. 4) and Kuhn (1962) highlighted four requirements of paradigms: (1) they are based on theories, (2) they have an impact on policies are articulated by a clearly determined group of scholars and (3) and (4) they exert an influence in practice. The appearance of changing forest development models, rooted in guiding development policies and strategies, and which have a strong link with practical implementation and experiences, fully cover these requirements. The interpretation of paradigm change by Kuhn (1962) was frequently criticized for being too vague and for having an insufficient grounding on empirical data (Lakatos 1970; Fuller 2003). Here it permits a valuable means of reflection and exploration, helping to fully understand and interpret development processes and, based on this, to design strategies for the future.

2.3 Paradigm Shifts in Rural Development and Forest Policy

2.3.1 Review of Existing Studies and Literature

Three research and publishing spheres are relevant for the interpretation of paradigm shifts in tropical forestry and rural development: (1) literature on development theories and rural development approaches, their genesis and implementation with a special focus on the tropics and subtropics; (2) literature on tropical forest policies and politics with a strong linkage to their historical development and (3) literature on the interaction between rural development and tropical forestry in theory and practice.

Abundant literature deals with general economic and social development models and their historical integration. The early models mainly focused on economic growth orientation (Boeke 1958; Hirschmann 1958; Hagen 1975; Hunt 1989; Thirlwall 1994; Todaro 1995, 1997; Cypher and Dietz 1997). Alternative studies focused on the origin of unequal exchange and dependencies (Mandel 1962; Perroux 1976; Frank 1977). After the Second World War, coinciding with the decolonization of many tropical countries, the Marshall Plan was viewed as a viable instrument to induce development. Preston (1996), among others, provides a comprehensive overview of the historical roots and the contents of the main development theories up to the 1990s. The development models of the 1960s and 1970s focused on capital accumulation, technology transfer and industrial and rural development, mostly assuming large production units. With the reorientation towards 'redistribution with growth' aspects such as basic needs, social justice and poverty alleviation gained ground (Hunt 1989). The negative consequences of pure market orientation were highlighted in dependence and dualism theories, which were rooted in the critical theory. Assuming that the pure market orientation had resulted in asymmetric power structures and structural dependencies with their historical determination (Bernstein 1973; Prebisch 1950, 1980; Furtado 1984), the main way out was seen in a fairer distribution of the means of production such as land and capital, fair incomes and in a sharing of the benefits accrued from the outputs.

Development models are a useful means to understand and evaluate development strategies and to formulate scenarios for future development. The theories facilitate an indicator-based analysis of their performance. Various authors have criticized the fact that all stages of the development discourse and the respective models have been determined by western countries, thereby inhibiting the generation of local solutions based on alternative development models (Cowen and Shenton 1996; Escobar 2012; Sen 1999).

A more profound investigation of the link to rural development and its genesis occurred from the 1960s onwards, when rural development experienced a fashionable period. Ruttan (1984) outlined the historical perspective of rural development programs between the 1960s and 1980s. In a very comprehensive review, Ashley and Maxwell (2001) discussed the sequence of 'narratives' in the rural development debate. They reviewed the approaches adopted between the 1960s and 1990s and identified shifts from state to market orientation and between social objectives to business orientation and back to a limited social orientation. In the synoptic overview it is assumed that rural poverty is an ongoing problem and that a more intensive analysis of the underlying reason for the failure to achieve development targets has to be undertaken (Chambers 1983). The synoptic overview is underpinned by Ellis and Biggs (2001, p. 439), who interpreted the rural development ideas more specifically in the context of development theories. They placed development models in a timeline, demonstrating a sequence of 'ideas' pertaining to rural development. Extensive literature on agricultural development in the tropics and its contribution to development is also available. Special emphasis is placed on the political economy of peasants (Bernstein and Byres 2001). There are various controversial opinions concerning the future of small farms (McMichael 2006; Hazell et al. 2007).

There have been few studies of tropical forestry development on a macroeconomic level to date. Comprehensive studies and publications on forest policy focus mostly on western countries (Ellefson 1992; Cubbage et al. 1993). Such studies are rare even for the large transition countries China, India and Brazil. In recent times there was a surge in forest policy studies addressing tropical countries, but in most cases these targeted specific aspects (Humphreys 1996, 2006; Palo and Mery 1996; Palo and Uusivuori 1999; Palo and Vanhanen 2000; Palo et al. 2001). The number of studies and publications with a focus on forest policy instruments implemented under the United Nations' conventions and addressing private sector activities has grown. Examples include studies of forest monitoring systems and payments for environmental services. Increasingly market orientation is the dominant issue and less emphasis is placed on the need for state intervention.

The analysis and modeling of the contribution of the forest sector to development has not been widely tackled and most of the studies do not permit either a systematic or a holistic understanding of macro forest policy structures and processes. Douglas (1983) presented an early review, using the situation in Bangladesh as an example. His study took place in the context of the 1980s and the basic needs approach, contrasting with the growth orientation of industrial plantations and large scale timber industries. Romm (1986) developed a framework for the integration between development and forest policy and Westoby (1987) revised critically his previous position on capital induced forestry development. Based on document analysis, rooted in the abstracts of the World Forest Congress and the FAO forestry journal UNASYLVA, in the 1980s Steinlin and Pretzsch (1984) presented a study on changing forestry development models. Tropical forestry development was structured in five stages. These stages correlated strongly with the latest and most prominent development models of the time. Towards the end of the 1990s the five stage approach was complemented by a sixth stage (Pretzsch 2003, 2005), which refers to the era of globalization. The as yet mainly pragmatically constructed genesis of tropical forestry development processes is now underpinned by a theoretical construct. The six stages approach is used as a foundation for the following detailed interpretation of the linkages between development policy and forest policy development. The differentiation of the six main discourses, identified here as 'paradigms', follows a discourse-oriented historical analysis. The delimitation is rooted in the rich literature on development models, in the practical, empirically observed action adopted in tropical forestry and in the rich discussions held with numerous tropical foresters and scientists. Following the claims to paradigms, the underlying indicators used to review the six stages are the theoretical foundation, the value systems and practice. Changes to the models are interpreted as paradigm change.

A more empirically embedded analysis of changing paradigms of forest plantation is being developed (Szulecka et al. 2013). Studies on tropical forest politics have been enriched by contributions from political ecology. The origins go back to the 1970s and one of the main initiators was Blaikie (1985), with his study on the political economy of watershed management. The approach was further developed subsequently, focusing on the relations between political, social and ecological factors as driving forces and the outcomes of environmental management (Peet and Watts 1996; Walker 2005, 2006). The publication by Bryant and Bailey (1997) provides an excellent overview of this approach, which was mainly developed by geographers and sociologists. Many of the recent studies focus on natural resource management in tropical regions. Intensive studies on forestry contribute profoundly to the discourse on the linkages between the pre-colonial, colonial and post-colonial development of forestry and forestry institutions in the tropics (Peluso 1992; Fairhead and Leach 1995; Bryant 1997, 1998; Pye 2005). A number of authors from India have worked on the British colonial influence on the post-colonial forest administration (Gadgil and Guha 1992; Gadgil and Berkes 1991). The positive influence of the European forest management paradigm has been called into question, because often its implementation was not adapted to the local cultural and political situation (Shiva 1989).

The following elaboration upon six tropical forestry development stages follows four steps: (1) a short classification of the historical stage and an introduction to the availability of corresponding data; (2) an interpretation and model of development; (3) a review of tropical forestry development and (4) synthesis in a paradigm.

2.3.2 *Stage I: Trees and Forests in the Tropics in Pre-colonial Times*

2.3.2.1 Classification of the Historical Stage and Data Availability

This stage covers the relationship between forestry and the general development of tropical countries before colonization by European countries. Information on the use of forests and the perception of trees and forests in pre-colonial societies is scarce. Written documents from testimonies are available only in a few countries. The oral transmission of information can barely bridge long periods of time. With increasing interest in the reconstruction of traditional knowledge, the research community grew considerably, employing epistemological and interdisciplinary methods (Warren et al. 1995; Agrawal 1995). A surge in the number of publications has taken place in recent years (Posey 1985; Berkes 2008; Parotta and Trosper 2012).

2.3.2.2 Interpretation and Model of Development

The complex nature and interdisciplinary approach needed to understand and interpret pre-colonial societies is documented by a variety of theoretical explanation models. The relationship between humans and their natural environment in the tropics was first depicted by anthropologists, stemming from the disciplines cultural ecology (Steward 1972) and cultural and human ecology (Bennett 1976). Steward (1972) investigated socio-economic patterns relevant for human habitat use. Cultural ecology explains the diversity of land use systems and its integration in farm household systems with a strong focus on subsistence strategies (Bennett 1976; Bargatzky 1986; Little 1999). Bennett (1976) presented a broader interpretation of cultural and human ecology. He applied an anthropological approach to explain transitions in rural societies. Most of the explanation models were formulated in western countries. In the 1960s, the Brazilian anthropologist Darcy Ribeiro structured the socio-cultural development of civilization in a stages approach (Ribeiro 1983). He described the alteration process from hunting practices to intensive land use, differentiating between sustainable and unsustainable, and even nature-destroying land use practices. Pre-colonial governance systems are characterized by great diversity, ranging from subsistence based local economies to highly developed state structures with far reaching market integration. Thus, governance systems differed between strong, hierarchically ordered societies, bottom-up, self-organized communities and even diffuse, almost anarchic organization in some tropical rain forests (Sigrist 1979). Examples of strong hierarchical organizations include most of the societies that have reached a high level of technological achievement such as the Incas and Mayas in Latin America and the traditional Chinese society. Typical bottom-up structures are represented by the social organization of African pygmy tribes and indigenous Amazon communities (Demesse

1992). Rather sporadic social organization is characteristic of sparsely populated tropical forest regions. Sigrist (1979) characterizes Western African forest people as living in a society nearly without hierarchies, which he defines as anarchic.

Widely discussed, and often still contradictory, are the reasons put forward for the clash of civilizations. Analyzes are based on assumptions, because the reconstruction of pre-colonial relations between civilization and ecosystems is only partly feasible (Ponting 1991; Diamond 2005). In addition to war, often ecological deterioration, overuse due to population pressure, overexploitation of the labor force and emergent diseases were drivers of the decline of cultures. Widely investigated examples include Easter Island, the early civilizations of the Euphrat and Tigris, the civilization of Angkor Wat in Cambodia (Masselos 2010) and the Maya and Inca civilizations.

2.3.2.3 Tropical Forestry Development

In pre-colonial times forest farmers availed of a large variety of forest goods and services. The short term provision of food and shelter was essential for the individual survival of the rural population. Household consumption products consisted of a broad portfolio of food items like vegetables, fruits, tubers and all types of animals, which covered the calorific and protein requirements of forest people. Up until today in India, the numerous traditional forest uses are denoted by the 'five or six f' referring to firewood, fruits, fodder, fiber, food and fertilizer. Cosmic and religious values have also played an important role. Often there were limits in terms of the market access and especially in tropical countries the storage of food was frequently difficult. The situation with regard to livelihoods differed profoundly between densely populated areas of intensive cultural development, such as civilizations in India and China, and societies located in rain forests with a low number of inhabitants and at the rural periphery. The variety in the consumption of forest products and services was complemented by complex institutional patterns of pre-colonial forest use, by which access, use and consumption was organized. The land tenure was mostly determined by common property, which Minsch (1992) denoted as *patrimonium*. Thus, the land was in the hands of the community under local chiefs or, like in many African countries, in the hands of the ancestors and future generations. Land users had no ownership rights and no land title; they merely possessed usufruct rights over land and vegetation, resulting in a stewardship or trustee relationship. The land had to be used in a sustainable way in order to guarantee that future generations would have the same land use options as the users of the present (Pretzsch 1986). With an increasing scarcity of forest products, and in order to ensure a continuous provision of food, some pre-colonial societies reacted by establishing rules, which were developed by the communities and chiefs. The rules contributed to reducing transaction costs. Land and resource access rights were restricted, the amount of resource use per capita was limited by quotas and communities agreed upon benefit sharing mechanisms. Often the institutional framework was complemented by religious and symbolic rules on an overarching level. The systems were often quite complex and efficient, including rules for reciprocity, benefit sharing and the redistribution of benefits. Numerous studies on the organization of traditional cultures demonstrate the high level of social capital (Pretzsch 1986, 1987b; Sponsel 1995).

The institutions regulating forestry were partly top-down in nature, such as in kingdoms or in communities governed by chiefs or the Indian rajas; they were determined by feudalistic superiors or religious leaders, and served to define the roles of the different segments in society by means of very strict rules. There are also many examples of bottom-up forest management, rooted in self-organization and constructed on the basis of local knowledge accumulated on farms and at village level. Elements of a moral economy were evidenced by the subsistence orientation and based on the principle of reciprocity. There are interesting cases of endogenous governance structures guiding sustainable resource management systems, albeit where local development was far from being automatically nature conserving (Gadgil and Berkes 1991).

Techniques and technologies were mainly handcraft-based, and some pre-colonial societies possessed sophisticated natural resource management systems. Traditional forest management practices were based on observation. Often these practices were copied from nature (Posey 1985; Neugebauer 1986).

2.3.2.4 Paradigm

Anthropological theories were identified as the main explanation models. Authority-driven societies with a strong top-down order were identified as were societies with bottom-up decision-making. The institutionalization of development processes led to technology development, thereby permitting an increasing number of inhabitants to co-exist with nature in a more or less balanced way. Bottom-up decision making was based on local experience and traditions, and led to the formulation of informal rules for forest management. The era was characterized by multiple forest uses for subsistence purposes and for trade in markets. Anthropological theories and the forest use practices can be integrated in a co-evolution model, which perfectly explains the interaction between people and forests and forms a core element of this paradigm.

2.3.3 Stage II: Forests and Trees in Colonial Times

2.3.3.1 Classification of the Historical Stage and Data Availability

The majority of tropical countries were colonized by European states. The era of European colonization extended from the end of the fifteenth century to the decolonization of the last tropical countries in the 1970s. Knowledge of colonial development and colonial forestry in particular are well documented. Many of the colonizing countries established precise documentation systems. Often the information collated was never assessed and to this very day is hosted in the colonial archives of the colonizing states. Documented information is biased for the view of the colonizers. Critical investigations of the philosophy and impacts of colonial forestry are still scarce (Shiva 1989; Blank 2006).

2.3.3.2 Interpretation and Model of Development

Europe's isolation within the bounds of the Atlantic Ocean to the west, Russia to the east, and the unknown African continent and the Islamic countries to the south and south east led to expansion activities in the fifteenth century in the form of expeditions and the installation of trade posts and markets. These were occupied as colonial territories in different stages (Fieldhouse 1965). The colonial occupation started with the 'voyage and discovery' of Christopher Columbus in 1492 and with Vasco da Gama's voyage to India in 1497–1498 (Preston 1996, p. 143; Fürstenberg 1966). In 1494 the colonial areas were split into Spanish and Portuguese domains under the Treaty of Tordesillas. The Aztec and the Maya territories were occupied and settlements were initiated around 1500. In Latin America colonization led to a far reaching destruction of local cultures. It was oriented towards the assimilation of the local administration and culture by the colonizing power. The establishment of agricultural plantations for the production of cash crops like sugarcane required a labor force. The native Indian population resisted and the transatlantic slave trade was initiated to meet the labor demand. With the colonizers struggling to produce agricultural crops in a reliable and efficient way the *encomienda* system was introduced, whereby local communities were forced to produce cash crops (Konetzke 1956). In 1564 the Spanish colonizers occupied territory of the Philippines; the first, and for some time the only, colonial territory in Asia. At that time the European states other than Spain and Portugal possessed only limited colonial territories. At the end of the eighteenth century French occupation was limited to but a few West Indian islands and bases in Africa. British colonies were established only in North America and in West Indian islands like Jamaica and Barbados, where, following the Brazilian model, plantation colonies were established. Further developments were accelerated by the foundation of the East Indian Trade Company, which was initially managed on a private basis. With the decline of the American colonies in the eighteenth century, colonial territories decreased before a new colonial era began in Africa and Asia (James 1998; Osterhammel 2009). Step by step the European capitalist system expanded to most tropical countries. Hobsbawn (1987) called this era "Age of Empire". Powerful administrations were established, undermining endogenous value systems and development paths (Preston 1996, p. 139; Escobar 2012).

The colonial systems differed in their strategies with respect to the implementation of local organizations. The French colonizers followed the direct rule system, which led to a complete substitution of the local administration by colonial staff (Pretzsch 1986, p. 21). Meanwhile, under the British indirect rule system, the local structure was partly taken over by the colonial ruler, integrating it within the colonial administration. All systems were based on unequal power structures and a one way flow of communication in a system of law and order (Hobsbawn 1987). The colonial occupation advanced hand in hand with the slave trade, which led to a far reaching brain drain and destruction in many topical countries (Loth 1981). Even more damaging than the physical power was the forced dissemination of European value systems. The existing social order, including the traditional resource management systems, were disrupted and replaced by external rules and practices. Imperialistic attitudes, continuous suppression and the extreme exploitation of the colonies led to increasing resistance against the colonizers, which finally resulted in struggles for independence and ultimately liberation.

2.3.3.3 Tropical Forestry Development

Huge parts of the colonial territories were covered by forests and the question of their use or conservation was a central issue. Forestry was greatly affected during the colonization process (Tucker and Richards 1983). Although each colonizer practiced its own specific ways of colonization, the final outcomes were similar. The early colonization of the Americas was much oriented towards the exploitation of timber for ship building, construction and a number of other uses. In spite of this economic importance attributed to the forests, the colonial legislation, the 'Leyes de los Indias', made no reference to forest use (Fürstenberg 1966, p. 158). With the establishment of the authoritarian colonial administration, traditional rights to forest use were negated and the traditional links between rural people and nature destroyed (Fürstenberg 1966). During the later colonization of Africa and Asia similar strategies led to comparable outcomes. Direct colonization by France resulted in the widespread destruction of local institutions and knowledge systems dealing with forest use (Pretzsch 1986). Although local institutions remained partly untouched under the British system of 'indirect rule', the rights of the local population to forest use were restricted. Most colonial interventions led to the alienation of local people from their forest environment. Accompanied by transatlantic slavery, the colonization process led to a far reaching destruction of local land use and management systems, often including the forest and tree components.

Forestry was subjugated to western philosophies and knowledge systems. In this way, the relatively complex and holistic human–nature interactions and the respective forest use systems were substituted by one-dimensional concepts of financial forest benefits and the creation of pure conservation areas. Early influence came from German academic forestry schools. The German Royal Forest Academy in Tharandt was one of the first academic institutions, transferring middle European scientific forestry models to tropical countries (Pretzsch 1994; Uibrig 2007). Much emphasis was placed on the development of mono-cyclical natural forest management systems. Especially under the French and the British colonial forest service, experimental plots were established with a number of silvicultural homogenization models (Dawkins and Philip 1998; Lamprecht 1990). Due to high labor inputs, low production per hectare and high complexity, most of the systems were far from operational (Pretzsch 1998, 2003). Adhering to the needs and interests of the colonizing countries, the 'modern' forestry referred to above made little account for subsistence production and was mostly oriented towards markets. The diversity of forest uses of pre-colonial times was replaced by an orientation towards the production of a small number of commercial crops for export, including coffee, cacao, oil palm and rubber. Forests were also seen as a land reserve for mainly plantation production. The forest use rights of the local population were drastically reduced. Following the Roman tradition, private property land titles were introduced. Minsch (1992) referred to this as *dominium*. Under French colonization local land users had to prove that the land had been under cultivation for at least 10 years. This was hardly the case in the traditional shifting cultivation systems, but the European settlers, working with permanent agriculture, got property rights (Pretzsch 1986). In this way a two-class society was created, consisting of people with and people without title to land. Other forest use rights such as the collection of non-timber forest products and firewood were placed under strict control.

The previous concept of 'ownership' over specific forest products and services was lost and often local people acted against colonial forest administrations (Blank 2006; Shiva 1989). The substitution of traditional rules by the paradigm of modern European forestry, which at that time was based on the maximization of the timber production, led to a partial rupture of the traditional relationship between people and their natural environment. The bureaucracy concept of Max Weber, with single line organizations, and the Prussian top-down approach and the respective law and order-based forest administration was often implemented (Rist 2002, p. 48 et seqq.). As Shiva (1989) illustrated, the post-colonial forest management paradigms barely referred to pre-colonial experiences but were instead dominated by western philosophy.

2.3.3.4 Paradigm

The theoretical foundations to explain this era are imperialism theories and analytical models explaining colonialism and related power structures. It is necessary to generalize as colonial expansion took place at different times and to different intensities in Latin America, Africa and Asia. The colonization practices of the colonizing country also varied. The basic principles and consequences were similar, however. Apart from direct forest destruction, colonization took place profoundly in the mind. Western value systems were imposed and traditional rules on how to treat and use the forest ecosystems were replaced by simple state control. Local people became alienated from traditional forest use.

2.3.4 *Stage III: Independence and Capital Formation*

2.3.4.1 Classification of the Historical Stages and Data Availability

The decolonization process extended from the time of independence of the Latin American countries until that of the African and Asian countries in the 1960s and 1970s. The era is characterized by remaining relicts of the strong bi-national linkages, which were either peaceful or characterized by post-colonial struggle. In the case of some countries it is difficult to obtain information for this era, because often the colonial administration's archives were taken over by the new national government, leading to a gap in the documental information.

2.3.4.2 Interpretation and Model of Development

The decolonization phase was still much dominated by value systems and structures implemented in colonial times. The colonial heritage led to a certain path dependence (Fieldhouse 1981) and it took time for own, endogenous development strategies to develop. Many of the higher government representatives had been educated in European countries, Soviet Union or United States of America. Their education was based on the latest scientific knowledge and belief systems of these countries. Their understanding of development strategies, and especially rural development and forestry, were primarily based on western development models (Rist 2002, p. 103). Most of the newly formed governments voted for a market oriented governance system, and integration in the world market. Development was almost exclusively measured by the level and the increase of the gross domestic product. After the Second World War the main economic orientation followed the assumption that capital accumulation is the main precondition for economic growth. Preston (1996, p. 154) summarized four main influencing factors that determined this era: "(1) the theoretical and practical impact of the Keynesian revolution, (2) increasing global influence of the USA in prescribing development strategies, (3) the reconstruction of western Europe with the Marshall Plan and (4) demand of new national states towards their own independent and nationalist development." The positive experiences of the Marshall Plan, a large investment made by the allied forces to reconstruct Germany after the war, were simply transferred to tropical countries (Nurkse 1953; Rosenstein-Rodan 1961). Development theories were formulated offering paths to overcome underdevelopment. Important in the context of understanding the spirit of this time is the Rostow model (Rostow 1960; Kuznets 1963). Similar to Marx's development stages, he proposed a sequence of five stages from traditional society, the preconditions for take-off, take-off, the drive to maturity and finally the age of high mass consumption (Rostow 1960, p. 4; Kuznets 1963; Thirlwall 1994, p. 61). It was assumed that the shift from traditional society to preconditions for take-off was determined by the availability of capital and entrepreneurial potential. This approach was the origin of the so called third world development theories (Rist 2002). In practice the model was difficult to implement because in the majority of post-colonial tropical countries the necessary capital was lacking. Productive industries, which might have permitted a nationally relevant formation of capital, did not exist in most countries. Savings and 'forced savings' in the form of taxes could barely be tapped. Capital, therefore, had to come either from outside the country or from nationally exploitable natural resources. Given the lack of capital, technologies and know-how, the resources were often exploited by international enterprises without any added value accruing within the country.

Only a small number of countries followed the path of import substitution, in most cases only temporarily, prioritizing the refinement of the available resources nationally and placing restrictions on imports. One such example is Ghana under Nkrunah. In socialist Tanzania under Julius Nyere, a concept of local community development was initiated with the objective of improving the local production structure. Similar concepts were realized in Indonesia and Brazil, as well as under the international trade boycott in Cuba. These years were an important time for nation building after decolonization (Cypher and Dietz 1997, p. 65). Euphoria arose in part at the idea of recovering development by means of rapid modernization (Preston 1996, p. 153 et seqq.). A number of neo-colonial imperialism theories appeared (Mommsen 1987), dealing with unequal center and periphery relations, power relation in the world system, decreasing terms of trade and income imbalances between industrial and developing countries and cultural imperialism.

2.3.4.3 Tropical Forestry Development

In the decolonization stage, tropical forestry strictly followed the general economic development models. The newly independent forest authorities were acquainted with western forest administration and forestry development models. Even the few academic institutions in which scientific forestry was taught had been founded by western countries and the contents of the curricula were entirely oriented towards European concepts of modern forestry. At the same time there was great pressure to generate income for the newly established National Forest Services and for national development in general. Under these circumstances Rostow's (1960) theory, and the respective European interpretation, served as a justification to accelerate tropical forest exploitation, driven mainly by European enterprises (Westoby 1962). At the sixth World Forest Congress in Madrid, Zivnuska (1966, p. 561) argued that under the conditions of initial growth and transition towards industrialization, the principle of sustainable forest management would be not applicable: "Here it must be recognized that, despite its central role in the doctrines of forest regulation, sustained yield is a concept which is economically justifiable only under a particular set of economic circumstances. The sustained yield concept is not applicable to the conditions of rapid growth". The liquidation of natural forest resources was seen as one of only a small number of options available to contribute to the formation of national capital and to induce development. It was assumed, that the massive exploitation of forest resources, a subsequent investment of the raised capital in timber industries, and a later re-investment in forest plantations, would lead to growth of the forestry sector and the wood industry (Zivnuska 1966). It was even argued that this might become a 'leading sector' in the national economy, similar to the experiences made in Scandinavian countries such as Finland and Sweden, and also in Canada (Palo 1988). As a consequence, natural forest exploitation for timber production increased rapidly in most tropical countries. As the state forest administrations possessed neither the proper logging machines, nor the know-how and the necessary financial capital, the forests were handed over to concessionaires for logging. They mostly worked with capital from Europe and later on from Japan, Russia and other industrialized countries, and engaged in forest exploitation with a short term orientation, maximizing the rent from forest stands. Critical scientifically based analyses of these practices only emerged belatedly (Pretzsch 1987a; Repetto and Gillis 1988).

In general, the management of state forest administrations was largely top-down, following a hierarchical line organization principle. As a colonial relict, it was similar to the Prussian type of forest organization propagated by Weber (Weber 1922). This opened the door to corruption and often the rights and interests of small forest farmers were ignored. The research into tropical forestry also mostly adhered to the old colonial practices, with the majority of the forestry research taking place in research stations and under experimental conditions. The findings were of limited applicability in practice and on a large scale (Lamprecht 1988). Most of the silvicultural models proved to be too complex, labor intensive and of little financial benefit (Pretzsch 1998, 2003).

2.3.4.4 Paradigm

The paradigm follows a rather simplified interpretation of development, as proposed by Rostow (1960). It was adopted by a whole generation of development economists from both the north and the south, and centered on the need to overuse forests for the formation of financial capital to initiate economic development (Zivnuska 1966). The implicit western model of growth orientation is still a part of many agendas today. The situation in the decolonized territories was rather different from the reality in Europe, however. The textbook ideas about sustainable forest management, which focused on western countries, did not fit with the situation in most tropical countries. After decolonization the new national states were de-capitalized and partly suffered from the brain drain caused by slavery. The traditional knowledge of the subsistence use of tropical forests had been negated and the scientific knowledge available with respect to tropical forest management was rather unsophisticated (Dawkins and Philip 1998).

2.3.5 *Stage IV: The Internationalization of Forestry*

2.3.5.1 Classification of the Historical Stage and Data Availability

The dissolution of bilateral colonial links happened at different speeds. The new nationalist governments struggled for independent and autonomous development, but experience with development models and their implementation in practice was lacking. In this situation the old colonial links were replaced in part by blue print concepts of western countries and international organizations. There was a boom in the number of publications on development, including forestry, with a focus on technologies, industrialization and economic growth.

2.3.5.2 Interpretation and Model of Development

It was an era of expanding international cooperation, based on relatively simple and linear blue print concepts related to the assumption that large scale rural development is feasible. The underlying thinking was still rooted in the prioritization of capital accumulation for industrialization, combined with the emerging modernization theories (Preston 1996). A strong dichotomy between traditional societies and modernization was assumed. Modernization was interpreted solely in the western sense, as entrepreneurship, and denied most of the historical facts experienced in developing countries. Still adhering to Rostow's stages of economic growth, the traditional stage had to be overcome using savings and with the formation of productive capital (Preston 1996, p. 167). This stage was accompanied by large scale investment in technologies in the primary sector and in industrialization and infrastructure. Modernization projects were often planned in a growth-oriented and technocratic way, with little consideration given to ecological and social impacts. The local stakeholders affected were in most cases not involved in decision making. Possible negative social consequences of large scale investments were invalidated using the argument of strong trickle down effects, which would reach and benefit the rural poor. Innovations were propagated following the linear technology transfer approach (Leeuwis and van den Ban 2004; Rogers 2003). Examples include the implementation of the green revolution based on agricultural inputs like fertilizer and improved seed in India, large scale colonization projects in the Amazon and the transmigration project in Indonesia. International agencies such as the United Nations' bodies and the World Bank strongly propagated this approach and implemented a huge number of projects (Goldman 2005). Land use planning was rooted in far reaching instrumental thinking, which led to the implementation of relatively uniform large scale projects with a technocrat orientation and a focus on hardware.

One of the consequences was the increasing indebtedness of many developing countries. The large scale projects acted as debt traps. Frequently only a part of the available funding was donated and many projects did not permit repayment of debt, composed of the principal and the interest. Payer (1974) reviewed the World Bank activities of that time and criticized the generally low long term efficiency of these studies.

2.3.5.3 Tropical Forestry Development

Tropical forestry adhered strongly to the modernization model. With the noticeable improvements to communication networks and the international trade infrastructure, the tropical timber trade channels diversified and timber production and the timber industries entered a stage of increasing internationalization. With the main focus remaining the initiation of national growth by means of capital formation, the bilateral colonial links where gradually replaced by multilateral market contacts. Colonial governance shifted to the domain of international organizations such as the United Nations bodies, the World Bank and the International Monetary Fund. The idea of the 'homogenization' of natural forests, following European experiences, was investigated in many research stations and propagated in many tropical countries.

Funds urgently needed by developing countries for improvement led to the continuous increase in the exploitation of natural forests. It became increasingly lucrative for private concessionaires, who assumed for the nation states the tasks of logging and selling timber. Forest concessions were rented on a short term basis, leading to a volatile capital investment. Due to high calculated logging costs and a corresponding underestimation of stumpage values, windfall profits were created and forest revenues for the national states were low. Also in the majority of the cases the national investment strategy failed because of the weak institutional structure, lack of know-how and political will and corrupt misuse of the financial capital generated. Thus, these concepts posed a severe threat to local people and their livelihoods. Due to deficiencies in the institutional framework, most of the capital released left the locality, the region and even the country, and incentives for local development were minimal (Pretzsch 1990; Vincent 1990; Vincent and Binkley 1992).

Once the natural forests had been destroyed by over-logging, the transformation to large scale forest plantations was suggested. The World Bank was a major proponent of this. The planning of these mostly large scale projects followed a strict top-down approach. Often the projects failed because of the adoption of techniques and technologies not adapted to the purpose, or a lack of acceptance by the local people resulting in resistance. In many cases plantations were established with the aid of low interest loans provided by international organizations. The failure of large scale forest plantation projects has contributed to the international debts of some tropical countries (Payer 1974; Pretzsch 1986; Goldman 2005).

Large scale and multinational enterprises appeared on the market and entered third world countries. Private investors established large scale forest plantations. The investment in the Jari Project in the Amazon by Daniel Ludwig in 1967 demonstrates the spirit of that time. He bought 1.6 million hectares of forest for an amount of $1.6 million and invested a total of $1 billion installing a pulp factory he had shipped from Japan. This gigantic project was largely a failure and he sold the whole consortium for $280 million in 1981 (Fearnside and Rankin 1982, 1985; Russel 1987; Lins 1994).

As in agriculture, most forestry projects followed the technology transfer concept, which granted high priority to exogenous technology development. Local knowledge was rarely availed of. The utopia of the 'homogenization' of natural forest similar to European forest experiences further dominated the thinking of tropical foresters. It was assumed that through the homogenization of the natural forests, the growth rates of commercially useable timber could be increased considerably. Many silviculture systems were developed based on the classical European system but in most cases the corresponding experiments took place in research stations and not in the field. Large scale implementation often failed because of financial and technological limitations. Meanwhile, the establishment of plantations led to technological innovations and private enterprises especially entered the charcoal, pulp and paper sectors.

The ongoing destruction of the forests led to the creation of an ad hoc working group, created as an output of the FAO Committee on Forest Development in the Tropics in 1983. Together with the World Bank and the World Resources Institute (WRI) a 5-year program was drafted in 1985, coined the Tropical Forestry Action Plan (TFAP). A corresponding organizational unit was set up within the FAO. Seventy countries had joined the TFAP by 1990. In that same year, however, it became obvious that deforestation in the tropics had not slowed down. The TFAP had followed a very technocratic and sector-oriented approach and its implementation was typical of the final stage of internationalization and large scale planning in tropical forestry. The procedure was strongly criticized by the WRI, citing the following deficits: (1) too restrictive top-down planning; (2) no participation of local actors and NGOs and (3) the lack of sector integration. Funding was stopped and a fundamental reorientation process started, which resulted in the creation of the National Forest Action Plans and later on the National Forest Programs (Winterbottom 1990; Humphreys 1996, p. 31 et seqq; Glück et al. 1999).

2.3.5.4 Paradigm

Although the traditional colonial links were still strong in some countries, guiding concepts for forestry development were now increasingly the domain of international organizations such as the Food and Agriculture Organization of the United Nations and the World Bank. Both international organizations and large forest enterprises followed one common paradigm: tropical forest management and plantation development are feasible within the framework of good planning and require above all technological skills. The modernization theory was applied for forestry. Following the strategy of the World Bank, natural tropical forests were now logged over on a large scale based on concessions. This very much favored international concession enterprises, which appropriated large parts of the forest rent. A certain euphoria towards large scale projects gained ground. Still following the lack of capital for growth model, it was argued that development incentives should not focus on the poor, because their first preference would be consumption. Consequently, they would not in a position to save money for investment. A certain level of support for the poor would be ensured by the trickle down from the leading growth sectors. All components of this concept led to a disregard for the livelihoods of local people.

2.3.6 Stage V: Polarization

2.3.6.1 Classification of the Historical Stage and Data Availability

The shift towards the diversification of thinking and models in development cannot be sharply determined within a particular timeframe. It was a gradual process starting in the mid of the 1970s and continuing up to the time of the Rio Conference in the 1992. The new era is characterized by a fundamental widening of the perceptions of rural development, including tropical forest management. This is manifested by a multiplication of the actors involved, the sources of information and the number of field projects (Chambers 1983). New economic development models such as the basic needs approach were tested and documented. Publications from emerging NGOs indicated increasing involvement and activities on the part of the civil society in projects and in national environmental policies (Lane 1995).

2.3.6.2 Interpretation and Model of Development

The majority of third world countries followed the growth models propagated by western countries. Although an overall rise in economic growth was registered, the distribution of income became increasingly unequal with a steadily rising Gini-coefficient. As was mentioned previously, large scale investment projects mostly failed or, like the green revolution in India, benefited only a part of society. Social and ecological problems increased dramatically. Many countries suffered under a policy of austerity, prescribed by the International Monetary Fund, which also served to put additional pressure on natural resources (Choussudovsky 1997). The origins of international debts were often the aforementioned large international projects, which were badly planned and failed to deliver the calculated returns (Payer 1974). The idea of a leading sector, which accelerates development across the entire national economy by means of a 'trickle down' effect to the poor segments of society, did not work in practice. At the same time, however, only very few practical examples of the implementation of endogenous development strategies, such as import substitution, existed. Often, these too were unsuccessful (Thirlwall 1994, p. 13).

In 1973, during the era of McNamara's presidency of the World Bank, the development policy shifted from a growth orientation to 'redistribution with growth' (Chenery et al. 1979). This reorientation led to the implementation of the basic needs approach (Streeten and Burki 1978; ILO 1977). Apart from taking care to cover the basic needs such as food, shelter, health and education of poor farmers and laborers, it focused on the implementation of labor-intensive technologies, which permitted a maximum provision of labor to the rural poor. The basic needs approach was justified partly on the basis of ethical arguments for poverty alleviation. It was also based on the insight that poor segments of society contribute to economic development, because they satisfy the demand of the poor for consumption goods like mopeds and televisions, although basic needs are not covered. Further reasons were the reduction of population growth through the satisfaction of basic needs and the assumption that an increasing number of poor households would lead to political instability and ecological destruction. The approach was assessed controversially. The general improvement to livelihood conditions contrasted with strong limitations in technology development (Hunt 1989). It was argued that because of the low inputs in terms of capital and technology, labor productivity might not increase significantly, which would lead to the perpetuation of a life of poverty (Hunt 1989). The approach was also criticized for being expensive and potentially leading to greater resort to credit and accumulation of debt in developing countries.

The shift from a growth orientation towards a more ecological and social thinking was manifested in an enlargement of the development community by civil society organizations such as interest groups and non-governmental organizations (Uphoff 1993). Criticisms of destructive government projects, funded using international money, spread in the NGO community. The partial failure of the modernization approach led to a rapid diversification of development models, approaches and discourses. The failure of the linear growth model initiated a critical discourse on how to 'induce' development. New ideas and concepts developed independently of western influences. An example is the Indian 'environmentalism', which is characterized by a high level of cultural identity and site-specific local value systems (Gadgil and Guha 1992, 1994; Poffenberger and Gean 1996). With the diversification of development philosophies and actor constellations the ground for the Rio Conference was prepared. The process led to the drafting of international conventions.

2.3.6.3 Tropical Forestry Development

Negative developments in tropical forestry led to a basic rethinking and reshaping of forest politics. The negative factors included the lost revenues from tropical forest exploitation, the ongoing deforestation linked to large scale forest conversion for agricultural purposes, and finally the failure of the Tropical Forest Action Plan as a powerful instrument for tropical forestry development. Belatedly it was demonstrated that the revenues from the tropical forest exploitation of the last 20 years had barely affected the national budget, as had been assumed in the Zivnuska model (1966). Studies by the World Resources Institute have shown that a large part of the stumpage value from tropical forests was transferred to western countries and Japan, where most of the large timber trade enterprises were based (Repetto 1988). These so called 'windfall profits' were a consequence of a lack of knowledge of the cost structure of local forest administrations. The foreign capital based enterprises overestimated the costs, which led to a rather low calculated rent for the resource. In many cases the conversion of natural forests to agricultural cash crop plantations like soybean, rubber and coffee, or to pasture for cattle grazing, was not successful in the long term. The ecological consequences were negative and often the local population was barely involved in land use decisions. Colonization and mining projects in tropical rain forest areas, like the Polonoroeste Project implemented by the World Bank and the Carajas Project in the Amazon, were assessed negatively and, in part, their implementation was stopped. In Asian countries, the green revolution led to the segregation of society into winners and losers. With the Transmigration Project in Indonesia, involving the transmigration of peasants from Java to Kalimantan, it became obvious that often large scale technocratic projects led to increased destruction and the social marginalization of a large segment of the rural population, as was also the case with the conversion of exploited forest concessions to large scale forest plantations in West Africa. The Tropical Forestry Action Plan, launched in the mid-1980s and quietly abandoned in the early 1990s, had followed the conventional technocratic forestry philosophy (Winterbottom 1990). Fundamental criticism of the conventional development models led to a rapid emergence of new social forces such as communities, interest groups and different types of grassroot and non-governmental organizations (Kolk 1996). Often these were more active and efficient than state forest services and governmental projects.

All of these negatively perceived factors led to resistance by the many new groups active in civil society and the partial blocking of development initiatives. One of the consequences was a reorientation and diversification of land use systems, including tropical forest management. There was a shift towards the use of more flexible planning systems, the integration of a multitude of other sectors relevant for forestry, the reestablishment of a link to local knowledge and the proactive inclusion of conflict management mechanisms. The negative consequences of tropical forest exploitation led to a boycott of tropical timber. European NGOs initiated this boycott, which was successfully adopted in public and municipal administrations in Germany and Britain. The private sector was much slower to react. In a further step, the Forest Stewardship Council (FSC) was founded and forest certification became a complementary instrument in the implementation of sustainable forest management. Forest certification represents a typical instrument of this era. It resides between civil society and private engagement, and represents the search for alternatives to government action (Meidinger et al. 2003).

More and more actor-oriented approaches appeared, explaining forest use based on actor interest. The myth of a linear link between population growth and forest destruction, following the Malthusian explanation model, was critically analyzed, indicating the importance of an sufficient number of active rural population per area for the initiation of land use innovations (Boserup 1965; Heilig 1996). An increasing number of political ecology studies focused on tropical forestry, including the analysis of power structures and conflicts between actors. It became more and more obvious that thinking in linear and large scale dimensions would not lead to success. The concept 'small is beautiful' was realized in micro-enterprises, which were developed to transform forest products (Schuhmacher 1973). The new national forestry program concept, and later on the national forest plans, related closely to local agendas, the UNCED process and finally to the outcomes of the Rio Conference.

The revival of local and traditional forestry knowledge from the mid-1980s onwards was also significant, as were the adoption of site-specific and historically rooted land use alternatives. Rich knowledge still exists within rural populations in tropical and subtropical countries, although in many places the knowledge systems are being gradually lost as the knowledge is only transferred orally. The guiding management principles often derived from natural cosmic visions (Chap. 7).

Many elements of the pre-colonial paradigm, site-specific local knowledge and the importance of community institutions formed the basis for the emerging paradigm of 'social forestry', which developed in parallel with the purely utilitarian paradigm of industrial forestry. The basic needs model and the livelihood approach contributed to the further development of community and social forestry, providing a third way between state and private activities.

In this polarization stage, a third path besides the industrial forestry and social forestry paths took shape, whereby the conservation activities introduced in a top-down manner during colonial times were broadened in their perspective, taking into account the needs of the local population and integrating these needs within the landscape.

2.3.6.4 Paradigm

With the shift away from the internalization of tropical forest policy to the polarization paradigm, forestry development strategies diversified in three directions. With a concentration in large international timber enterprises, the focus on forest-based industrialization gained further ground. This development was embedded in conventional growth theories and the theorem of competitive advantage in the world market. In many rural areas of the tropics, locally oriented rural development, in which forestry plays an important role in livelihood development, the alleviation of rural poverty and local income generation, replaced the linear development models and the industrialization strategy. Various development models, such as the basic needs approach, dualism and dependency theories, laid the groundwork for the expanding strategies of social and community forestry. Within local development strategies, instruments for a 'third way' of development, based on the creation of social capital, evolved. The discourse on tropical forest conservation intensified.

2.3.7 *Stage VI: Globalization*

2.3.7.1 Classification of the Historical Stage and Data Availability

The historical review reveals the expansion of international market networks, rapid multiplication of multinational enterprises and increasing acculturation, following western, Islamic or eastern Chinese philosophies. A qualitative advance appeared with the end of the 'cold war' at the beginning of the 1990s. Market forces dominate financial transactions, labor markets and the organization of commodity and service markets even in the public sphere. The availability of data has improved, although often the data do not portray the increasingly complex reality. Dense and unfathomable inter-sector relations, the multiplicity of actors and the volatile information policy of large multinational enterprises make data collection difficult and often diffuse. In many sectors, and specifically in tropical forestry, case study research is commonplace and most of the tropical forestry activities, excluding the private enterprise sector, are well documented. However, often the synthesis of case studies towards an overall theoretical explanation model is lacking.

2.3.7.2 Interpretation and Model of Development

Globalization is a multi-facetted process, which has strong cultural, social and economic dimensions and impacts. The interaction between various spheres makes it difficult to apply only one analytical tool (Beck 1986, 2007; Rehbein and Schwengel 2012). From the viewpoint of development policy, and focusing preferentially on economic targets, the globalization era is characterized by deregulation. The predominance of market mechanisms over state interventions, increasing power of large multinational enterprises and banks result in an antagonism between international and global plans and strategies and their implementation on a local level. Methodological definitions of the new global development paradigm are still not very precise. A generalization under the heading 'neo-liberal' has caused misunderstanding (Rapley 2004; Thorsen 2011). The term neo-liberal is generally used with an overall negative connotation, and without enough theoretical abstraction. Examples of populist autocratic governments like Chile under Pinochet, influenced by M. Friedman, or China under Deng Xiaoping, which were partly interpreted as neo-liberal, do not contribute to a theory-oriented clarification either (Nonini 2008). From the perspective of development theory, two controversial schools are relevant. An extreme neo-liberal school focuses on the preference for a self-regulating market with an extremely high level of entrepreneurial freedom and absolutely minimal state interference. It is characterized by far reaching deregulation; state activities are limited to just a few areas such as national defense and police (Hayek 1960, 1988). Meanwhile, the moderate liberal approach represented by the Freiburg school of national economy, deals with the ordo-liberal principles of regulating and constituting functions of the state. It is assumed that market forces need to be directed and limited by a comprehensive system of rules and regulations (Eucken 1990). The neo-liberal policy is one practical component of the present implementation of globalization, but it is poorly theoretically embedded. Globalization may also advance based on other economic foundations (Giddens 1998). Although many authors see the globalization effects as most positive (Stiglitz 2006); the most dramatic and negative consequence of the current neo-liberal policy is the increasing imbalance and inequality between social classes and nations (Rapley 2004). The global economic community does not take into account the fact that the starting positions of different nations are unequal as a result of historically determined events and politics, including colonialism, the slave trade and austerity policies dictated by the International Monetary Fund (Sen 1999). From the "postdevelopment" view after the Second World War development processes in the South were completely determined by the North (Escobar 1995; Goldman 2005). Increasingly, part of the developing countries is placed in a dependent and subordinate position, with a rapid decrease in the development options available. Economic growth is only feasible at the expense of social justice and subject to the short term overuse of natural resources. There are few opportunities to strive towards sustainable development targets, taking into account growth and distributional justice as well as a non-destructive management of natural resources. The underlying, unrealistic targets were mostly formulated by western countries. A discourse on this injustice, its socially and ecologically destructive consequences, and the need to reformulate these targets, is necessary in order to find a compromise in negotiations over common global goods like water, climate, biodiversity and forests (Ziegler 2002, 2005). The simple focus on market mechanisms and the espousing of the sustainable development principle as a superficial dogma can be interpreted as being the end of any development theory. An increasing number of studies focus on the limits of growth on a global level. In the meantime, however, creative, innovative and proactive theoretical models on the integration of developing countries in the future world economy are poorly developed (Preston 1996, p. 254). China and other former developing countries offer development aid in Africa, partly following rather mechanistic models (Michel and Beuret 2008). The changing power constellations on a global level will result in rather differing development options and chances for the future (Ferguson 2012). A respective discourse is necessary, including Eastern and Oriental thinking (Said 1978). Strategies on a green economy focus mostly on development in Western countries and barely address the consequences for developing countries (Jackson 2009; Fücks 2013).

The lack of adequate development models makes it difficult to implement social and ecological targets in rural development sustainably. The emerging paradigm has a new dimension, namely the supranational exchange of the production factors labor and capital, the run on limited land resources as well as increasing knowledge and commodity exchange. Mechanisms of self-compliance, corporate social responsibility and the drafting of a code of conduct under the global compact are not convincing if there are no legally binding rules (Humphreys 2009).

2.3.7.3 Tropical Forestry Development

After the Rio Conference 1992 the attention granted to tropical forestry issues gradually waned. Negotiations on the forest convention were without success because of controversial positions adopted with regard to the nature of tropical forests as a private or a common good. Although global deforestation declined, in most tropical countries forests were still being destroyed for alternative land uses, land speculation or for urbanization and large scale projects like dams. Poverty remained an important driving force of forest destruction and degradation. This negative development was overcome unexpectedly and rapidly with the increasing recognition of the importance of tropical forestry in climate change mitigation and in the context of the overall scarcity of and rising demand for forest products and services, which is related to the increasing world population. The substitution of fossil fuels by renewable sources of energy created an additional demand for wood as a source of energy, resulting in an increase in prices. Shifts can be detected in the organization of tropical forestry. Embedded in the institutional triangle between state, private actor and communities, a shift towards private actors and community or forest user group organization can be partly explained by the aforementioned neo-liberal influences. Many state forest administrations failed to implement forest policies benefiting rural and national development. They were not able to regulate access to forest products and services by the local population or to integrate international enterprises in forest management strategies ensuring adequate, long term revenues for the state. State forest administrations were gradually replaced by independent state forest enterprises operating according to a full cost calculation and controlling relatively independent budgets.

Mechanisms of deregulation and privatization have led to the outsourcing of many activities to private service enterprises, and in some cases to a partial delegation or even total privatization of forestry activities. The main objectives of these activities are cost reduction, and in line with this the reduction in the number of permanent employees. The negative consequences are lower social standards and poorly trained workers with greater incidences of accidents (Poschen 1997, 2001). In many post-socialist countries land reforms, which were often supported by the World Bank and other international organizations, led to the redistribution of former state forest. It was divided into small plots and handed over to small landowners. This adheres fully to the neo-liberal philosophy, but often has negative impacts for the quality and productivity of the forest resources. It is difficult to formulate management plans for small farm forest plots and to conduct sustainable forest management. The liberalization of access rights to land and forests on an international level has led to the growing establishment of large international forest enterprises in tropical countries. In a few model cases, relative transparency is assured by means of certification and open access to comprehensive documentation, as is at least partly the case for the Precious Woods shareholder company (Frattini 2006). The great majority of large forest enterprises working in the tropics, however, are rarely audited with respect to their management plans and destructive practices are widespread. In some countries use rights are delegated for the time span of one or two forest rotation cycles.

In parallel, rules for the use of forest under common property use regimes were developed further (Schmithüsen 1986). This 'third way' of granting rights to forest use to local communities is deemed to be a viable alternative to privatization. The important social function and, above all, the contribution towards poverty alleviation can be satisfied in conjunction with timber production and other market functions (FAO/DFID 2001). The corresponding devolution of forest management rights to local communities and the increasing participation of civil society in forestry decisions, having started with examples such as the forest user groups in Nepal, is extending to many tropical countries. This process can lead to the empowerment of local communities, but may also be counterbalanced by negative effects. Transition processes are often organized too superficially, decentralized forests become degraded and the need to initiate training and to plan the necessary funds are often overlooked. The decentralization of any form of forest use requires prior training and education, so that the communities or households affected are well prepared and informed with regard to the respective rules. The land or forest in question must also be in a state that permits the extraction of benefits sufficient to cover at least the cost of organization and management. Ostrom (1990) referred to the principle of congruence in this point; the activity must generate more outputs than the inputs required in their generation, usually in the form of labor and small investments. New forms of forest organization such as partnership agreements like joint forest management and leasehold forestry are important tools for rural development. A number of new instruments have emerged, most of which are rooted in market mechanisms. Examples are the mechanisms under the United Nations Convention on Climate Change, the clean development mechanism (CDM), reduced deforestation and degradation (REDD) and payment for environmental services (PES) schemes related to biodiversity, water and many forest products. These mechanisms are designed to implement the increasingly complex forest use and conservation systems in tandem. The strong orientation towards market forces puts an emphasis on products and services with market values. As a consequence, the increasing 'commoditization' of forestry may lead to an unbalanced valuation of forest services (Kosoy and Corbera 2010). Furthermore, recent experiences with PES demonstrate that because of the public good or open access character of many forest goods and services, the options to operate within market mechanisms are rather limited. Most of the payments for forest services are still based on state budget transactions (Chap. 9).

From the perspective of cultural determination and epistemology, forestry is extremely site-specific and locally embedded. With colonial expansion, valuable forest management traditions disappeared. Part of this traditional knowledge is being restored today. With the further expansion of global knowledge systems, the risk is the establishment of one dominant western scientific school of modern forestry and sustainability, which will again come to suppress local, experience-based knowledge.

2.3.7.4 Paradigm

Since the end of the cold war priority has been given to market mechanisms. A guiding theoretical foundation was neo-liberalism, which had a considerable impact on tropical forestry, with a trend towards privatization, the outsourcing of certain activities by forest enterprises to service providers and the improvement of accounting systems within state forest enterprises. The relationship between forestry and development suffers from a concentration of development thinking on market forces and related instruments. Overarching theoretical constructs are lacking and there is no solution as to how to proceed with international negotiations on global environmental problems. The dysfunctions between hypothetical neo-liberal market mechanisms and the reality of dependence and power relations has led to failures in international negotiations.

2.3.8 Lessons from the Paradigms and Their Changes

The six stages discussed may overlap and strict timeframes cannot always be assigned. They have the character of discourses and storylines and cannot always be exactly proved scientifically. Apart from this fuzziness, they follow entirely the retirements for paradigms outlined by Hunt (1989, p. 3): (1) each of the stages is backed by theoretical models explaining socio-economic development, more or less interrelated with tropical forestry development concepts; (2) all six stages are characterized by beliefs and perceptions of the empirical reality; (3) they are articulated by a clearly determined group of scholars and (4) the respective models have a strong link to practice in the field.

The transition between the stages discussed involves paradigm shifts, which characterize dynamics in theory development and demonstrate the linkages between theoretical approaches and practice, their strengths and weaknesses. The following interactions between the six historical stages and the corresponding paradigms demonstrate the reflective character of tropical forestry development. In the pre-colonial stage subsistence production was most common in rural areas, being linked to the use of multiple forest products and services. In colonial times imperialist policies dominated land use through control over land ownership, use rights, crop selection and forced labor. Forest production was geared towards the interests of the western colonizers and employed their techniques and technologies, although often these did not fit. Site-specific local knowledge was negated and lost. Tropical forest management mostly followed Western silvicultural models. With decolonization tropical forests were generally perceived as a financial resource to build up national economies. They were overexploited and destroyed on a large scale. The examples of many tropical countries in South East Asia and Africa demonstrate that the appropriation of natural resources by western countries extended beyond colonial times. The models on the non-applicability of the sustainability approach at this stage of economic growth proposed by Rostow and Zivnuka were propagated by international organizations and influenced a whole generation of tropical foresters. This influence is still present in many discourses taking place in tropical countries and contrasts with the sustainable development principle propagated today. Similarly, during the growth stage in most western industrial countries capital formation from natural resources was ranked higher than sustainable yield principles. In the 1970s the stage of modernization thinking followed, accompanied by large scale technocratic projects and top-down planning mechanisms. Tropical forestry was increasingly driven by international organizations. Multilateral international business employing large wood transformation units such as pulp, paper and plywood factories spread. Increasing negative ecological and social consequences of large scale interventions like colonization and land conversion projects led to the initiation of the stage of polarization. This stage is characterized by new power constellations, the gradual involvement of civil society actors in decisions on development and forest management together with the incubation of new ideas, institutions and technologies. With a strong reflective link to pre-colonial forestry, social forestry advanced and complemented industrial forestry by re-discovering traditional practices and local knowledge. The many new ideas found their expression in the Rio Conference, which also represented a turning point in terms of the importance of tropical forestry because agreement on a global forest convention was not reached and funding sources decreased. Many innovations from this era form a toolbox for future tropical forestry development: experience of public participation, ideas on community development as a third way, new production systems such as extractive reserves, joint forest management and forest user groups (Chap. 5). During the globalization stage the broad spectrum of development models is gradually replaced by a unique market model, following neo-liberal policies. In this stage a critical review of the modernization paradigm may contribute to finding adequate ways to address the bio-economy and the related large scale forest plantations. In the current globalization stage tropical forestry is undergoing a basic structural change, initiated by the polarization stage, under which it is now being re-oriented towards market mechanisms and partly supplemented by the further development of decentralized community and association structures. This current stage is characterized by a lack of new and innovative development theories. It is assumed that only theories which are embedded in the endogenous empirical reality permit the necessary discourse on development options. Theories are necessary because the complexity is reduced and the causal relationships are visible.

In summary, the following outcomes may be derived from the paradigm-based historical analysis:

Learning from paradigms

The six paradigms facilitate 'learning' in relation to the diversity of interpretations of tropical forestry development and the partly dysfunctional linkages between tropical forestry and development policies. Learning and understanding lay the groundwork for the development of future strategies and action. Experiences gleaned from different historical stages about the relationship between state and people, the consequences of different types of interventions, the success or failure of institution building, the involvement of different actor constellations and the overall contribution to rural development are transferable to other contexts. Often they are relevant for current decision-making, because similar situations and instruments appear. The review of past development stages provides insights into the consequences of theoretical approaches to development and their implementation.

Deficits in theory and model building

The transferability of experiences can only be ensured with a fundamental theory orientation (Weber 2012). Theories and models permit abstraction from a specific practical case. Tropical forestry was characterized by action based on site specific practices, which were, as demonstrated previously, embedded in development theories. The current situation is again characterized by the implementation of practical instruments but the theoretical discourse on an overall sector integration and on forestry development strategies is still weak. In the context of the present globalization stage this is especially critical because recent development discourses have focused primarily on market orientation, relying on simple market mechanisms. Tropical forestry has sought to adapt to this, but without a proper theory and merely adhering to instruments such as reduced deforestation and degradation and payments for environmental services steps ahead are hardly feasible. It is unrealistic to assume that the main driving mechanism for tropical forest management might solely be market-based, without a strong emphasis on the constituting and regulating principles of the state and communities.

External determination of tropical forestry

Most of the techniques and value systems in tropical forestry were introduced from western industrialized countries and often they followed a rather simple, linear development model (Fals-Borda and Mora-Osejo 2003). Much emphasis was placed on the principle of sustainable forest management. The applied perspective was very technical, with the European way determined by a strong preference for economic growth in line with sustainable forest management. This concept was even defended by international organizations in the 1960s. The simplified adoption of this European model of sustainability led to the imposition of forestry interventions not at all beneficial for local forest people in the tropics.

Short term waves and fashions determine the discourse on tropical forestry

The western domination of the discourse on tropical forestry is underlined by short term 'fashions', which dominate international cooperation and often bear little relation to local practice. New tropical forest management instruments emerge when the last instruments fall out of fashion (Pretzsch 2005). Most of these instruments are imposed on tropical countries from the outside. This may be justified to some extent, as institutional change needs to be induced from outside to overcome innovation barriers. Too often, however, local target groups are not involved in instrument development. Often they 'identify' with these instruments only a form of strategic behavior adopted to receive international funding.

Forest policies do not reach the people

The discourses initiated by western countries and adopted by international organizations often fail to reach local communities. This can be observed in relation to UN conventions such as the UNCBD and the UNFCCC especially. The objective of the policy level is to reach international environmental targets; the corresponding discourse takes place during international conferences and occupies staff at research institutions. Frequently, however, often it does not really contribute to the development of rural areas in the long term. Much of the recent discussion on sustainable development follows this rationale and does not tackle the problems facing the rural majority in many counties. Often the instruments are not integrated in an overall global forest development model.

2.4 Paths to the Future

It was demonstrated that in the past, tropical forestry mainly followed general development models and there was a lack of pro-active theory building. Many development models served to enrich tropical forestry, such as the dualism, basic needs and dependency approaches. Today the situation is more difficult. Since the 1990s the absolute preference for market orientation has led to a deficit in the generation of development models beyond market instruments. Learning from the past, and a reflective integration of the large variety of models practiced, may open the door to the development of approaches for the future; including some not adhering to the neoliberal mainstreams of current development policies.

The development of future concepts is challenging because the complexity of the factors influencing tropical forestry has increased considerably. The global population is rising towards about ten billion. The demand for food, energy and other nature-based services, like recreation, is increasing dramatically; especially to supply the growing urban population with forest products and services. With the need for the substitution of fossil sources of energy, there is a rising competition for land for use in the production of food, renewable sources of energy such oil palm and fast growing trees, timber and other industrial raw materials like rubber and gum. This competition between different products and services will continue to increase steadily and trade-offs may lead to greater conflict. The ecological challenges associated with climate change, biodiversity conservation and man-made catastrophes render ecological and social systems more vulnerable. As a consequence, future development will be much less predictable and even within the long planning horizons of the forestry sector flexible, adaptive structures will be required (Williams 2011). A future development strategy for tropical forestry must adhere to the principles of flexibility, keeping future options open and integrating all members of society and be based on rather different development models. Much greater emphasis must be placed on endogenous development strategies with a high level of local ownership (Escobar 2012).

A strong emphasis must also be placed on the involvement of all three of the actor groups that have proved to be important for the development of tropical forestry, namely (1) state institutions, (2) private, market-oriented actors and (3) organizations of the civil society.

1. The strong criticism of inefficient state administration and policy making has been outlined in detail. Nevertheless, state participation in tropical forestry is essential. First of all strong rules are required; rules that cannot be left solely to certification procedures and self-compliance. The state must implement strong constituting and regulating principles, especially with respect to tropical forestry (Eucken 1990; North 1990; 1991). These must be innovative and formulated in discussion with different actor groups. The national forest programs have shown that a dialogue concerning target setting in tropical forestry involving various actors from society is feasible in many situations. The multiple services provided by tropical forestry, especially in the field of poverty alleviation, in supporting health by providing medicinal products and also recreation should not be placed in the market domain and should be readily accessible to all members of society. State forestry should also be in a position to control price structures in natural forest logging and plantation management. Experience has shown that in the tropics the private sector has profited from extremely high windfall profits in the past, because state bureaucrats were unable to estimate stumpage values and rents. Practical experiences gleaned by the state in relation to tropical forestry are extremely important for strategic planning and agenda setting. Strong control over land use is necessary in order to avoid land grabbing and large scale bio-economy projects with little contribution to long term rural development. The lessons learnt from the plantation boom in 1970s and its far reaching failures are important.
2. In many rural areas of the tropics, markets do not work in the way assumed by the economic theory. There is no transparency, competition is lacking and transaction costs are extremely high. Furthermore, because of the nature of traditional property rights, often land is not transferable. In this situation it is important that the local actors engage in the market and learn about its functioning. Endogenous value chain development can contribute to the formation of local capital and the establishment of adapted transformation units. The engagement of private tropical forestry concerns should focus on local strategies and a corresponding empowerment of local market actors. Small farms continue to play an important role in tropical countries (Killick 2001; Hazell et al. 2007). Often forestry and trees are part of the farm and, compared with plantations managed by large enterprises, the biodiversity is high. With the ongoing globalization processes and the lowering of international trade barriers the future for these farms is uncertain. Although the subsistence component is often considerable, these farms still host a large part of the world food production and provide products and services for the local economy especially. New forms of social capital need to be developed as a coping strategy for the survival of these units. This capital might be based on a further development of traditional cooperatives and user groups to allow rural areas to compete with urban areas in economic terms. The current tendency towards an increasing influence of the global market leads to the destruction of local institutions and social structures,

because they cannot compete with the external market power. On the other hand, there are huge market potentials in tropical forestry, which can only be developed with integration in international markets.
3. A significant part of the global forest area is managed by communities under the traditional common property regime, the patrimonium. Over a long period of time the state sought to integrate this type of management within the conventional state administration. In most countries where this has occurred it has not produced any solutions; often the actors felt suppressed or controlled and the motivation to participate in sustainable forest management was rather limited. In some countries, like India, cooperation models between the state and communities were implemented (Poffenberger qne McGean 1996). Increasingly theories relating to rules and structures for community development were developed (Oakerson 1992). Experiences are rapidly growing about this third way of development with the roots in the works of Ostrom (1990) and their further extension in different networks like the program on Collective Action and Property Rights (CAPRI; www.capri.cigiar.org). In forestry the rather specific acquisition rights towards land, land cover, products and services have to be taken in account (Agrarwal and Ostrom 2001).

The following lessons should also be taken on board for future strategy development:

Endogenous and integrative tropical forest management strategies

In most stages of tropical forestry development local people were not involved in decision-making to any significant degree. In future-oriented forestry development strategies co-management with all of the groups involved is essential. The focus must be put on local knowledge and strategies so as to avoid the multiple failures associated with external interventions (Fals-Borda and Mora-Osejo 2003; Escobar 2012). The diversity of the local techniques and technologies available represents a valuable foundation for improvements with the highest possible degree of flexibility. The interaction between state, private and community forestry must be extended further in the form of partnerships and associations.

Reduce the gap between global policies and local people

In forestry communication between public administrations and local people has always been difficult and characterized by an asymmetry in the flow of information, different understandings and interpretations of messages and limitations in the diffusion of information. The gap between global discourses and conventions, and their implementation on a local level, continues to increase. On the local level better cooperation instruments need to be developed; these must permit the integration of local narratives, images and positions (Roe 1994). The creation of socio-economic field laboratories can contribute to a better interaction between the two levels (Lindner and Pretzsch 2013). First of all, however, the at times utopian global discourses must be brought down to a tangible level.

Theory building

Tropical forestry suffers from a lack of theoretical orientation. This often results in merely instrumental approaches that are not integrated in overarching rural development strategies and models. Given the current market orientation, the lack of theory building leads to a 'reductionist' view of rural development and as a result needs are barely addressed at a practical level. Prominent example is the dominance of market approaches in the implementation of the United Nations' Framework Convention on Climate Change. Innovative theory building integrating different development models and especially different types of communities as actor groups is necessary.

Pro-active agenda setting

Although forests cover around 30 % of the global land area, in many cases the development of forest management strategies was a reaction to the emergence of framework conditions such as development policies or followed on from short term trends initiated by international organizations or resulting from policies. Few important impulses, and hardly any agenda setting, were initiated by the forestry community. Pro-active and constructivist engagement in model building for problem solving is an urgent task for the future (Berger and Luckman 1967).

Flexible planning and adaptive management

Lessons learned from the six tropical forestry paradigms have shown that often tropical forestry was planned in a technocratic way over long periods of time. Although rich experiences have been made in the domain of social forestry, this is only now slowly also becoming the case for other sectors and in state forest administration. The negative experiences of large scale planning in the 1970s must be evaluated carefully, especially as they relate to similar modern initiatives such as bio-economy projects. This inflexible form of planning is currently evident in rubber plantation projects in South East Asia, for example. An important objective of flexible and adaptive planning in tropical forestry is to maintain development options for the future (Williams 2011).

Building social capital

The complexity of forest management is increasing while human behavior is becoming ever more individualistic. This is partly due to the emergence of the information society, which substitutes traditional social family links (Sennet 2012). On the other hand strategies are needed to cope with the increasing vulnerability of small production units, brought about by market volatility or climate change effects. Individual forest farms hardly will be able adapt individually; collective action requires the formation of new forms of social capital.

Integration in a green economy

The task for the future is to initiate a next step in theory building and in practice, towards the integration of tropical forestry in a green economy (Jackson 2009; UNEP 2011). Theory building is important given the emergent interactions of the forestry sector with medicine and health, energy and recreation. Forestry provides an increasing variety of raw materials for different industries and the development of corresponding value chains can prompt rural development. Furthermore, the dynamic relationship between the rural and the urban requires further theoretical foundations, which must be integrated in development models. Cultural practices and experiences enrich forestry and rural development especially on a local, site specific level. Experiences from the

East as well as from the orient require more attendance. Green economies have to be developed in a "pluriverse" way (Escobar 2012), based on the huge amount of development options, available on a local, regional and national level.

References

Agrarwal A, Ostrom E (2001) Collective action, property rights and the devolution of forest and protected area management. In: Meinzen-Dick R, Knox A, Di Gregorio M (eds) Collective action, property rights and devolution of natural resource management. Exchange of knowledge and implications for policy. Proceedings of the international conference held from 21–25 June in Puerto Azul. DSE/ZEL Feldafing

Agrawal A (1995) Dismantling the divide between indigenous and scientific knowledge. Develop Change 26:413–439

Ashley C, Maxwell S (2001) Rethinking rural development. Dev Policy Rev 19(4):395–425

Bargatzky T (1986) Einführung in die Kulturökologie. Umwelt, Kultur und Gesellschaft. Reimer, Berlin

Beck U (1986) Risikogesellschaft. Auf dem Weg in eine andere Moderne. Suhrkamp, Frankfurt

Beck U (2007) Weltrisikogesellschaft. Auf der Suche nach der verlorenen Sicherheit. Suhrkamp, Frankfurt

Bennett JW (1976) The ecological transition: cultural anthropology and human adaptation. Pergamon, New York

Berger PL, Luckman T (1967) The social construction of reality. A treatise in the sociology of knowledge. Doubleday and Middlesex, Garden City

Berkes F (2008) Sacred ecology, 2nd edn. Routledge, New York/London

Bernstein H (ed) (1973) Underdevelopment and development. The third world today. Penguin, Harmondsworth

Bernstein H, Byres T (2001) From peasant studies to agrarian change. J Agrar Change 1(1):1–56

Bizikova L (2011) Understanding the contribution of the environment to human well-being: a review of literature. IISD Report, International Institute for Sustainable development, Winnipeg

Blaikie P (1985) The political economy of soil erosion in developing countries. Longman, London

Blank W (2006) Die kolonialwirtschaftliche Forstnutzung der ehemaligen Kolonie Deutsch-Ostafrika. Masterarbeit Technische Universität Dresden; Institut für Internationale Forst- und Holzwirtschaft, Dresden

Boeke JH (1958) Economics and economic policy in dual societies. Institute of Pacific Relations, New York

Boserup E (1965) The conditions of agricultural growth. Aldine, Chicago

Bryant RL (1997) The political ecology of forestry in Burma 1824–1994. Hurst, London

Bryant RL (1998) Power, knowledge and political ecology in the third world: a review. Prog Phys Geogr 22(1):79–94

Bryant RL, Bailey S (1997) Third world political ecology. Routledge, London

Bryant D, Nielsen D, Tengley L (1997) The last frontier forests. Ecosystems and economies on the edge. World Resources Institute, Washington

Carlowitz CV (1713) Sylviculture Oeconomica oder haußwithliche nachricht und Naturgemäßige Anweisung zur Wilden Baum-Zucht, Johann Friedrich Braun 2. Leipzig

Carney D (2003) Sustainable livelihoods approaches: progress and possibilities for change. Finesse Print, Toronto

Chambers R (1983) Rural development: putting the last first. Longman, London/Lagos/New York

Chenery HB (1979) Structural change and development policy. Oxford University Press, New York/Oxford

Choussudovsky M (1997) The globalisation of poverty: impacts of IMF and World Bank reforms. Third World Network, Penang
Cooper F, Packard R (2005) The history and politics of development knowledge. In: The anthropology of development and globalization. From classical political economy to contemporary neoliberalism. Blackwell, Malden/Oxford/Carlton, pp 126–139
Cowen MP, Shenton RW (1996) Doctrines of development. Routledge, London/New York
Cubbage FW, O'Laughlin J, Bullock CS III (1993) Forest resource policy. Wiley, New York
Cypher JM, Dietz JL (1997) The process of economic development. Routledge, London
Dawkins HC, Philip MS (1998) Tropical moist forest silviculture and management. A history of success and failure. CAB International, Oxon/New York
De Groot RS (1992) Functions of nature. Evaluation of nature in environmental planning, management and decision making. Wolters-Nordhoff, Groningen
Demesse L (1992) Technique et économie des Pygmées Babinga. Insitut d'Ethnologie, Musée de l'Homme, Paris
Detten RV (2011) Sustainability as a guideline for strategic planning? The problem of long-term forest management in the face of uncertainty. Eur J Forest Res 130:451–465
Diamond J (2005) Collapse. How societies choose to fail or to succeed. Viking, New York
Douglas JJ (1983) A re-appraisal of forestry development in developing countries, Forestry sciences 8. Martinus Nijhoff/Dr. Junk Publishers. The Hague
Douglas J, Simula M (2010) The future of the world's forests. Ideas vs. ideologies. Springer, Dordrecht
Ellefson PE (1992) Forest resources policy. Process participants and programs. McGraw-Hill, New York
Ellis F (1998) Household strategies and rural livelihood diversification. J Dev Stud 35:1–38
Ellis F, Biggs S (2001) Evolving themes in rural development 1950s–2000s. Dev Policy Rev 19 (4):437–448
Escobar A (1988) Power and visibility: development and the intervention and management of the third world. Cult Anthropol 3(4):428–443
Escobar A (2012) Encountering development, 2nd edn. Princeton University Press, Princeton and Oxford
Eucken W (1990) Grundsätze der Wirtschaftspolitik, 6th edn. J.C.B. Mohr, Tübingen
Fairhead J, Leach M (1995) False forest history, complicit social analysis: rethinking some West African environmental narratives. World Dev 23(6):1023–1035
Fals-Borda O, Mora-Osejo LE (2003) Context and diffusion of knowledge. A critique of eurocentrism. Act Res 1(1):29–37
FAO (1989) Forestry and food security, FAO forestry paper 90. FAO, Rome
FAO (2001) Global forest resources assessment 2000. Main report, FAO forestry paper 140. Rome
FAO (2010) Global forest resources assessment 2010. Main report, FAO forestry paper 163. Rome
FAO/DFID (2001) How forests can reduce poverty. FAO, Rome
Farah HI, Vasapollo L, coordinadores (2011) Vivir bien: Paradigma no capitalista? CIDES/UMSA, La Paz
Fearnside PM, Rankin JM (1982) The new Jari: risks and prospects of a major Amazonian development. Interciencia 7(6):329–339
Fearnside PM, Rankin JM (1985) Jari revisisted: changes and the outlook for sustainability in Amazonia's largest silvicultural Estate. Interciencia 10(3):121–129
Ferguson N (2012) Civilization. The six killer apps of Western power. Penguin Books, London
Fieldhouse DK (1965) Die Kolonialreiche seit dem 18. Jahrhundert. Fischer Weltgeschichte, Bd 29. Fischer, Frankfurt a. M.
Frank AG (1977) L'accumulación mondiale 1500–1800. Calmann-Lévy, Paris
Frattini M (2006) Dauerhafte Naturwaldbewirtschaftung in Amazonien: Perspektive für Holz verarbeitende Unternehmen? Schriftenreihen des Instituts für Internationale Forst- und Holzwirtschaft 12, Technische Universität Dresden, Tharandt
Fücks R (2013) Intelligent Wachsen. Die Grüne Revolution. Hanser, München

Fuller S (2003) The struggle for the soul of science. Kuhn vs. Popper. Icon, Cambridge/Toronto
Fürstenberg P von (1966) Formen der Waldnutzung im spanischen Nord- und Mittelamerika während der Kolonialzeit (16. – 18. Jahrhundert): dargestellt an Beispielen aus dem ehemaligen Vizekönigreich Neu-Spanien einschließlich der Audiencias Guatemala und Santo Domingo. Diss., Universität Göttingen
Furtado C (1984) Cultura e desenvolvimento em época de crise. Paz e Terra, Rio de Janeiro
Gadgil M, Berkes F (1991) Traditional resource management systems. Resour Manage Optim 8 (3/4):127–141
Gadgil M, Guha R (1992) This fissured land. An ecological history of India. Oxford University Press, New Delhi
Gadgil M, Guha R (1994) Ecological conflicts and the environmental movement in India. In: Ghai D (ed) Development and environment sustaining people and nature. UNRISD, Oxford, pp 101–136
Galston W, Baehler HJ (1995) Rural development in the United States. Connecting theory, practice, and possibilities. Covelo, Washington, DC
Galtung J, O'Brien P, Preiswerk R (eds) (1980) Self-reliance: a strategy for development. Institute of Development Studies, Geneva
Giddens A (1998) The third way. The renewal of social democracy. Polity Press, Cornwall
Glück P, Oesten G, Schanz H, Volz K-R (1999) Formulation and implementation of national forest programmes. Vol. I: theoretical aspects; Vol: II: state of the art in Europe; vol. III: international experiences; EFI Proceedings No. 30, European Forest Institute, Joensuu
Goldman M (2005) Imperial nature. The World Bank and struggles for social justice in the age of globalization. Yale University Press, New Haven/London
Habermas J (1988) Theorie des kommunikativen Handelns, Bd. 1 and 2. Suhrkamp, Frankfurt
Hagen EE (1975) The economics of development. R.D. Irwin, Homewood, rev. Ed
Hanewinkel M (2011) Multifunktionalität des Waldes. In: Eidgenössische Forschungsanstalt WSL, Birmensdorf, Forum für Wissen S. 7–14
Hardin G (1968) The tragedy of the commons. Science 162:1243–1248
Hayek F (1960) The constitution of liberty. Routledge & Kegan Paul, London
Hayek F (1988) The fatal conceit: the errors of socialism. Routledge, London
Hazell P, Poulton C, Wiggins S, Dorward A (2007) The future of small farms for poverty reduction and growth. International Food Policy Research Institute, 2020 Discussion paper 42, Washington, DC
Heilig GK (1996) How many people can be fed on earth? In: Lutz W (ed), The future population of the world. What can we assume today? Revised edition. Earthscan, London, pp 196–249
Hirschman AO (1958) The strategy of economic development. Yale University Press, New Haven
Hobsbawm EJ (1987) The age of empire 1875–1914. Weidenfeld and Nicolson, London
Hoggart K (1990) Let's do away with rural. J Rural Stud 6(3):245–257
Humphreys D (1996) Forest politics. The evolution of international cooperation. Earthscan, London
Humphreys D (2006) Logjam. Deforestation and the crisis of global governance. Earthscan, London
Humphreys D (2009) From corporate social responsibility to the democratic regulation of transnational corporations. Int J Environ Cult Econ Soc Sust 5(4):208–217
Hunt D (1989) Economic theories of development. An analysis of competing paradigms. Harvester Wheatsheaf, Hertfordshire
ILO International Labour Organisation (1977) The basic needs approach to development. International Labour Office, Geneva
Jackson T (2009) Prosperity without growth? The transition to a sustainable economy. Sustainable Development Commission, London
James L (1998) The rise and fall of the British empire. (First published 1994). Abacus, London
Kay C (2008) Reflections on Latin American rural studies in the neoliberal globalization period: a new rurality? Dev Change 39(6):915–943

Killick T (2001) Globalization and the rural poor. Dev Policy Rev 19(2):155–180
Kolk A (1996) Forests in international environmental politics. Environmental Organizations, NGOs and the Brazilian Amazon. International Books, Utrecht
Konetzke R (1956) Süd- und Mittelamerika I. Die Indianerkulturen Altamerikas und die spanisch-portugiesische Kolonialherrschaft. Fischer Weltgeschichte, Band 22 (Edition 1986). Fischer Taschenbuch verlag, Frankfurt a. M
Kosoy N, Corbera E (2010) Payment for ecosystem services as commodity fetichism. Ecol Econ 69:1228–1236
Kuhn TS (1962) The structure of scientific revolutions. University of Chicago Press, Chicago
Kuznets S (1963) Notes on the take-off. In: The economics of take-off into sustained growth. Macmillan, London, pp 22–43
Lakatos I (1970) Falsification and the methodology of scientific research programmes. In: Lakatos I, Musgrave AP (eds) Criticism and the growth of knowledge. Proceedings of the international colloquium in the philosophy of science, London, 1965, vol 4. Cambridge University Press, pp 91–196
Lamprecht H (1988) Über Grenzen und Möglichkeiten forstlicher Entwicklungshilfe in den Feuchttropen, Allg. Forst- u. J.-Ztg 159(6):103–107
Lamprecht H (1990) Silviculture in the tropics, Gesellschaft für Technische Zusammenarbeit. TZ Verlag, Rossdorf
Lane J (1995) Non-governmental organizations and participatory development: the concept in theory versus the concept in practice. In: Wright S, Nelson N (eds) Power and participatory development: theory and practice. London, pp 181–191
Leeuwis C, van den Ban A (2004) Communication for rural innovation: rethinking agricultural extension, 3rd edn. Blackwell Science, Oxford, pp 22–48; 55–56; 247–259; 292–301
Lindner A, Pretzsch J (2013) An international network on climate change impacts on small farmers in the tropical Andes – global conventions from a local perspective. Sust Agr Res 2(2):92–98
Lins C (1994) Jari. 70 Anos de História, 2nd Edn. Dataforma, Rio de Janeiro
Little PE (1999) Environments and environmentalisms in anthropological research. Annu Rev Anthropol 28:253–284
Ljungman L, Nair CTS (2001) Changing perceptions on technical assistance in support of sustainable forest management. In: Workshop-paper on financing sustainable forest management. Oslo, 22–25 Jan 2001, available (accessed Nov. 2013) http:/www.cifor.cgiar.org/fsfm/Papers/01Ljungman.pdf
Loth H (1981) Sklaverei. Die Geschichte des Sklavenhandels zwischen Afrika und Amerika. Hammer, Wuppertal
Mandel E (1962) Traité d'economie marxiste. Julliard, Paris
Masselos JE (2010) Imperien Asiens. Von den alten Khmer bis zu den Meiji. Theiss, Stuttgart
McCarthy J, Prudham S (2003) Neoliberal nature and the nature of neoliberalism. Geoforum, http://dx.doi.org/10.1016/J.geoforum.2003.07.003
McMichael P (2006) Peasant prospects in the neoliberal age. New Polit Econ 11(3):407–417
Meidinger E, Elliott C, Oesten G (eds) (2003) Social and political dimensions of forest certification. Kessel, Remagen-Oberwinter
Michel S, Beuret M (2008) La Chinafrique Pékin à la conquête du continent noir. Grasset, Paris
Millenium Ecosystem Assessment (2005) Ecosystems and human well-being: biodiversity synthesis. World Resources Institute, Washington, DC
Miller B (1992) Collective action and rational choice: place, community, and the limits to individual self-interest. Econ Geogr 68:22–42
Minsch J (1992) Ethik der Ressourcennutzung – ökonomische Aspekte. In: Schweizer Zeitung für das gesammelte Forstwesen. 143:893–907
Mommsen WJ (1987) Imperialismustheorien, 3rd edn. Vandenhoeck & Ruprecht, Göttingen
Moseley MJ (2003) Rural development. Principles and practice. Sage, London/Thousand Oaks/New Dehli

Nederween Pieterse J (2001) Development theory. Deconstructions/reconstructions. Sage, London/Thousand Oaks/New Dehli
Neugebauer B (1986) Der Wandel kleinbäuerlicher Landnutzung in Oxkutzcab, Yucatán. Todtmoos
Nonini DM (2008) Is China becoming neoliberal? Crit Anthropol 28:145–170
Norgaard RB (1994) Development betrayed the end of progress and a coevolutionary revisioning of the future. Routledge, London/New York
North DC (1990) Institutions, institutional change and economic performance. Cambridge University Press, Cambridge
North DC (1991) Institutions. J Econ Perspect 5(1):97–112
Nurkse R (1953) Problems of capital formation in underdeveloped countries. Blackwell, Oxford
Oakerson JR (1992) Analysing the commons. A framework. In: Bromley DW (ed) Making the commons work. Theory, practice and policy. ICS Press, San Francisco, pp 41–59
Osterhammel J (1997) Colonialism: a theoretical overview. Markus Wiener Publishers, Princeton
Osterhammel J (2009) Die Verwandlung der Welt. Eine Geschichte des 19. Jahrhunderts. Beck, München
Ostrom E (1990) Governing the commons: the evolution of institutions for collective action. Cambridge University Press, New York
Palo M (1988) The forest-based development theory revisited with a case study of Finland and prospects for developing countries. In: Deforestation or development in the third world? Volume II. Finnish Forest Research Institute, Helsinki, pp 13–157
Palo M, Mery G (eds) (1996) Sustainable forest challenges for developing countries. Kluwer, Boston/London/Dordrecht
Palo M, Uusivuori J (eds) (1999) World forests, society and environment, vol I, World forests. Kluwer, Dordrecht/Boston/London
Palo M, Vanhanen H (eds) (2000) World forests from deforestation to transition? vol II, World forests. Kluwer, Dordrecht/Boston/London
Palo M, Uusivuori J, Mery G (eds) (2001) World forests. Markets and policies, vol III, World forests. Kluwer, Dordrecht/Boston/London
Parotta JA, Trosper RL (2012) Traditional forest-related knowledge: sustaining communities, ecosystems and biocultural diversity, vol 12, World Forest. Springer, Heidelberg
Payer C (1991) Lent and lost. Foreign credit and third world development. Zed Books, London/New Jersey
Peet R, Watts M (1993) Introduction: development theory and environment in an age of market triumphalism. Econ Geogr 69(3):227–235
Peet R, Watts M (1996) Liberation ecologies. Environment, development, social movements. Routledge, London/New York
Peluso N (1992) Rich forests, poor people. Resource control and resistance in Java. Berkley, Los Angeles/London
Perroux F (1976) Les enterprises transnacionales et le nouvel ordre économique du monde. Institut des Sciences Sociales Appliquées, Lyon, p 112
Poffenberger M, McGean B (eds) (1996) Village voices, forest choices. Joint forest management in India. Oxford University Press, Oxford
Polanyi K (1944) The great transformation. Octagon Books, New York
Ponting C (1991) A green history of the world. The environment and the collapse of great civilisations. Penguin, New York
Poschen P (1997) 'Forests and employment – much more than meets the eye'. In: Invited paper XI world forestry congress, Antalya, 13–22 Oct 1997, proceedings vol 4, p 61–79
Poschen P (2001) 'Globalisation and sustainability – the forestry and wood industries on the move' with Mattias Lövgren. ILO, Geneva
Posey D (1985) Indigenous management of tropical forest ecosystems: the case of the Kayapo Indians of the Brazilian Amazon. Agroforest Syst 3:139–158

Prebisch R (1950) The economic development of Latin America and its principal problems. Lake Success, New York
Prebisch R (1980) Towards a theory of change. CEPAL Rev 10:155–208
Preston PW (1996) Development theory, an introduction. Blackwell Publishers, Oxford
Preston PW (2011; first published 1987) Rethinking development. Essays on development and Southeast Asia. Routledge library editions: development, vol 109, Abingdon
Pretzsch J (1986) Traditionelle Bodenbewirtschaftung, weltmarktorientierte Plantagenproduktion und Tropenwaldzerstörung in der Republik Elfenbeinküste. Eine Untersuchung über die Folgewirkungen einseitig exportorientierter Landnutzung auf lokale Gesellschaftsstrukturen, Produktionsverhältnisse und Ökosysteme. Schriftenreihe des Instituts für Landespflege, Heft 8, Freiburg
Pretzsch J (1987a) Die Entwicklungsbeiträge von Holzexploitation und Holzindustrie in Ländern der feuchten Tropen. Dargestellt am Beispiel der Elfenbeinküste. Schriftenreihe des Instituts für Landespflege, Heft 11, Freiburg
Pretzsch J (1987b) Sichtweise und Nutzung des tropischen Feuchtwaldes in traditionellen Kulturen, unter Kolonialverwaltung und nach der Unabhängigkeit. Forschungsauftrag der FESt, Heidelberg unpublished manuscript
Pretzsch J (1990) Contributions to the development of timber exploitation and the wood-processing industry in countries of the humid tropics. The case of Ivory Coast. In: Natural resources and development 31:7–21
Pretzsch J (1994) Kontinuität und Bruch. Tropenforstliche Ausbildung und Forschung in Tharandt. Der Wald Berlin 44(5):148–152
Pretzsch J (1998) Chances and limits of tropical forest management from the socio-economic view: Status quo and perspectives. In: Plant research and development 47/48:88–97
Pretzsch (2003) Methodological aspects of tropical forest management research. Contribution to the XII world forestry congress, Quebec City, http://www.fao.org/DOCREP/Article/WFC/XII/0791-C4.HTM. Last access 5 Jan 2013
Pretzsch J (2005) Forest related rural livelihood strategies in national and global development. Forest Trees Livelihoods 15:115–127
Pye O (2005) Khor Jor Kor. Forest politics in Thailand. White Lotus Press, Bangkok
Rapley J (2004) Globalization and inequality. Neoliberalism's downward spiral. Lynne Rienner, Boulder/London
Rehbein B, Schwengel H (2012) Theorien der Globalisierung, 2. Aufl. UTB, Konstanz
Repetto R (1988) The forest for the trees? Government policies and the misuse of forest resources. World Resources Institute, Washington, DC
Repetto R, Gillis M (eds) (1988) Public policies and the misuse of forest resources. Cambridge University Press, Cambridge
Ribeiro D (1983) Der zivilisatorische Prozeß. Suhrkamp, Frankfurt
Rist G (2002) The history of development from Western origins to global faith, 2nd edn. Zed Books, London/New York
Ritter E, Dauksta D (eds) (2011) New perspectives on people and forests. Springer, Heidelberg/London/New York/Dordrecht
Roe E (1994) Narrative policy analysis. Theory and practice. Duke University Press, Durham/London
Rogers EM (2003) Diffusion of innovations, 5th edn. Free Press, New York
Romm J (1986) Forest policy and development policy. J World Forest Resour Manage 2:85–103
Rosenstein-Rodan PN (1961) International aid for underdeveloped countries. In: Review of economics and statistics XLIII:107–138
Rostow WW (1960) The stages of economic growth: a non-communist manifesto. Cambridge University Press, Cambridge
Rudel TK, Coomes OT, Moran E, Achard F, Angelsen A, Xu J, Lambin E (2005) Forest transitions: towards a global understanding of land use change. Glob Environ Change 15:23–31

Russel CE (1987) Plantation forestry. In: Jordan CF (ed) Amazonian rain forests. Springer Verlag, New York, pp 76–89
Ruttan VW (1984) Integrated rural development programmes: a historical perspective. World Dev 12(4):393–401
Sachs J (2006) The end of poverty. Economic possibilities for our time. Penguin Press, New York
Said EW (1978) Orientalism. Routledge & Kegan Paul, London
Schmithüsen F (1986) The impact of land-tenure on rural forestry development. In: Schmithüsen F (ed) Forestry legislation. Report of the IUFRO working party S4.06.-04, Forstwissenschaftliche Beiträge. ETH-Zentrum, Zürich, pp 121–140
Schuhmacher EF (1973) Small is beautiful. A study of economics as if people mattered. Blond and Briggs, London
Scott JC (1985) Weapons of the weak. Everyday forms of peasant resistance. Yale University Press, New Haven/London
Seers D (1979) The meaning of development. In: Lehmann D (ed) Development theory – four critical studies. Bournemouth, pp 9–33
Seligson MA, Passé-Smith JT (eds) (2003) Development and underdevelopment. The political economy of global inequality. Lynne Rienner Publishers, Boulder
Sen A (1999) Development as freedom. Oxford university press, Oxford
Sennet R (2012) Zusammenarbeit. Was unsere Gesellschaft zusammenhält. Hanser Berlin, Berlin
Shiva V (1989) Staying alive: women, ecology and development. Zed Books, London
Sigrist C (1979) Regulierte Anarchie. Untersuchungen zum Fehlen und zur Entstehung politischer Herrschaft in segmentären Gesellschaften Afrikas. Syndicat, Frankfurt
Silva DE (1997) Why institutional reforms in forestry? Lessons from international experience. Nat Res Forum 21:51–60
Sponsel L (ed) (1995) Indigenous peoples and the future of Amazonia. An ecological anthropology of an endangered world. University of Arizona Press, Tucson/London
Steinlin H, Pretzsch J (1984) Der tropische Feuchtwald in der internationalen Forstpolitik. In: Holz-Zentralblatt 110(138):20–20
Steward JH (1972) Theory of culture change: the methodology of multilinear evolution. University of Illinois Press, Urbana
Stiglitz JE (2006) Making globalization work. New York/London
Streeten P, Burki SJ (1978) Basic needs: some issues. World Bank Reprint. Series 53. World Development 6, Washington, DC
Sunderlin WD, Dewi S, Puntodewo A, Müller D, Angelsen A, Epprecht M (2008) Why forests are important for global poverty alleviation: a spatial explanation. Ecol Soc 13(2):24 (online, last access Nov. 2013) :http:/www.ecologyandsociety.org/vol13/iss2/art24/
Szulecka J, Pretzsch J, Secco L (2013) Paradigms in tropical forest plantations: a critical reflection on historical shifts in plantation approaches. International Forestry review, Vol. 15 (4) in press
Thirlwall AP (1994) Growth and development, 5th ed. MacMillan Press TTD, London
Thorsen DE (2011) The politics of freedom. A study of the political thought of Isaiah Berlin and Karl Popper, and the challenges of Neoliberalism. Ph.D. thesis, Dept. of Political Science, University of Oslo, Oslo
Thünen v JH (1966) Thünen's isolated state (trans: Hall P). Pergamon, London (first German edition 1826)
Todaro MP (1995) Economics for a developing world. An introduction to principles, problems and policies for development, 3rd edn. London/New York, Longman
Todaro MP (1997) Economic development, 6th edn. New York, Longman
Toynbee A (1998) Menschheit und Mutter Erde. Die Geschichte der großen Zivilisationen. Ullstein, Berloin
Tucker RP, Richards JF (eds) (1983) Global deforestation and the nineteenth-century world economy. Duke University Press, Durham
Uibrig H (2007) Chronik des Tharandter Tropenforst- und Holzwirtschaftlichen Insitutes 1963–2007, Schriftenreihe des Institutes für Internationale Forst- und Holzwirtschaft. Tharandt

UNEP (2011) Green Economy Report. Towards a green economy: pathways to sustainable development and poverty eradiction. http://www.unep.org/greeneconomy/GreenEconomyReport. Pree release 2011, last access October 2013
Uphoff N (1993) Grassroots organizations and NGOs in rural development: Opportunities with diminishing states and expanding markets. World Dev 21(4):607–622
Vincent JR (1990) Rent capture and the feasibility of tropical forest management. Land Econ 66 (2):212–223
Vincent JR, Binkley CS (1992) Forest-based industrialisation: a dynamic perspective. In: Sharma NP (Hrs) Managing the world's forests: looking for balance between conservation and development. Kendall/Hunt Publishing Company, Dubuque, pp 93–175
Walker PA (2005) Political ecology: where is the ecology? Prog Hum Geogr 29:73–82
Walker PA (2006) Political ecology: where is the policy? Prog Hum Geogr 30:382–395
Warren DM, Slikkerveer LJ, Brokensha D (1995) The cultural dimension of development. Indigenous knowledge systems. Intermediate Technology Publications, London
Weber M (1922) Wirtschaft und Gesellschaft. Verlag. J. C. B. Mohr, Tübingen
Weber N (2012) Reflections on theories in forestry policy: testing, combining or building? Forest Policy Econ 16:102–108
Weischet W, Caviedes CN (1993) The persisting ecological constraints of tropical agriculture. Longman Scientific & Technical, Essex
Westoby JC (1962) Forest industries in the attack on economic underdevelopment. UNASYLVA 16(4):168–201
Westoby (1987) The purpose of forests. Follies of development. Basil Blackwell Inc. New York, Oxford
Wiggins S, Proctor S (2001) How special are rural areas? The economic implications of location for rural development. Dev Policy Rev 19(4):427–436
Williams BK (2011) Adaptive management of natural resources – framework and issues. J Environ Manage 92:1346–1353
Winterbottom R (1990) Taking stock: the tropical forestry action plan after five years. World Resources Institute, Washington, DC
Wunder S (2001) Poverty alleviation and tropical forests – what scope for synergies? World Dev 29:1–17
Ziegler J (2002) Les nouveaux Maitres du Monde et ceux qui leur resistante. Fayard, Paris
Ziegler J (2005) L'Empire de la honte. Fayard, Paris
Zivnuska JA (1966) The integration of forest development plans and national development plans. How to make the forestry case at the national level. In: Proceedings of the 6th World Forest Congress, Madrid

Chapter 3
The Challenges Facing Forest-Based Rural Development in the Tropics and Subtropics

Dietrich Darr, Marolyn Vidaurre, Holm Uibrig, Andre Lindner, Eckhard Auch, and Klaus Ackermann

Abstract In spite of decades of unprecedented exploitation of resources, rural areas in the tropics and subtropics still possess vast human, biological, mineral, land, forest and other natural resources that potentially represent a primary source of human and economic development. However, the benefits produced by past development programs and policies often bypassed rural areas. The apparent failure of societies to develop rural areas to the same extent as their urban and metropolitan centers has largely been attributed to a complex set of factors at the level of the natural and the social sub-system at local, national and global scales. In this chapter the authors attempt to provide a comprehensive review of the manifold dimensions of the rural development challenge in forest areas of tropical and subtropical countries. In so doing, the authors draw heavily on empirical evidence published in the pertinent scientific literature.

Keywords Environmental factors • Deforestation • Demographic trends • Infrastructure development • Forest governance

---

D. Darr (✉)
Faculty of Life Sciences, Hochschule Rhein-Waal, Marie-Curie-Str. 1, Kleve 47533, Germany
e-mail: dietrich.darr@hochschule-rhein-waal.de

M. Vidaurre • H. Uibrig • A. Lindner • E. Auch
Technische Universität Dresden, Institute of International Forestry and Forest Products, 1117, Tharandt 01735, Germany
e-mail: marolynvida@gmail.com; holm.uibrig@forst.tu-dresden.de; andre.lindner1@mailbox.tu-dresden.de; eckhard.auch@tu-dresden.de

K. Ackermann
Gesellschaft für Internationale Zusammenarbeit (GIZ) GmbH, Godesbergerallee 119, Bonn 53175, Germany
e-mail: klaus.ackermann@giz.de

J. Pretzsch et al. (eds.), *Forests and Rural Development*, Tropical Forestry 9,
DOI 10.1007/978-3-642-41404-6_3, © Springer-Verlag Berlin Heidelberg 2014

3.1 Introduction

Approximately one third of the world's land surface is located in the tropics and subtropics. More than 90 % of the land area of tropical and subtropical countries can be classified as rural, and – even though the proportion of urban populations is increasing globally – on average in excess of half of the populations of these countries continue to live in rural areas (WRI 2011). In spite of decades of unprecedented exploitation of resources, rural areas in the tropics and subtropics still possess vast human, biological, mineral, land, forest and other natural resources. Notwithstanding the growing importance of 'soft' factors such as knowledge, information and creativity for the economic success of modern societies, these rural resources still constitute – and likely continue to do so – one primary source of overall human and economic development.

However, the benefits of past development programs and policies often bypassed rural areas. The greatest proportion of value creation derived from primary resources typically ensues outside the rural areas of developing countries. The positive welfare effects stemming from the utilization of the many rural resources usually accrue at other locations and are inadequately redistributed to rural regions. At the same time, these areas must often bear the social and environmental costs linked to the exploitation of natural resources. As a consequence, rural areas lag behind urban centers in terms of standards of living, life expectancy, education and many other basic indicators of human development.

The apparent failure of societies to develop rural areas to the same extent as their urban and metropolitan centers has largely been attributed to a complex set of factors at the level of the natural and the social sub-system at local, national and global scales. Differences in the biophysical endowment prevailing at a particular site, such as the geology, topography and climate, and the demographic, socio-economic, political and cultural conditions, have resulted in a wide array of specific challenges to development that are not amenable to quick-fix, one-size-fits-all solutions. The relationship between these factors and socio-economic development is usually multi-dimensional: resource degradation, lack of physical and social infrastructure, unfavorable demographic dynamics and other challenges are a major cause, but simultaneously also a direct consequence of the underdevelopment of rural areas in the tropics and subtropics.

In this chapter a comprehensive review of the manifold dimensions of the rural development challenge that tropical and subtropical countries currently face is provided. The authors draw heavily on empirical evidence published in the pertinent scientific literature. Although the objective is to provide a balanced account in terms of the specific impediments to forest-based rural development in various regions, due to the highly location-specific character of many of these factors, it is also necessary to generalize and to neglect the specificities of each individual case.

3.2 Biophysical Conditions

3.2.1 Climate, Hydrology and Soil

Tropical and subtropical climates are diverse and variable, depending on altitude and geographic location. Generally, the tropical climate is continuously warm and frost free, with daily fluctuations of temperature exceeding the annual variation. The subtropical climate is typically characterized by an average annual temperature of above 20°C, with an average temperature of below 20°C during the coldest month.

Due to the prevalence of air convergence near the equator, the amount of annual precipitation typically decreases with increasing latitude. The areas around the equator are dominated by a continuously wet regime with persistent rainfall ranging from 1,800 to 5,000 mm/year. Outside the intertropical convergence zone various forms of seasonal humid climates prevail, such as climates with two distinct wet seasons with high rainfall, two wet seasons with lower rainfall, a monsoon or just one distinct rainy season. The total rainfall in these areas ranges from 800 to 1,800 mm/year with decreasing precipitation and duration of the rainy season with increasing latitude. Landscapes subject to less than 800 mm but more than 250 mm/year are referred to as semi-arid and denote the transition to tropical and subtropical deserts with precipitation of less than 250 mm/year (Schultz 2005). There exists a negative correlation between annual precipitation and the inter-annual distribution of rainfall (Kellman and Tackaberry 1997). Tropical areas like north-eastern Brazil or the Horn of Africa are seriously affected by low annual precipitation, which is further aggravated by the large variability of rainfall events. These rainfall patterns are likely to become even more unpredictable and unreliable due to climate change (Corlett 2012).

Hydrological regimes reflect seasonal variations. The surface flow of rivers diverges into distinctive dry- and wet-season modes with an increase in the seasonality of rainfall. This seasonality impacts on the composition and functioning of natural ecosystems, and on the management of natural resources (Naeem 2010). For example, along large streams like the Amazon and the Orinoco, high wet-season rainfall leads to extensive flooding of the surrounding floodplains for up to several months. Highly specialized ecosystems (e.g., Igapo and Varzea) are able to withstand such flooding (Junk 1997). Due to reduced infiltration capacity, deforestation can result as a consequence of increased flood peaks (Bathurst et al. 2011).

Tropical soils are generally poor and acidic, and characterized by high iron and aluminum oxide contents. These soils are typically infertile and phosphorous deficient. Their capacity to maintain lush forest vegetation in spite of this stems from efficient nutrient cycling and the stability of the soil organic matter (SOM). A reservoir of carbon, nutrients and energy, SOM sustains the fertility, productivity and quality of tropical soils. Disturbances to the nutrient cycle can lead to a drastic decline in soil productivity (Juo and Franzluebbers 2003). This is also the reason why tropical forest ecosystems cannot readily be converted into arable land, in spite of their high biodiversity and rates of primary production. Any change from a natural forest ecosystem to an agri-ecosystem is accompanied by a loss of SOM due to altered carbon inputs and turnover rates, which drastically affects water retention capacity and other soil properties. If not sufficiently supplemented with fertilizers (Craswell and Lefroy 2001), the SOM serves temporarily as the source of nutrients until its stock is depleted and soils are left degraded.

3.2.2 *Main Tropical Forest Types*

Tropical and subtropical ecosystems comprise a high diversity of life forms and a complex network of biological interactions, which make their biota unique (Levin 2007). While favorable climatic conditions allow for biological activity throughout the entire year in most parts of the tropics, temperature and/or hydrological regimes limit biological activity during some part of the year in the semi-arid and arid regions of the tropics and the subtropics.

Alongside savannas, forests are the most important natural plant formations in tropical and subtropical areas. Whilst these forests share certain functional characteristics, their structures and compositions differ considerably (Whitmore 2009).

A variety of forest types can be distinguished within the entire tropical forest formation, with evergreen lowland tropical rainforest changing along environmental gradients such as precipitation, altitude and soil properties. Together with moist and rain forests, tropical dry forests and semi-evergreen forests are two additional major tropical forest types (Table 3.1). Dry forests are typically characterized by a distinct annual dry period and, therefore, lower rates of net primary production. Seasonality is less pronounced in semi-evergreen rain forests, whereas wet and rain forests, by contrast, show no patterns of seasonality.

Woody plants are the predominant components of tropical forest ecosystems. Trees tend to be rather tall, and branching is concentrated in the higher strata of the canopy. The canopy itself is the main regulator of the microclimate within tropical evergreen forests. With increasing altitude there is a continuous change in forest structure and composition, which has many regional variations. At higher altitudes tree size and species richness decrease (Whitmore 2009).

Forest ecosystem diversity and tree stature also change along a gradient of decreasing precipitation. The most obvious adaptation strategy of trees to drought is deciduousness. Deciduous forests regularly occur as patches in savannas, restricted to sites that are less susceptible to fire (Kellman and Tackaberry 1997; Schultz 2005).

Table 3.1 Extent of the main tropical forest types and their respective productivity expressed as net primary production (NPP) of above-ground biomass (AGB)

	Proportion of total forested tropical and subtropical landmass (percent)	NPP of AGB (Mg/ha*year)
Wet and rain forests	25	12–22
Semi-evergreen forests	33	8–14
Dry forests	42	2–10

Source: Murphy and Lugo (1986), Clark et al. (2001), modified

3.2.3 Forest Functions

All over the tropics and subtropics, forests are the source of a vast number of tangible and intangible benefits. While historically the provision of natural products and raw materials has been the predominant function of forests (Fuhrer 2000), nowadays ecosystem services are defined more broadly, encompassing provisioning, regulating, cultural and supportive services (CBD 2010). A tripartite set of functions can generally be identified with regard to forests: the productive function, i.e. the ability of forests to deliver resources such as timber and non-timber products; a set of protective functions such as the beneficial effects of forests on soil, water and climate; and the social functions of forests, referring to their importance for recreation and culture (Pistorius et al. 2011).

More than half of the Earth's forests serve the provision of wood and non-wood products (FAO 2010). Close to 1.2 billion ha, or 30 % of the world's total forests, are primarily dedicated to the productive function, while an additional 949 million hectares (24 %) are designated for multiple uses, including in most cases the production of timber and forest products. In 2005 the total global harvest of wood amounted to 3.4 billion cubic meters annually, about half of this being fuelwood (ibid.). Considering that illegally harvested timber and much of the wood used for subsistence purposes is not usually recorded, the actual quantity of wood harvested is unquestionably considerably higher. The value of the wood harvested was placed at just over USD 100 billion annually, whereas the reported value of the non-wood forest products harvested amounted to USD 19 billion (ibid.). However, this latter value probably covers only a fraction of the true total value of harvested non-wood forest products given the difficulties associated with quantifying and valuing produce harvested for subsistence use.

Less tangible benefits of forests are mainly made up of regulating ecosystem functions, for example, the conservation of biodiversity, watershed protection, erosion control, climate protection, pollination and pest control (Pistorius et al. 2011). Approximately 12 % of the total global forest area, corresponding to more than 460 million hectares, is designated for the conservation of biological diversity (FAO 2010). Most but not all of these forests are located within protected areas. The main argument for preserving biodiversity is that losses of diversity may impair life-supporting processes required by humans, such as primary production, carbon storage, water retention and the provision of clean water (Bengtsson et al. 2000). The primary management objective of an additional 8 % of the world's forests is soil and water conservation. The global area of protective forests increased by 59 million hectares between 1990 and 2010, with the biggest contribution coming from China's large-scale tree planting and reforestation programs aimed at desertification control and soil and water conservation (FAO 2010).

Cultural services provided by forests refer to non-material benefits people obtain through spiritual and religious enrichment, educational uses and cognitive development, reflection, tourism and recreation, and aesthetic appreciation or cultural heritage (MEA 2005; Pistorius et al. 2011). The variety of cultural and symbolic forest functions are as numerous and diverse as the human communities and their cultures, and comprise all aspects of culture such as language, history, art, religion and medicine. Approximately 4 % of the world's forests are designated for the provision of social and cultural services globally, with significant regional variations (FAO 2010). Whereas in East Asia 3 % and in Europe 2 % of the total forest area is primarily dedicated to the provision of social services, more than 20 % of the national forest area in Brazil serves the protection of the culture and way of life of forest-dependent people.

3.2.4 Biophysical Conditions in Flux

Natural processes and human activity are constantly altering the biophysical and biogeochemical conditions prevailing in any given ecosystem. While natural environmental shifts typically occur at the time scale of millennia (e.g., Marchant and Hooghiemstra 2004), human induced environmental change is a more recent and rapid phenomenon. The combustion of fossil fuels, the release of chemical substances into the environment and the alteration of land cover rank among the most significant human interventions on the planet (Keys and McConnell 2005). A growing human population and an increasing per capita consumption of goods and services are assumed to be the two greatest drivers of anthropogenic global change (Tilman et al. 2001). Tropical forests are at the centre of these environmental changes, both affecting and being affected by other parts of the global ecosystem.

3.2.4.1 Soil Degradation

Effecting a long-term decline in ecosystem function and productivity, soil degradation is increasing in severity and extent in many parts of the world. It has been estimated that more than 20 % of all agricultural land, 30 % of forest land and approximately 1.5 billion people are currently affected by the effects of soil degradation (Bai et al. 2008). Soil and land/forest degradation is typically driven by a complex and context-specific set of ecological, biophysical, socio-economic and institutional factors, and is not amenable to unambiguous cause-effect linkages of universal validity (cf. Ananda and Herath 2003; Upadhyay et al. 2006; Warren 2002). Reduced soil fertility and decreasing agronomic productivity, changing species composition and loss of biodiversity, lower levels of food security, increased human pressure on neighboring ecosystems, migration and deteriorating quality of life are among the direct and indirect consequences of land degradation (e.g., Eswaran et al. 2001; Lal 2010; Warner et al. 2010; Ye and van Ranst 2009). While soil erosion is the most widespread process of soil degradation, other important forms include the loss of nutrients and organic matter, salinization, soil acidification, pollution, compaction and waterlogging, leaching and subsidence (e.g., Asio et al. 2009; Eswaran et al. 2001). Tropical soils are particularly prone to degradation because of their relatively unfavorable chemical and physical properties, and the prevalent climatic conditions.

3.2.4.2 Climate Change

A warming of the climate system is unequivocal, as is now evident from observations of increases in global average air and ocean temperatures, the widespread melting of snow and ice and a rising global average sea level. Most of the observed increase in the global average temperatures since the mid-twentieth century is very likely an effect of the increase in anthropogenic greenhouse gas emissions (IPCC 2007a). Above- and below-ground woody biomass and forest soils represent major global carbon stocks (e.g., Saatchi et al. 2011; Don et al. 2011). Deforestation and forest degradation, therefore, are significant contributors to human-induced carbon emissions (Houghton 2010; Baccini et al. 2012).

While the ultimate magnitude and timing of climate change impacts are still uncertain, the current scientific understanding of the consequences of climate change is that (IPCC 2007b):

- Water availability is projected to increase by 10–40 % at high latitudes and in some wet tropical areas, and to decrease by 10–30 % at mid-latitudes and in the dry tropics.
- Approximately 20–30 % of plant and animal species face an increased risk of extinction if the global average temperature rise exceeds 1.5–2.5°C.
- Ecosystem structure and functions are likely to change with predominantly negative consequences for biodiversity, water and food supply if the global average temperature increase exceeds 1.5–2.5°C.
- Crop productivity is projected to increase slightly at mid to high latitudes, and to decrease at lower latitudes, especially in seasonally dry and tropical regions, which would increase the risk of hunger.
- Increases in the frequency of droughts and floods are projected to affect local crop production negatively, especially in the subsistence sectors in the tropics and subtropics.

- Global commercial timber productivity will rise modestly in the short- to medium-term, with large regional variability.

Tropical forest ecosystems are among the most vulnerable to changes in temperature and rainfall (Nkem et al. 2009; Corlett 2012). Although certain modern forestry practices can contribute to either maintaining or enhancing the adaptive capacity of natural and planted tropical forests to global climate change, little progress has been made to date in the adoption of such management practices (Guariguata et al. 2008; Bonan 2008).

Climate change will potentially have significant impacts not only on ecosystems but also on the social, economic, industrial and enterprise-level structures that are reliant upon them (Linnenluecke et al. 2011). Populations in the developing world are likely to be the most seriously affected (e.g., Nelson et al. 2009). The particular vulnerability of smallholders and subsistence farmers stems from their being located predominantly in the tropics, being heavily dependent on forests for their livelihoods, and from various other unfavorable socio-economic, demographic and policy trends limiting their capacity to adapt (e.g., Morton 2007).

3.2.4.3 Loss of Biodiversity

The functioning of global ecosystems, and the wide range of material and non-material services they provide to human societies, depend on their biological diversity (UNEP 2007). While the production of timber and food is seemingly most efficient in less biologically diverse systems (MEA 2005), it is generally accepted that when biodiversity is lost, ecosystems become less resilient and more vulnerable to external disturbances, and thus most of their regulating, supporting and cultural services become threatened (CBD 2010).

Tropical ecosystems – and the tropical moist forests in particular – are characterized by high levels of biodiversity, including genetic diversity, species richness, family richness, species endemism and ecosystem richness (MEA 2005; Schultz 2005; Whitmore 2009). It has been estimated that between 50 % and 90 % of all global terrestrial species live in tropical forests (WRI 1992). Deforestation, forest fragmentation, introduction of non-native species and other forms of habitat destruction are, therefore, major drivers of global species extinction and biodiversity loss (e.g., Dirzo and Raven 2003; Ghazoul and Sheil 2010; Sodhi et al. 2010; Norris et al. 2010).

Although the negative impact of biodiversity loss on ecosystem functioning is largely uncontested, the precise effect is highly site-specific and not well known. Three main hypothetical relationships can be differentiated (Fig. 3.1). The redundancy hypothesis proposes that organisms are functionally redundant, so that the loss of some species may be compensated by other species. Although increasing richness may lead to increased function, such relationships are asymptotic and may saturate at fairly low levels of diversity; more quickly when there is high redundancy. The diversity-stability hypothesis, by contrast, suggests that each species has

Fig. 3.1 Relationships between biodiversity and ecosystem functioning: (**a**) redundancy hypothesis, (**b**) diversity-stability hypothesis; (**c**) idiosyncratic response hypothesis (Source: Naeem 1998, modified)

a unique role in an ecological system, and that the loss of a species contributes to the collapse of the system, therefore diminishing its function. Finally, the idiosyncratic response hypothesis suggests biodiversity-function relationships that are not monotonic (Loreau et al. 2007).

It has been demonstrated that loss of biodiversity is closely linked to the problem of poverty (e.g., Adams et al. 2004). The rural poor depend most directly upon local ecosystem services for their livelihoods and are often the least able to access or afford substitutes when these become degraded (CBD 2006). While conserving biological diversity is, therefore, an important component of long-term poverty alleviation, poverty at the same time can significantly compromise efforts towards biodiversity conservation (Fisher and Christopher 2007).

Areas of high biodiversity are often simultaneously the most diverse in terms of human culture (e.g., Moore et al. 2002; Loh and Harmon 2005). It has been shown that the loss of biological and ethno-linguistic diversity is correlated (e.g., Harmon 1996; Sutherland 2003; Maffi 2005). The involvement of indigenous communities and their traditional ecological knowledge and cultural values is, therefore, increasingly recognized as a major strategy in biodiversity conservation (Oviedo and Maffi 2000).

3.3 Human Population

Approximately half of the global human population currently lives in rural areas. At 60 % and more, the proportion assumed by the rural population is particularly high in tropical Oceania, sub-Saharan Africa and Asia (WRI 2011). Despite the enormous progress in terms of human development that has been achieved globally in recent decades (UNDP 2010), living standards and levels of socio-economic and human development lag far behind in the rural areas of most tropical countries. For example:

- Fertility rates are typically above the national average in rural areas. A lower average age of marriage, lower levels of education of rural women, lower levels of contraception use, high infant mortality and food insecurity are some of the factors that contribute to larger family sizes and higher reproduction rates in rural areas (e.g., Mekonnen and Worku 2011; White and Speizer 2007).
- In most developing countries, rural areas are disadvantaged compared to urban areas with regard to the provision of basic social infrastructure, such as clean drinking water and improved sanitation (e.g., Wolf 2009). In addition to other factors, inadequate water supply and sanitation can adversely affect health (Hunter et al. 2010). Poor health conditions can lead to declining household productivity and poverty (Asenso-Okyere et al. 2011).
- Life expectancy and active life expectancy of rural populations are typically below that of urban residents. Better access to healthcare services in urban centers, higher socio-economic status, but also differences in social support and health behavior contribute to the rural–urban divide (e.g., Thompson et al. 2003; Zimmer et al. 2010).
- Rural areas are typically characterized by comparatively low levels of literacy, formal education and school enrollment. Often national education policies assign priority to urban areas (e.g., Fu 2005). A lack of or inadequate resources and professionally qualified teachers impair the quality of education in rural areas (e.g., Mtahabwa and Rao 2010). As in the developing world, the return to education is low for rural compared to urban activities (Knight et al. 2008), rural parents often spend less on the primary education of their children than urban households in absolute and relative terms (e.g., Mussa 2010).
- In many tropical and subtropical countries, there exist significant disparities with respect to income, consumption and wealth between urban and rural residents (e.g., Shi and Chuliang 2010; Sicular et al. 2006; Zacharias and Vakulabharanam 2011). These gaps result largely from comparatively low levels of labor productivity in rural areas and the predominance of the primary economic sector (e.g., Cheng 2010). On the other hand, many rural households have a more diversified portfolio of income compared to their urban counterparts, making them less vulnerable to sudden changes in their environment (e.g., Ersado 2006).

As sources of provisioning, regulating, cultural and supporting services, tropical and subtropical forests are of direct importance for human well-being. Between 0.96 and 1.46 billion people worldwide depend directly on forests for their livelihoods and daily sustenance, including 60 million indigenous hunters, gatherers and shifting cultivators, 350 million rural people living at the forest margins, 45 million employees in formal and informal forest-based enterprises, and 0.5–1 billion smallholder farmers growing farm trees or managing remnant forests for subsistence and income (Calibre Consultants and SSC 2000; Krishnaswamy and Hanson 1999). Increasing levels of human and socio-economic development have often been accompanied, or even rendered possible, by exploitative forest use and the conversion of forest land. Further human population growth and migration of people into pristine forest areas will continue to shape the appearance and distribution of tropical and subtropical forests globally (e.g., Carr 2009; DeFries et al. 2010; Jorgenson and Burns 2007). Yet, demographic processes are just one of several social, political, economic and ecological driving forces behind tropical deforestation (Carr et al. 2005).

Despite the fact that the proportion of the global population living in urban areas recently passed the 50 % mark, most of the population growth projected until 2030, and possibly beyond, will still take place in rural areas, mainly in sub-Saharan Africa and Asia (e.g., Bilsborrow 2002; Sudhira and Gururaja 2012; Potts 2012). Cincotta et al. (2000) estimated that in 1995 nearly 75 million people were living within the three major areas of tropical forest, namely the Congo, the Amazon and the Borneo-Mekong forest basins. While the average population density in these areas is still rather low, they are experiencing a rapid population rise, with growth rates more than twice the global average (ibid.). A growing population typically reduces the amount of agricultural land available per capita, thereby leading to smaller farm sizes, reduced fallow periods, declining soil fertility and agricultural yields. In situ agricultural intensification, the expansion of cropping into forest lands, temporary and permanent migration and changes to reproductive behavior are some of the potential responses to the growing environmental and socio-economic pressures resulting from population increase (e.g., Bilsborrow 1987, 2002; Boserup 1965). Given the fact that a clear positive relationship between increasing populations and deforestation has been observed, particularly in areas of low population density (Rosero-Bixby and Palloni 1998), it is likely that the pressure on the remaining forests and other natural resources is going to further intensify.

Whereas the impact of natural population growth on forests typically takes between one and several generations to fully unfold, migration-related demographic changes take place more rapidly and often result in much more sudden impacts on forest areas. Mobility of people, especially migration to cities, is deemed to be a precondition of economic development (World Bank 2009). Yet, contrary to widely held beliefs, rural-rural migration is still more prevalent than rural–urban population flows in many developing countries (Bilsborrow 2002). Migration streams depend on the perceived opportunities at the new destination. For example, these streams tend to be larger to destinations where the public share of forest land and the road system are larger (Amacher et al. 1998). The propensity to migrate is highest among young adults and falls steadily with increasing age, reflecting the combined effect of cultural influences, norms, traditions and economic opportunities that determine migration decisions (Bell and Muhidin 2009). Due to the fact that migrants often originate from destinations located in other agri-ecological zones and, therefore, lack knowledge of locally adapted farm and forest management practices, they plan in shorter time horizons or simply because they are burdened by higher levels of impoverishment, migrant settlers often tend to adopt more land-intensive and/or land-degrading management practices (Codjoe and Bilsborrow 2012). The fact that settlers still obtain formal or informal land titles by clearing forest land (e.g., Casse et al. 2004; Fearnside 2001; Unruh et al. 2005; Peres and Schneider 2012) contributes to forest conversion in many regions of the tropics and subtropics. Bilsborrow (2002) also pointed out that migration of people often results in negative impacts on the areas of their origin, such as depleted labor supply, disrupted communities and family lives, and the fact that those to migrate are often the most talented, educated and motivated people.

At the same time, the expected mid- to long-term decline of rural populations in tropical and subtropical countries as a consequence of their progressing demographic transition has the potential to contribute significantly to the improvement of rural living conditions (cf. ibid.). Increased land availability per rural household combined with a growing demand for food leading to higher output prices will in turn lead to higher rural employment, higher wages, higher farm incomes and ultimately higher overall rural standards of living. Yet, the timing and magnitude of this demographic dividend are highly uncertain.

3.4 Land Use

Terra firma constitutes approximately one third of the Earth's surface. While forests originally represented the natural cover of about half of all land, recent estimates of the global forest cover lie at 31 % (FAO 2010). Closely related to its physical attributes such as soil, relief, location, distance to settlements and markets, forest land has been subjected to transformation to other land use types such as agriculture since ancient times (Angelsen 2007). For example, as cited in Carr et al. (2005), forest clearance from the most fertile and most favorable agricultural land in tropical Latin America occurred centuries ago, confining recent conversions of forest to agricultural land to less favorable and marginal sites. While the specific drivers and causes of tropical deforestation might differ in varying geographical and historical contexts, there is clearly no single-factor explanation. Public and individual decisions that respond to changing, national- to global-scale economic opportunities and/or policies, mediated by local-scale institutional factors have led to regionally distinct modes of agricultural expansion, wood extraction and infrastructure extension that prevailed in causing deforestation (Geist and Lambin 2002; Goers et al. 2012). Changes to forest cover, therefore, are just one facet in a dynamic response of the land use mosaic to general livelihood transitions driven by diverse and contrasting rural development policies, changing markets, price fluctuations and extreme climatic events (Ribeiro Palacios et al. 2013).

In spite of the fact that the global annual deforestation rate has gone down slightly in recent times (FAO 2011a), tropical forests are still disappearing at alarming rates in numerous tropical countries. On the other hand, in countries like China, Vietnam and Chile, for example, the forest area has stabilized or even begun to increase due to afforestation, landscape restoration and natural expansion (FAO 2010). This shift from a shrinking to an expanding forest area is often referred to as the forest transition (Mather and Needle 1998; Gardner-Outlaw and Engelman 1999; Mather 2001). Proponents of this theory claim that the forest cover dynamics observed in Europe and North America over the past two centuries may shed light

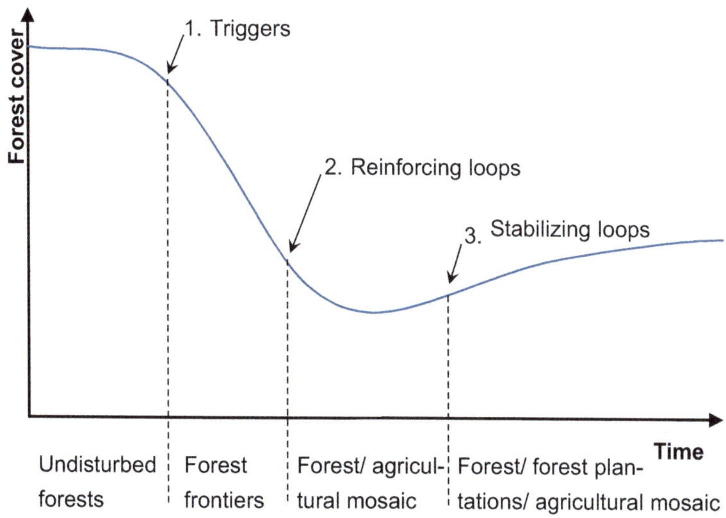

Fig. 3.2 Stages and main drivers in the forest transition (Source: adapted from Angelsen 2007: 32)

on the likely future direction of changes to the forest areas in tropical countries (Fig. 3.2).

According to Angelsen (2007, 2009), four stages of the forest transition can be identified. Initially, in the undisturbed forests stage, countries have a high and relatively stable forest cover. Deforestation is oftentimes triggered by colonization programs and the establishment of roads in previously inaccessible frontier forests, which open up these areas for markets and create demand for forest and agricultural products. Further infrastructure development, population increase, rising incomes and capital accumulation reinforce the pressure on the forest resources, accelerate deforestation, and significantly alter the prevailing land cover into a forest/agricultural mosaic. As the environmental consequences of continued deforestation become more apparent and, simultaneously, socio-economic development comes to depend less on primary production, deforestation rates ultimately slow down and the forest cover stabilizes or begins to recover.

The greater part of tropical forest conversion can be attributed to the expansion of subsistence agriculture by small farmers (Carr 2009). Taking all historical land cover changes together, agriculture has already cleared or converted 70 % of the original grasslands, 50 % of the savannas, 45 % of the temperate deciduous forests, and 27 % of the tropical forests that once existed (Foley et al. 2011). Despite the fact that improvements in agricultural productivity have significantly boosted agricultural yields and global food production over the last decades, cropland expansion remains a common phenomenon in sub-Saharan Africa, Latin America and the Caribbean, and Southeast Asia (Smith et al. 2010). A growing human population, changing consumer diets and a corresponding increase in demand for meat and animal fodder, and the growing competition between the production of food and biofuels are likely to add pressure to the forest and agricultural land resources globally (ibid.). It has been estimated that global food demand will increase by 50–100 % up to the year 2050 (e.g., The Royal Society 2009). The optimization of land use decisions with respect to production, rural development, environmental, social justice and food consumption considerations becomes increasingly relevant in this context (Pretty et al. 2010).

Simultaneously, studies have revealed significant global reserves of land that are currently under- or unutilized. Deininger and Byerlee (2011), for example, estimate that some 445 million hectares of uncultivated, non-forest land would be ecologically suitable for rainfed cultivation. Most of this land is located in sparsely populated areas of sub-Saharan Africa and other lower-income countries far from any infrastructure. Voluntary land transfers to large-scale private sector investors are proposed as one option to put this land to economic use, thereby increasing global food production (ibid.). The acquisition of large tracts of land by international agribusiness, investment funds and foreign government agencies, a development colloquially also referred to as 'land grabbling', has already encompassed a total area of approximately 60 million hectares in developing countries over recent years (Cotula 2012). As these investors commonly target the best available land in terms of water availability, soil fertility, proximity to markets etc., and because at least some of this seemingly unutilized land was previously used for shifting cultivation, dry season grazing and similar low-intensity uses, this development may have profound implications for local smallholders and for the future of the entire world agriculture system (ibid., Cotula et al. 2009).

In addition to the conversion of forest land for agricultural use, forest areas are also being cleared for urban and industrial uses, such as settlements, infrastructure development and mining. While the resultant forest clearings are often confined and comparatively small in size, they are associated with a number of negative indirect consequences that compound the impact of clearance on the remaining forests, such as emissions, increased forest exploitation due to improved access, increased fragmentation, the isolation of genetic resources, an increased incidence of forest fire and subsequent forest degradation (e.g., Ahrends et al. 2010; Saha and Padhy 2011; Tang et al. 2011, also see section 3.5).

Changes to the forest cover refer not only to a loss of natural forest, but potentially also to increased areas of planted forest. In 2010, the total area of forest plantations was estimated at 264 million hectares, of which three quarters were grown for productive purposes and one quarter for protective purposes, such as the rehabilitation of degraded land, combating desertification, soil and water protection, and carbon sequestration (FAO 2010). The area of planted forest has grown significantly in recent years. Between 2005 and 2010 the average annual increase was approximately five million hectares. Given this trend, an increase of up to 300 million hectares can be expected in the near future (ibid.). Aside from their substantial impact on biodiversity, forest plantations contribute to the global timber supply and also provide opportunities for significant rural employment and the diversification of the local economy (Kulbhushan and Singh 2007; Nair and Rutt 2009).

Table 3.2 Forest area transition matrix at pan-tropical level for the years 1990–2000

(Million ha) Land cover in 1990	Land cover in 2000: Closed forest	Open forest	Long fallow	Fragmented forest	Scrub	Short fallow	Other land cover	Water	Plantation	**Total 1990**	% of total land area
Closed forest	1131.6	1.2	5.7	9.4	1.3	9.8	43.1	1.1	1.9	**1205.1**	39.3
Open forest	0.2	287.3	0.5	6.8	0.7	2.2	6.6	0.1	0.0	**304.5**	9.9
Long fallow	1.1	0.1	63.2	0.2	0.0	4.8	4.7	0.0	0.2	**74.4**	2.4
Fragmented forest	0.5	0.4	0.2	202.1	0.5	2.2	11.2	0.1	0.2	**217.5**	7.1
Scrub	0.1	0.1	0.0	0.1	143.5	0.6	9.7	1.8	0.1	**155.9**	5.1
Short fallow	1.0	0.3	1.2	1.5	0.2	122.7	11.6	0.2	0.4	**139.0**	4.5
Other land cover	0.6	0.5	0.5	2.3	3.7	4.9	928.4	1.3	2.3	**944.4**	30.8
Water	0.2	0.0	0.0	0.0	0.8	0.0	1.2	5.6	0.0	**7.8**	0.3
Plantation	0.0	0.0	0.0	0.0	0.0	0.0	1.1	0.0	18.0	**19.3**	0.6
Total 2000	**1135.2**	**290.0**	**71.5**	**222.5**	**150.6**	**147.3**	**1017.6**	**10.2**	**23.2**	**3068.0**	
% of total land area	37.0	9.5	2.3	7.3	4.9	4.8	33.2	0.3	0.8		100.0

Source: Schoene et al. (2007: 4) after FAO (2001)

The magnitude and direction of land cover changes have been documented in a transition matrix (FAO 2001, Table 3.2). In the most recent version available, the spectrum and scale of area transitions occurring in the pan-tropics between 1990 and 2000 are illustrated. The matrix compares the land cover in 1990 with the corresponding totals in 2000, with cells reflecting specific transitions[1] (Schoene et al. 2007). Adhering to the FAO definition of forests, agroforestry and urban forests are subsumed under the heading ‘other land cover’.

Facing a continuing trend of deforestation and forest degradation, national legislation and international agreements have been enacted to bring to a halt both processes. Large tracts of forest land meanwhile possess formal protection statuses of varying intensity. Approaches to forest protection vary from strict conservation to selective management restrictions and land use planning (Box 3.1).

Despite the continuing depletion of forest resources in many tropical and subtropical countries, a decreasing rate of global deforestation and decelerating rates of forest destruction in all three major tropical forest basins (FAO 2011b) give hope that tropical deforestation and forest degradation can be reversed through combined global and local initiatives targeting increased reforestation, forest restoration and sustainable forest management.

The development of sustainable forestry practices helps to address this problem at the local level. Tree harvesting approaches emulating gap phase dynamics have

[1] For example, between the years 1990 and 2000, 1.2 million hectares of closed forest underwent a transition to open forest. In the same period, 0.2 million hectares of open forest reverted to closed forest.

been put forward as one means to naturally manage forests while facilitating continued timber supply (Pinard and Putz 1996). The implementation of alternative practices during harvesting (e.g., directional felling, skid trails, tree marking) further reduces the impacts of management practices on the remaining stand. Such reduced-impact logging (RIL) can minimize tree mortality after harvesting in natural forests significantly. In general, these approaches improve forest growth and quality, retention of forest biomass and thus increase the long-term value of the timber resources. The successful implementation of such practices remains one of the challenges with respect to the management of tropical forests (Putz et al. 2008).

Box 3.1 Approaches to Protecting Forest Areas: The Situation in Madagascar

Madagascar is famous for its natural resources, featuring a high level of endemic fauna and flora, most of which depend on forests as habitat. Although estimates of the original forest cover vary, there is no doubt that the overall forest area has decreased considerably. Between 1990 and 2005, 1.2 million hectares of forest were destroyed, corresponding to about 12 % of the forest area.

Despite a slight decrease in the annual deforestation rate from 0.8 % to 0.5 % between 1990 and 2005, the country's forest area remains under pressure. Important driving forces behind forest degradation and deforestation are shifting cultivation, charcoal production, illegal logging and recurrent forest fires. Weather variations and climate change are also serving to increase the vulnerability of the forest ecosystems. Additionally, the political crisis that began in January 2009 has weakened forest governance at all levels and has, for example, led to increased illegal exploitation of wood (e.g., rosewood exploitation for exportation and charcoal making) in forests within and outside national parks.

To reverse the trend of forest degradation and deforestation, sustainable forest management has been promoted under the national forest policy since the 1990s. Key elements include:

Forest zoning: Based on the priorities of forest management at national, regional and local level, forest areas are divided into different management zones, such as conservation, rehabilitation, afforestation and sustainable utilization. While forest zoning can in principle help to establish a coherent and spatially balanced distribution of the different forest management options, it has not been implemented across the whole country as yet.

The *transfer of forest management* to local user groups based on agreed forest management contracts has been promoted under various laws since 1996. The active participation of local forest users in forest management shall lead to more sustainable management outcomes. By 2007, approximately

(continued)

Box 3.1 (continued)

500,000 ha corresponding to 6 % of the forests outside protected areas had been transferred to local user groups.

In recent years, new *protected areas* have been created. These correspond mainly to classes III, V and VI of the IUCN protected area management categories.[2] While the specific regulations for forest management may vary, forest management in these protected areas can generally be delegated to private operators, NGOs and local populations. In total, approximately 4.7 million hectares of new protected areas have been created.

Land use planning is now being promoted at national and regional level under the national land use policy. While regional land use plans are not yet available for all regions, land use planning at municipality level is currently still in an experimental state.

Despite these policy efforts, forest degradation still represents a major national challenge. Weak enforcement of existing regulations, insufficient capacity of the forest administration, the lack of a multi-sectoral approach to land use planning that sufficiently integrates forest zoning with the strategic interests of other industries, and local forest management agreements that often fail to acknowledge the needs and livelihood strategies of local populations sufficiently while focusing too much on forest conservation are among the factors limiting the effectiveness of forest protection in Madagascar.

Source: Ackermann (2003), Dufils (2008), Ferguson (2007), Finoana (2008), Hannah et al. (2008), Innes (2010), Jariala (2009), Raik (2007), REBIOMA (2012), Schuurman and Lowry (2009), Tadross et al. (2008a, b)

3.5 Infrastructure

A defining feature of rural areas in the tropics and subtropics is their remote location, lack of or poorly developed basic infrastructure, and thus their limited integration in the urban and metropolitan sphere. Rural infrastructure such as roads, warehouses and market places, wells and irrigation systems, electricity and telecommunication services, medical centers and schools, as well as access to credit are essential for many economic activities, a prerequisite for agriculture and forestry supply chains and also determine the quality of life of rural communities.

Despite massive public and private sector investment in infrastructure projects in recent decades, a huge gap remains between current funding levels and that which is still needed in most developing countries. The World Bank has estimated that, on

[2] Class III represents a natural monument or feature, class V a protected landscape or seascape, and class VI a protected area with sustainable use of natural resources (Dudley 2008).

average, developing countries actually invest about 3–4 % of their GDP on infrastructure annually, whereas they should be spending 7–9 % on new investment projects and the maintenance of existing infrastructure, if broader economic growth and poverty reduction goals are to be achieved (cf. UNCTAD 2008). The overall performance of infrastructure and logistics services in low and lower middle income countries is considerably below the global average. Sub-Saharan Africa and southern Asia are the regions with the lowest performing infrastructure and logistics sectors (World Bank 2012a). Access to infrastructure and credit in rural areas is usually even lower than in urban areas, even in countries where urban coverage is already low by international standards (Foster and Briceño-Garmendia 2010).

The challenge facing rural infrastructure services is that their provision typically requires significant capital investment in both establishment and maintenance. The national budgets of developing countries are often overburdened by these demands. Although growing, private sector engagement in rural infrastructure remains relatively limited – particularly so as these services are widely regarded as quasi-public goods that should be affordable for the rural populations and so can rarely be provided cost-effectively. Small-scale and community-based rural infrastructure projects are seen as an alternative to conventional infrastructure development projects that can help to overcome this dilemma (e.g., Kariuki and Schwartz 2005; Fujiie et al. 2011; Kirubi et al. 2009; Galperin 2005).

A bidirectional relationship between infrastructure and economic development undoubtedly exists; i.e., high quality infrastructure services are simultaneously a precondition and a consequence of economic development. Yet, recent research has clearly isolated and quantified the contribution of infrastructure to economic development (e.g., Canning and Pedroni 2004; Calderon and Serven 2010; Cook 2011; Foster and Briceño-Garmendia 2010; Gruber and Koutroumpis 2011). Among other things it has been shown that:

- Transportation infrastructure links rural areas to regional and national markets and helps to integrate them into the global economy. Rural roads lead to lower input and transportation costs, higher agricultural output prices at local village markets, higher average household labor supply and higher agricultural productivity (Khandker et al. 2009). Lack of and poorly maintained rural infrastructure can explain productivity differences between regions (Fan and Zhang 2004).
- Improved rural roads facilitate the development of local markets and the emergence of a flourishing service sector, particularly so in poor communities (Mu and van de Walle 2011). Access to rural roads and electricity supply creates opportunities for non-farm employment and helps increase and diversify household incomes (Gibson and Olivia 2010). The establishment of rural tourism, for example, depends to a large degree on the availability of rural infrastructure (Wilson et al. 2001).
- Road improvements increase the availability of social services and lead, for example, to higher school enrollment and completion rates (Mu and van de Walle 2011; Fahy Bryceson et al. 2008). In addition, improvements in transport

infrastructure reduce travel time, make more convenient modes of travel and shipment available, and also influence land values (UNCTAD 2008).

- Improved household and agricultural water supply can contribute to improved farm production and productivity, enhance farm and non-farm employment and income, encourage the utilization of fertilizer, allow diversification into high value products, and improve nutritional status, health and societal equity (Namara et al. 2010).
- Market infrastructure, such as storage, grading and processing facilities, help farmers to obtain higher prices for their agricultural products by decoupling the sale of output from the time of harvest, aggregation of sub-scale quantities, and greater local added value. A lack of agricultural logistics services greatly restricts the competitiveness and economic development of a region (Xu 2011).
- Information and communication technology has the potential to support and enhance sustainable rural development, particularly in remote areas (Chapman and Slaymaker 2002; Bandias and Vemuri 2005). The availability of telephones leads to lower transaction costs and increased information and market transparency. Through mobile phones, extension services become more widely available and access to finance and insurance products expands in rural areas (Qiang et al. 2011). The rural poor usually benefit most from telephone-based services (Bayes 2001).
- Rural banks and microfinance institutions provide rural credit for the expansion and development of household activities and rural enterprises, thereby contributing to private sector development and the creation of productive employment (Mpuga 2010; Rijkers et al. 2010). It has also been shown that informal credit institutions can contribute to rural development (e.g., Guirkinger 2008; Panda and Atibudhi 2010).

Infrastructure projects can also have significant negative direct or indirect impacts on the natural and social systems in rural areas, however. Large-scale infrastructure projects such as dams typically lead to greater health risks (e.g., Lerer and Scudder 1999), changes in the rural economy and employment structure, social disruption, and impacts on gender relations and cultural aspects of life as a consequence of involuntary resettlement programs (e.g., Fearnside 1999; Tilt et al. 2009; Wilmsen et al. 2011). The clearance of tropical forest for roads, power lines and other linear infrastructures leads to increased physical disturbance of the ecosystem and its wildlife, higher levels of chemical pollution, edge and barrier effects affecting species composition, and invasion by exotic species and humans (Laurance et al. 2009). Rural roads often induce uncontrolled encroachment, exploitation and land conversion in areas of previously intact forest cover – despite the fact that some studies find no direct – or even a positive – impact of new roads on forest cover (e.g., Andersen et al. 2002; Deng et al. 2011; Deininger and Minten 2002). By providing links to agricultural and timber markets, lowering transportation costs and increasing the value of forest land, rural roads and tracks provide the incentives that often induce forest degradation and deforestation (e.g., Chomitz and Gray 1996; Cropper et al. 2001; Freitas et al. 2010; Geist and Lambin 2001; Pfaff et al. 2007).

3.6 Institutional Framework

The configurations of legal framework, informal regulations, organizations, institutional responsibilities, political and power relations at the local, national and international levels shape the opportunities for sustainable forest management and rural development. These institutional framework conditions are relatively unfavorable in many regions of the tropics and subtropics, further aggravating the many challenges facing forest-based rural development.

In spite of the fact that community-based natural resource management projects do not guarantee sustainable and equitable management outcomes per se (e.g., Kellert et al. 2000), the potential of community forest management to contribute to a more sustainable management of tropical and subtropical forests has been widely recognized (e.g., Ellis and Porter-Bolland 2008; Pagdee et al. 2006; Bowler et al. 2010). However, forest ownership in the tropics and subtropics is still dominated by the state, particularly so in Africa. While private ownership of forests is most significant in Latin America, communities hold remarkable shares in Latin America and Asia (Table 3.3). Although the proportion of forest area under communal management has increased significantly in recent decades, and currently accounts for 27 % globally (Larson et al. 2010), in many countries forest legislation and tenure arrangements do not yet sufficiently recognize the rights of local communities to own, use and access their forests, and represent a major impediment to rural livelihoods (Colchester et al. 2006).

Ownership of land and forest resources is associated with the full set of rights: extraction, management, exclusion and alienation rights (Agrawal and Ostrom 2001). Where formal property rights do not exist, or institutions and enforcement are weak, informal property arrangements typically emerge. These are often grounded in customs and traditions and developed by communities as self-organizing structures that correspond to local livelihood needs (Pacheco et al. 2008). However, local people often suffer from the tenure insecurity inherent in these arrangements as long as they are not formally recognized or endorsed by the authorities. It has been clearly shown that land tenure insecurity is associated with higher levels of tropical deforestation (Robinson et al. 2011) and that it also negatively influences the decision by rural households to grow trees and invest in other soil-improving conservation measures (Mekonnen 2009; Boulay et al. 2012; Abdulai et al. 2011).

The quality of governance, in particular the ability and willingness of the political stakeholders and institutional system to enforce the pertinent legal norms and to protect the rights of tree and land owners against unauthorized claims, directly determine the extent and quality of the national forest resources. Failing and failed states, warfare and corruption are conditions that are clearly detrimental to sustainable forest management (e.g., Irland 2008; Karsenty and Ongolo 2012). An increase in governance quality, on the other hand, tends to be associated with a decrease in deforestation rates (Umemiya et al. 2010) and an increase in plantation establishment (Smith et al. 2003).

Table 3.3 Forest land ownership in the tropics and subtropics

Region	State (%)	Communal (designated) (%)	Communal (owned) (%)	Private (%)
Africa	97.9	1.6	0.1	0.4
Latin America	36.1	7.3	24.6	31.9
Asia and the Pacific[a]	67.95	2.98	23.37	5.7

Source: adapted from Dahal (2011)
[a]Data for 11 countries covering 80 % of the regions' forests

At the international level, global agreements and initiatives – or a lack thereof – have significant impacts on tropical and subtropical forest resources. Driven by the concerns on the part of national governments that they would confer too much decision-making power over their environmental resources to supranational bodies, participants of the 1992 World Summit on Sustainable Development in Rio de Janeiro failed to agree on a legally binding international convention on forests. Subsequent forest-related global policy initiatives and bodies have had only limited impact and have largely failed to effectively halt deforestation and foster sustainable forest management (Levin et al. 2008). In view of the slow progress towards a binding global forest regime, various voluntary bilateral and private sector instruments have been developed and a variety of stakeholders with forest related mandates and significant overlaps have emerged (Reischl 2012). These, together with the legally binding instruments of the Convention on Biological Diversity, currently form the distinct international regime on forests (Humphreys 2006). REDD + constitutes the latest forest instrument being promoted globally (Gupta 2012). However, in light of the continuing absence of binding carbon emissions targets, there is a risk that this initiative will ultimately remain largely ineffective.

3.7 Summary

Summarizing the previous sections, presented in Table 3.4 is an overview of the major dimensions of the rural development challenge facing the forest areas of tropical and subtropical countries. The relevance of these dimensions in each of the major global geographic regions was assessed using public global statistics and pertinent publications. Certainly the realities facing individual rural communities and their particular development challenges are much more complex than any such table can ever indicate, regardless of how elaborate it may be. However, even at the scale of these global regions, differences in the structure and extent of the many barriers to forest-based rural development are evident. These differences are even more pronounced for smaller units of assessment at more granular scales of analysis. As diverse as the particular facets of the daily problems in a distinct area are, so highly customized and adapted must any proposed solutions to these problems be in order to successfully develop rural areas in the tropics and subtropics.

Table 3.4 Dimensions of the challenges facing forest-based rural development in the tropics and subtropics

Regions[a] dimensions	Asia	Oceania	North Africa/ ME	SSA	Central America	South America	Indicators	Sources
Local level								
Soil quality	++	+	+	++	+	++	Nutrient availability	Fischer et al. (2008)
Soil degradation[b]	+++	++	n/a	+++	+++	++	Human-induced soil degradation (percent of land surface)	GLASOD (1991)
Abundance of forest resources	++	++	+	+++	++	+++	Forest and wooded land (% of land area) and proportion of global forest reserves, 2010	FAO (2010)
Land use change	+	++	+	+++	++	+++	Annual forest cover change (ha), 1990–2010	FAO (2010)
Reforestation	+++	+	++	+	+	+	Forest plantations (% of forest area) and proportion of global plantation area, 2010	FAO (2010)
Sustainable forest management	+++	+	+++	++	++	+++	Predominance of SFM (2010 status) and progress 1990–2010	FAO (2010)
Biodiversity loss	+++	++	++	++	++	+++	Species and habitat loss in terrestrial ecosystems, 1975–2010	Gibson et al. (2011), UNEP (2012), CBD (2010)
Availability of infrastructure	+++	+	++	+	++	+	Road network density (m per km^2 land) and condition (% paved), 2006 or latest available	WRI (2011), World Bank (2010)
Population density	+++	++	++	++	++	+	Population density in rural areas, 2010	WRI (2011)
Population growth	+++	+++	+++	+++	++	+	Rural population growth (%), 1950–2010	WRI (2011)
Migration	+++	+	+	+	++	+	Average annual net international migrants, 2005–2010	WRI (2011)
International land deals	+++	+	+	++	+	+	Land area affected (ha), 2000–2011	ILC (2012)
National level								
Land use planning	+	+	+	+	+	+	Effectiveness of forest-related land use planning	Authors' assessment

Community involvement[c]	++	+++	n/a	+	+++	++	Share of CFM (designated and owned) in total forest area, 2008	Sunderlin et al. (2008)
Land and tree tenure security	+	++	+	+	++	+	Insecure land tenure and property rights, 2005	USAID and ARD (2008)
Quality of governance	++	+	++	+	+++	++	Government effectiveness, 2011	World Bank (2012b)
Global level								
Climate change	++	+	+++	++	++	+++	Projection of consecutive dry days and soil moisture anomalies, 2046–2100	IPCC (2012)
Global forest regime	+	+	+	+	+	+	Strength of legally binding instruments, 2012	Authors' assessment

Status: + low, +++ high, n/a data not available

[a]Classification of regions based on WRI (2011). Asia excl. Japan; Oceania excl. Australia and New Zealand; ME = Middle East (excl. Israel), SSA = sub-Saharan Africa (excl. South Africa)

[b]Classification of regions based on original source

[c]Based on the assessment of 37 countries globally covering 85 % of the global forest area

References

Abduali A, Owusu V, Goetz R (2011) Land tenure differences and investment in land improvement measures: theoretical and empirical analyses. J Dev Econ 96(1):66–78

Ackermann K (2003) Sozioökonomische Bedeutung und ökologische Potentiale von Sekundärwäldern in Nordmadagaskar: Analyse und Ableitung von Bewirtschaftungsempfehlungen. Technische Universität Dresden, Schriftenreihe des Institutes für Internationale Forst- und Holzwirtschaft 9

Adams WM, Aveling R, Brockington D, Dickson B, Elliott J, Hutton J, Roe D, Vira B, Wolmer W (2004) Biodiversity conservation and the eradication of poverty. Science 306 (5699):1146–1149

Agrawal A, Ostrom E (2001) Collective action, property rights and decentralization in resource use in India and Nepal. Polit Soc 29(4):485–514

Ahrends A, Burgess ND, Milledge SAH, Bulling MT, Fisher B, Smart JCR, Clarke GP, Mhoro BE, Lewis SL (2010) Predictable waves of sequential forest degradation and biodiversity loss spreading from an African city. Proc Natl Acad Sci 107(33):14556–14561

Amacher GS, Cruz W, Grebner D, Hyde WF (1998) Environmental motivations for migration: population pressure, poverty, and deforestation in the Philippines. Land Econ 74(1):92–101

Ananda J, Herath G (2003) Soil erosion in developing countries: a socio-economic appraisal. J Environ Manag 68(4):343–353

Andersen LE, Granger CWJ, Reis EJ, Weinhold D, Wunder S (2002) The dynamics of deforestation and economic growth in the Brazilian Amazon. Press Syndicate of the University of Cambridge/Cambridge University Press, Cambridge/New York

Angelsen A (2007) Forest cover change in space and time: combining the von Thünen and forest transition theories. World bank policy research working paper 4117

Angelsen A (ed) (2009) Realising REDD+: national strategy and policy options. CIFOR, Bogor

Asenso-Okyere K, Chiang C, Thangata P, Andam K, Mekonnen DA (2011) Understanding the interaction between farm labor productivity, and health and nutrition: a survey of the evidence. J Dev Agric Econ 3(3):80–90

Asio V, Jahn R, Perez FO, Navarrete IA, Abit SM Jr (2009) A review of soil degradation in the Philippines. Ann Trop Res 31(2):69–94

Baccini A, Goetz SJ, Walker WS, Laporte NT, Sun M, Sulla-Menashe D, Hackler J, Beck PSA, Dubayah R, Friedl MA, Samanta S, Houghton RA (2012) Estimated carbon dioxide emissions from tropical deforestation improved by carbon-density maps. Nat Clim Change. Advance online publication. doi:10.1038/nclimate1354

Bai ZG, Dent DL, Olsson L, Schaepman ME (2008) Global assessment of land degradation and improvement. 1: identification by remote sensing. Report 2008/01, ISRIC – World Soil Information, Wageningen

Bandias S, Vemuri SR (2005) Telecommunications infrastructure facilitating sustainable development of rural and remote communities in Northern Australia. Telecommun Policy 29 (2–3):237–249

Bathurst JC, Iroumé A, Cisneros F, Fallas J, Iturraspe R, Novillo MG et al (2011) Forest impact on floods due to extreme rainfall and snowmelt in four Latin American environments 1: field data analysis. J Hydrol 400(3–4):281–291

Bayes A (2001) Infrastructure and rural development: insights from a Grameen Bank village phone initiative in Bangladesh. Agric Econ 25(2–3):261–272

Bell M, Muhidin S (2009) Cross-national comparison of internal migration. Human development research paper 2009/30, United Nations Development Program (UNDP), New York

Bengtsson J, Nilsson SG, Franc A, Menozzi P (2000) Biodiversity, disturbances, ecosystem function and management of European forests. For Ecol Manag 132(1):39–50

Bilsborrow RE (1987) Population pressure and agricultural development in developing countries: a conceptual framework and recent evidence. World Dev 15(2):183–203

Bilsborrow RE (2002) Migration, population change, and the rural environment. ECSP report 8. Environmental Change and Security Program (ECSP), Washington, DC
Bonan GB (2008) Forests and climate change: forcings, feedbacks, and the climate benefits of forests. Science 320:1444–1449
Boserup E (1965) The conditions of agricultural growth: the economics of agrarian change under population pressure. Aldine, Chicago
Boulay A, Tacconi L, Kanowski P (2012) Drivers of adoption of eucalypt tree farming by smallholders in Thailand. Agrofor Syst 84(2):179–189
Bowler D, Buyung-Ali L, Healey JR, Jones JPG, Knight T, Pullin AS (2010) The evidence base for community forest management as a mechanism for supplying global environmental benefits and improving local welfare: a STAP advisory document. Prepared on behalf of the Scientific and Technical Advisory Panel (STAP) of the Global Environment Facility (GEF). Centre for Evidence-Based Conservation, Bangor University, Bangor
Calderon C, Serven L (2010) Infrastructure and economic development in sub-Saharan Africa. J African Econ 19 (Suppl. 1):i13–i87
Calibre Consultants, SSC (2000) Number of forest-dependent people: a feasibility study for DFID's Forestry Research Programme. University of Reading, Reading
Canning D, Pedroni P (2004) The effect of infrastructure on long run economic growth. Department of Economics working papers. Department of Economics, Rice University, Houston
Carr DL (2009) Population and deforestation: why rural migration matters. Prog Hum Geogr 33 (3):355–378
Carr DL, Suter L, Barbieri A (2005) Population dynamics and tropical deforestation: state of the debate and conceptual challenges. Popul Environ 27(1):89–113
Casse T, Milhøj A, Ranaivosion S, Randriamanarivo JR (2004) Causes of deforestation in southwestern Madagascar: what do we know? For Policy Econ 6(1):33–48
CBD (2006) Global biodiversity outlook 2. Secretariat of the Convention on Biological Diversity, Montreal, p 81
CBD (2010) Global biodiversity outlook 3. Secretariat of the Convention on Biological Diversity, Montréal, p 94
Chapman R, Slaymaker T (2002) ICTs and rural development: review of the literature, current interventions and opportunities for action. Working paper, ODI, London
Cheng H (2010) Gap between urban and rural economic development in less developed areas – a case from Northern Jiangsu area. China Asian Agric Res 2(11):4–8
Chomitz KM, Gray DA (1996) Roads, land use, and deforestation: a spatial model applied to Belize. World Bank Econ Rev 10(3):487–512
Cincotta RP, Wisnewski J, Engelmann R (2000) Human population in the biodiversity hotspots. Nature 404:990–992
Clark DA, Brown S, Kicklighter DW, Chambers JQ, Thomlinson JR, Ni J, Holland EA (2001) Net primary production in tropical forests: an evaluation and synthesis of existing field data. Ecol Appl 11(2):371–384
Codjoe SNI, Bilsborrow RE (2012) Are migrants exceptional resource degraders? A study of agricultural households in Ghana. Geo J 77(5):681–694
Colchester M, Boscolo M, Contreras-Hermosilla A, Del Gatto F, Dempsey J, Lescuyer G, Obidzinski K, Pommier D, Richards M, Sembiring SN, Tacconi L, Vargas Rios MT, Wells A (2006) Justice in the forest: rural livelihoods and forest law enforcement. Forest perspectives 3. Center for International Forestry Research (CIFOR), Bogor
Cook P (2011) Infrastructure, rural electrification and development. Energy Sustain Dev 15 (3):304–313
Corlett RT (2012) Climate change in the tropics: the end of the world as we know it? Biol Conserv 151(1):22–25
Cotula L (2012) The international political economy of the global land rush: a critical appraisal of trends, scale, geography and drivers. J Peasant Stud 39(3–4):649–680

Cotula L, Vermeulen S, Leonard R, Keeley J (2009) Land grab or development opportunity? Agricultural investment and international land deals in Africa. FAO/IIED/IFAD, London/ Rome

Craswell ET, Lefroy RDB (2001) The role and function of organic matter in tropical soils. Nutr Cycl Agroecosyst 61(1/2):7–18

Cropper M, Puri J, Griffiths C (2001) Predicting the location of deforestation: the role of roads and protected areas in North Thailand. Land Econ 77(2):172–186

Dahal GR (2011) Forest tenure in Asia. ITTO Trop For Update 20(4):8–9

DeFries RS, Rudel T, Uriarte M, Hansen M (2010) Deforestation driven by urban population growth and agricultural trade in the twenty-first century. Nat Geosci 3:178–181

Deininger K, Byerlee D (2011) Rising global interest in farmland – can it yield sustainable and equitable benefits? Agriculture and rural development. World Bank, Washington, DC

Deininger K, Minten B (2002) Determinants of deforestation and the economics of protection: an application to Mexico. Am J Agric Econ 84(4):943–960

Deng X, Huang J, Uchida E, Rozelle S, Gibson J (2011) Pressure cookers or pressure valves: do roads lead to deforestation in China? J Environ Econ Manag 61(1):79–94

Dirzo R, Raven PH (2003) Global state of biodiversity and loss. Annu Rev Environ Resour 28:137–167

Don A, Schumacher J, Freibauer A (2011) Impact of tropical land-use change on soil organic carbon stocks – a meta-analysis. Glob Chang Biol 17(4):1658–1670

Dudley N (ed) (2008) Guidelines for applying protected area management categories. International Union for Conservation of Nature (IUCN), Gland

Dufils JM (2008) Couvert forestier restant. In: Goodman SM (ed) Paysages naturels et biodiversité de Madagascar. Musée National d'Histoire Naturelle, Paris, pp 67–79

Ellis EA, Porter-Bolland L (2008) Is community-based forest management more effective than protected areas? A comparison of land use/land cover change in two neighboring study areas of the Central Yucatan Peninsula, Mexico. Forest Ecol Manag 256(11):1971–1983

Ersado L (2006) Income diversification in Zimbabwe: welfare implications from urban and rural areas. Policy research working paper 3964. World Bank, Washington, DC

Eswaran H, Lal R, Reich PF (2001) Land degradation: an overview. In: Bridges EM, Hannam ID, Oldeman LR, Penning de Vries FWT, Scherr SJ, Sompatpanit S (eds) Responses to land degradation. Proceedings of the 2nd international conference on land degradation and desertification, Khon Kaen, Thailand. Oxford Press, New Delhi

Fahy Bryceson D, Bradbury A, Bradbury T (2008) Roads to poverty reduction? Exploring rural roads' impact on mobility in Africa and Asia. Dev Policy Rev 26(4):459–482

Fan S, Zhang X (2004) Infrastructure and regional economic development in rural China. China Econ Rev 15(2):203–214

FAO (2001) Global forest resources assessment 2000 – main report. FAO forestry paper no 140. Food and Agriculture Organization of the United Nations, Rome

FAO (2010) Global forest resources assessment, main report. FAO forest paper 163. Food and Agriculture Organization of the United Nations, Rome

FAO (2011a) State of the world's forests 2011. Food and Agriculture Organization of the United Nations, Rome

FAO (2011b) The state of forests in the Amazon basin, Congo basin and Southeast Asia. A report prepared for the summit of the three basins. Brazzaville. Food and Agriculture Organization of the United Nations, Rome, 31 May–3 June 2011

Fearnside PM (1999) Social impacts of Brazil's Tucuruí Dam. Environ Manag 24(4):483–495

Fearnside PM (2001) Land-tenure issues as factors in environmental destruction in Brazilian Amazonia: the case of southern Pará. World Dev 29(8):1361–1372

Ferguson B (2007) Community forest management in southern Madagascar: local institutions and the policy process. Procedural paper. University of East Anglia, School of Development Studies, Norwich

Finoana H (2008) Etat des lieux sur les transfert de gestion. Document d'études. Ministère de l'Environnement, des Forets et du Tourisme, Programme Gestion Durable des Ressources Naturelles (PGDRN), Deutsche Gesellschaft für Internationale Zusammenarbeit (GIZ) GmbH, Bonn

Fischer G, Nachtergaele F, Prieler S, van Velthuizen HT, Verelst L, Wiberg D (2008) Global Agro-ecological zones assessment for agriculture. IIASA, FAO, Rome

Fisher B, Christopher T (2007) Poverty and biodiversity: measuring the overlap of human poverty and the biodiversity hotspots. Ecol Econ 62(1):93–101

Foley JA, Ramankutty N, Brauman KA, Cassidy ES, Gerber JS, Johnston M, Mueller ND, O'Connell C, Ray DK, West PC, Balzer C, Bennett EM, Carpenter SR, Hill J, Monfreda C, Polasky S, Rockström J, Sheehan J, Siebert S, Tilman D, Zaks DPM (2011) Solutions for a cultivated planet. Nature 478:337–342

Foster V, Briceño-Garmendia C (2010) Africa's infrastructure. A time for transformation. Agence Française de Développement/World Bank, Washington, DC

Freitas SR, Hawbaker TJ, Metzger JP (2010) Effects of roads, topography, and land use on forest cover dynamics in the Brazilian Atlantic Forest. Forest Ecol Manag 259(3):410–417

Fu TM (2005) Unequal primary education opportunities in rural and urban China. China Perspect [online], 60. Available online at http://chinaperspectives.revues.org/500. Accessed 6 Jan 2013

Fuhrer E (2000) Forest functions, ecosystem stability and management. Forest Ecol Manag 132 (1):29–38

Fujiie H, Maruyama A, Fujiie M, Takagaki M, Merrey DJ, Kikuchi M (2011) Why invest in minor projects in Sub-Saharan Africa? An exploration of the scale economy and diseconomy of irrigation projects. Irrig Drain Syst 25(1):39–60

Galperin H (2005) Wireless networks and rural development: opportunities for Latin America. Inf Technol Int Dev 2(3):47–56

Gardner-Outlaw T, Engelman R (1999) Forest, population, consumption and wood resources futures. Publication Action International. Colorcraft of Virginia, Washington, DC

Geist H, Lambin E (2001) What drives tropical deforestation? A meta-analysis of proximate and underlying causes of deforestation based on subnational case study evidence. LUCC report series; 4. LUCC International Project Office, Brussels

Geist HJ, Lambin EF (2002) Proximate causes and underlying driving forces of tropical deforestatio. BioScience 52(2):143–150

Ghazoul J, Sheil D (2010) Tropical rain forest ecology: diversity and conservation. Oxford University Press, Oxford

Gibson J, Olivia S (2010) The effect of infrastructure access and quality on non-farm enterprises in rural Indonesia. World Dev 38(5):717–726

Gibson L, Lee TM, Koh LP, Brook BW, Gardner TA, Barlow J, Peres CA, Bradshaw CJA, Laurance WF, Lovejoy TE, Sodhi NS (2011) Primary forests are irreplaceable for sustaining tropical biodiversity. Nature 478:378–381

GLASOD (1991) World map of the status of human-induced soil degradation – an explanatory note. Global Assessment of Soil Degradation. International Soil Reference and Information Centre, Wageningen; United Nations Environment Programme, Nairobi

Goers L, Lawson J, Garen E (2012) Economic drivers of tropical deforestation for agriculture. In: Ashton M, Tyrrell M, Spalding D, Gentry B (eds) Managing forest carbon in a changing climate. Springer, Dordrecht/Heidelberg/London/New York

Gruber H, Koutroumpis P (2011) Mobile telecommunications and the impact on economic development. Econ Policy 26(67):387–426

Guariguata MR, Cornelius JP, Locatelli B, Forner C, Sánchez-Azofeifa GA (2008) Mitigation needs adaptation: tropical forestry and climate change. Mitig Adapt Strateg Glob Change 13 (8):793–808

Guirkinger C (2008) Understanding the coexistence of formal and informal credit markets in Piura, Peru. World Dev 36(8):1436–1452

Gupta J (2012) Glocal forest and REDD + governance: win-win or lose-lose? Curr Opin Environ Sustain 4(6):620–627

Hannah L, Dave R, Lowry PP II, Andelman S et al (2008) Climate change adaptation for conservation in Madagascar. Biol Lett 4:590–594

Harmon D (1996) Losing species, losing languages: connections between biological and linguistic diversity. Southwest J Ling 15:89–108

Houghton RA (2010) How well do we know the flux of CO_2 from land-use change? Tellus B 62 (5):337–351

Humphreys D (2006) Logjam: deforestation and the crisis of global governance. Earthscan, London

Hunter PR, MacDonald AM, Carter RC (2010) Water supply and health. PLoS Med 7(11): e1000361. doi:10.1371/journal.pmed.1000361

ILC (2012) Land matrix. International Land Coalition, Rome. Available online at http://landportal.info/landmatrix. Accessed 3 Jan 2013

Innes JL (2010) Madagascar rosewood, illegal logging and the tropical timber trade. Madag Conserv Dev 5(1):6–10

IPCC (2007a) Climate change 2007: synthesis report. Contribution of working groups I, II and III to the Fourth Assessment Report of the Intergovernmental Panel on Climate Change [core writing team, Pachauri RK, Reisinger A (eds)]. IPCC, Geneva

IPCC (2007b) Climate change 2007: impacts, adaptation and vulnerability. Contribution of working group II to the Forth Assessment Report of the Intergovernmental Panel on Climate Change [Parry ML, Canziani OF, Palutikof JP, van der Linden PJ, Hanson CE (eds)]. IPCC, Geneva

IPCC (2012) Managing the risks of extreme events and disasters to advance climate change adaptation. A special report of working groups I and II of the Intergovernmental panel on Climate Change [Field CB, Baross V, Stocker TF, Qin D, Dokken DJ, Ebi KL, Mastrandrea MD, Mach KJ, Plattner GK, Allen SK, Tignor M, Midgley PM (eds)]. Cambridge University Press, Cambridge/New York

Irland LC (2008) State failure, corruption, and warfare: challenges for forest policy. J Sustain For 27(3):189–223

Jariala (2009) Evolution de la couverture des forêts naturelles à Madagascar 1990-2000-2005. Ministère de l'Environnement, des Forets et du Tourisme, USAID, Conservation International

Jorgenson AK, Burns TJ (2007) Effects of rural and urban population dynamics and national development on deforestation in less-developed countries, 1990–2000. Sociol Inq 77 (3):460–482

Junk WJ (1997) The central Amazon floodplain. Ecology of a pulsing system. Ecological studies 126. Springer, Berlin

Juo ASR, Franzluebbers K (2003) Tropical soils: properties and management for sustainable agriculture. Oxford University Press, Oxford

Kariuki M, Schwartz J (2005) Small-scale private service providers of water supply and electricity. A review of incidence, structure, pricing and operating characteristics. Policy research working paper 3727. World Bank, Washington, DC

Karsenty A, Ongolo S (2012) Can "fragile states" decide to reduce their deforestation? The inappropriate use of the theory of incentives with respect to the REDD mechanism. Forest Policy Econ 18(1):38–45

Kellert SR, Mehta JN, Ebbin SA, Lichtenfeld LL (2000) Community natural resource management: promise, rhetoric, and reality. Soc Nat Resour 13(8):705–715

Kellman MC, Tackaberry R (1997) Tropical environments. The functioning and management of tropical ecosystems, Routledge physical environment series. Routledge, London

Keys E, McConnell WJ (2005) Global change and the intensification of agriculture in the tropics. Glob Environ Chang 15(4):320–337

Khandker SR, Bakht Z, Koolwal GB (2009) The poverty impact of rural roads: evidence from Bangladesh. Econ Dev Cult Change 57(4):685–722

Kirubi C, Jacobson A, Kammen DM, Mills A (2009) Community-based electric micro-grids can contribute to rural development: evidence from Kenya. World Dev 37(7):1208–1221
Knight J, Shi L, Quheng D (2008) Education and the poverty trap in rural China. CSAE working paper 2008–02. Center for the Studies of African economies, University of Oxford, Oxford
Krishnaswamy A, Hanson A (1999) Our forests, our future. Summary report World Commission on Forests and Sustainable Development. Cambridge University Press, Cambridge
Kulbhushan B, Singh K (2007) Prospects and problems of afforestation of wasteland in India: a synthesis of macro and micro perspectives. Geoforum 38:1276–1289
Lal R (2010) Soil degradation as a reason for inadequate human nutrition. Food Sec 1(1):45–57
Larson AM, Barry D, Ram Dahal G (2010) New rights for forest-based communities? Understanding processes of forest tenure reform. Int For Rev 12(1):78–96
Laurance WF, Goosem M, Laurance SG (2009) Impacts of roads and linear clearings on tropical forests. Trends Ecol Evol 24(12):659–669
Lerer LB, Scudder T (1999) Health impacts of large dams. Environ Impact Assess Rev 19 (2):113–123
Levin SA (2007) Encyclopedia of biodiversity. Academic Press, Waltham/MA, USA
Levin K, McDermott C, Cashore B (2008) The climate regime as global forest governance: can reduced emissions from deforestation and forest degradation (REDD) initiatives pass a 'dual effectiveness' test? Int For Rev 10(3):538–549
Linnenluecke MK, Stathakis A, Griffiths A (2011) Firm relocation as adaptive response to climate change and weather extremes. Glob Environ Change 21(1):123–133
Loh J, Harmon D (2005) A global index of biocultural diversity. Ecol Indic 5(3):231–241
Loreau M, Naeem S, Inchausti P (2007) Biodiversity and ecosystem functioning. Synthesis and perspectives. Reprinted (with corr.). Oxford University Press, Oxford
Maffi L (2005) Linguistic, cultural, and biological diversity. Annu Rev Anthropol 34:599–617
Marchant R, Hooghiemstra H (2004) Rapid environmental change in African and South American tropics around 4000 years before present: a review. Earth Sci Rev 66(3–4):217–260
Mather A (2001) The transition from deforestation to reforestation in Europe. In: Angelsen A, Kaimowitz D (eds) Agricultural technologies and tropical deforestation. CAB International, Wellingford
Mather AS, Needle CL (1998) The forest transition: a theoretical basis. Area 30:117–124
MEA (2005) Ecosystems and human well-being – current states and trends assessment, Millennium ecosystem assessment. Island Press, Washington, DC
Mekonnen A (2009) Tenure security, resource endowments, and tree growing: evidence from the Amhara region of Ethiopia. Land Econ 85(2):292–307
Mekonnen W, Worku A (2011) Determinants of fertility in rural Ethiopia: the case of Butajira Demographic Surveillance System (DSS). BMC Public Health 11:782
Moore JL, Manne L, Brooks T, Burgess ND, Davies R, Rahbek C, Williams P, Balmford A (2002) The distribution of cultural and biological diversity in Africa. Proc R Soc Lond B 269 (1501):1645–1653
Morton JF (2007) The impact of climate change on smallholder and subsistence agriculture. Proc Natl Acad Sci 104(50):19680–19685
Mpuga P (2010) Constraints in access to and demand for rural credit: evidence from Uganda. Afr Dev Rev 22(1):115–148
Mtahabwa L, Rao N (2010) Pre-primary education in Tanzania: observations from urban and rural classrooms. Int J Educ Dev 30(3):227–235
Mu R, van de Walle D (2011) Rural roads and local market development in Vietnam. J Dev Stud 47(5):709–734
Murphy PG, Lugo AE (1986) Ecology of tropical dry forest. Annu Rev Ecol Syst 17:67–88
Mussa R (2010) Rural–urban differences in parental spending on children's primary education in Malawi. A Southern Africa Labour and Development Research Unit working paper number 49. SALDRU, University of Cape Town, Cape Town
Naeem S (1998) Species redundancy and ecosystem reliability. Conserv Biol 12(1):39–45

Naeem S (2010) Biodiversity, ecosystem functioning, and human wellbeing. An ecological and economic perspective. Reprint. Oxford University Press, Oxford

Nair CTS, Rutt R (2009) Creating forestry jobs to boost the economy and build a green future. Unasylva 60(233):3–10

Namara RE, Hanjra MA, Castillo GE, Ravnborg HM, Smith L, van Koppen B (2010) Agricultural water management and poverty linkages. Agric Water Manag 97(4):520–527

Nelson GC, Rosegrant MW, Koo J, Robertson R, Sulser T, Zhu T, Ringler C, Msangi S, Palazzo A, Batka M, Magalhaes M, Valmonte-Santos R, Ewing M, Lee D (2009) Climate change: impact on agriculture and costs of adaptation. IFPRI, Washington, DC

Nkem J, Santoso H, Idinoba M, Saborio J, Herawati H (2009) Tropical forests and climate change adaptation: criteria and indicators for adaptive management for reduced vulnerability and long-term sustainability. Final technical report. CIFOR, CATIE, Bogor

Norris K, Asase A, Collen B, Gockowksi J, Mason J, Phalan B, Wade A (2010) Biodiversity in a forest-agriculture mosaic – the changing face of West African rainforests. Biol Conserv 143 (10):2341–2350

Oviedo G, Maffi L (2000) Indigenous and traditional peoples of the world and ecoregion conservation. An integrated approach to conserving the world's biological and cultural diversity. WWF International, Gland

Pacheco P, Barry D, Cronkleton P, Larson AM (2008) The role of informal institutions in the use of forest resources in Latin America. Forests and Governance Programme no 15/2008. CIFOR, Bogor

Pagdee A, Kimb Y, Daugherty PJ (2006) What makes community forest management successful: a meta-study from community forests throughout the world. Soc Nat Resour 19(1):33–52

Panda DK, Atibudhi H (2010) Impact of group-based microfinance on rural household income: evidence from an Indian state. J Rural Coop 38(2):173–186

Peres CA, Schneider M (2012) Subsidized agricultural resettlements as drivers of tropical deforestation. Biol Conserv 151(1):65–68

Pfaff A, Robalino J, Walker R, Aldrich S, Caldas M, Reis E, Perz S, Bohrer C, Arima E, Laurance W, Kirby K (2007) Road investments, spatial spillovers, and deforestation in the Brazilian Amazon. J Reg Sci 47(1):109–123

Pinard MA, Putz PE (1996) Retaining biomass by reducing logging damage. Biotropica 28:278–295

Pistorius T, Schaich H, Winkel G, Plieninger T, Bieling C, Konold W, Volz KR (2011) Lessons for REDD 2011plus: a comparative analysis of the German discourse on forest functions and the global ecosystem services debate. Forest Policy Econ. doi:10.1016/j.forpol.2011.09.001

Potts D (2012) Challenging the myths of urban dynamics in Sub-Saharan Africa: the evidence from Nigeria. World Dev 40(7):1382–1393

Pretty J, Sutherland WJ, Ashby J, Auburn J, Baulcombe D, Bell M, Bentley J, Bickersteth S, Brown K, Burke J, Campbell H, Chen K, Crowley E, Crute I, Dobbelaere D, Edwards-Jones G, Funes-Monzote F, Godfray HCJ, Griffon M, Gypmantisiri P, Haddad L, Halavatau S, Herren H, Holderness M, Izac AM, Jones M, Koohafkan P, Lal R, Lang T, McNeely J, Mueller A, Nisbett N, Noble A, Pingali P, Pinto Y, Rabbinge R, Ravindranath NH, Rola A, Roling N, Sage C, Settle W, Sha JM, Shiming L, Simons T, Smith P, Strzepeck K, Swaine H, Terry E, Tomich TP, Toulmin C, Trigo E, Twomlow S, Vis JK, Wilson J, Pilgrim S (2010) The top 100 questions of importance to the future of global agriculture. Int J Agric Sustain 8 (4):219–236

Putz FE, Sist P, Fredericksen T, Dykstra D (2008) Reduced-impact logging: challenges and opportunities. Forest Ecol Manag 256(7):1427–1433

Qiang CZ, Kuek SC, Dymond A, Esselaar S (2011) Mobile applications for agriculture and rural development. World Bank, Washington, DC

Raik D (2007) Forest management in Madagascar: an historical overview. Madag Conserv Dev 4 (2):98–102

REBIOMA (2012) Unpublished statistical information. Reseau de la Biodiversité de Madagascar, Antananarivo

Reischl G (2012) Designing institutions for governing planetary boundaries – lessons from global forest governance. Ecol Econ 81:33–40

Ribeiro Palacios M, Huber-Sannwald E, García Barrios L, Pena de Paz F, Carrera Hernández J, Galindo Mendosa MG (2013) Landscape diversity in a rural territory: emerging land use mosaics coupled to livelihood diversification. Land Use Policy 30(1):814–824

Rijkers B, Söderbom M, Loening JL (2010) A rural–urban comparison of manufacturing enterprise performance in Ethiopia. World Dev 38(9):1278–1296

Robinson BE, Holland MB, Naughton-Treves L (2011) Does secure land tenure save forests? A review of the relationship between land tenure and tropical deforestation. CCAFS working paper no 7. CGIAR research program on Climate Change, Agriculture and Food Security (CCAFS), Copenhagen

Rosero-Bixby L, Palloni A (1998) Population and deforestation in Costa Rica. Popul Environ 20 (2):149–185

Saatchi SS, Harris NL, Brown S, Lefsky M, Mitchard ETA, Salas W, Zutta BA, Buermann W, Lewis SL, Hagen S, Petrova S, White L, Silman M, Morel A (2011) Benchmark map of forest carbon stocks in tropical regions across three continents. Proc Natl Acad Sci 108 (24):9899–9904

Saha DC, Padhy PK (2011) Effect of air and noise pollution on species diversity and population density of forest birds at Lalpahari, West Bengal, India. Sci Total Environ 409(24):5328–5336

Schoene D, Killmann W, von Lüpke H, LoycheWilkie M (2007) Definitional issues related to reducing emissions from deforestation in developing countries. Forests and Climate Change working paper 5. FAO, Rome

Schultz J (2005) The ecozones of the world. The ecological divisions of the geosphere, 2nd edn. Springer, Berlin

Schuurman D, Lowry P (2009) The Madagascar rosewood massacre. Madag Conserv Dev 2 (1):5–10

Shi L, Chuliang L (2010) Re-estimating the income gap between urban and rural households in China. Procedia Soc Behav Sci 2(5):7151–7163

Sicular T, Yue X, Gustafsson B, Li S (2006) The urban–rural income gap and inequality in China. Research paper no 2006/135. World Institute for Development Economics Research, United Nations University, New York

Smith RJ, Muir RDJ, Walpole MJ, Balmford A, Leader-Williams N (2003) Governance and the loss of biodiversity. Nature 426:67–70

Smith P, Gregory PJ, van Vuuren D, Obersteiner M, Havlík P, Rounsevell M, Woods J, Stehfest E, Bellarby J (2010) Competition for land. Phil Trans R Soc (B) 365:2941–2957

Sodhi NS, Wilcove DS, Lee TM, Sekercioglu CH, Subaraj R, Bernard H, Yong DL, Lim SLH, Prawiradilaga DM, Brook BW (2010) Deforestation and avian extinction on tropical landbridge islands. Conserv Biol 24(5):1290–1298

Sudhira HS, Gururaja KV (2012) Population crunch in India: is it urban or still rural? Curr Sci 103 (1):37–40

Sunderlin WD, Hatcher J, Liddle M (2008) From exclusion to ownership? Challenges and opportunities in advancing forest tenure reform. Rights and Resources Initiative, Washington, DC

Sutherland WJ (2003) Parallel extinction risk and global distribution of languages and species. Nature 423(6937):276–279

Tadross M, Rakotondrafara ML, Randriamarolaza L, Rabefitia Z, Yip Z (2008a) Le changement climatique à Madagascar. Direction Générale de la Météorologie

Tadross M, Randriamarolaza L, Rabefitia Z, Yip Z (2008b) Climate change in Madagascar; recent past and future. Climate Systems Analysis Group, South Africa and National Meteorological Office, Madagascar

Tang CQ, He LY, Gao Z, Zhao XF, Sun WB, Ohsawa M (2011) Habitat fragmentation, degradation, and population status of endangered Michelia coriacea in southeastern Yunnan, China. Mt Res Dev 31(4):343–350

The Royal Society (2009) Reaping the benefits: science and the sustainable intensification of global agriculture. RS policy document 11/09. The Royal Society, London

Thompson R, Miller N, Witter S (2003) Health-seeking behavior and rural/urban variation in Kazakhstan. Health Econ 12(7):553–564

Tilman D, Fargione J, Wolff B, D'Antonio C, Dobson A, Howarth R, Schindler D, Schlesinger WH, Simberloff D, Swackhamer D (2001) Forecasting agriculturally driven global environmental change. Science 292(5515):281–284

Tilt B, Braun Y, He D (2009) Social impacts of large dam projects: a comparison of international case studies and implications for best practice. J Environ Manag 90(Supp 3):S249–S257

Umemiya C, Rametsteiner E, Kraxner F (2010) Quantifying the impacts of the quality of governance on deforestation. Environ Sci Policy 13(8):695–701

UNCTAD (2008) World investment report 2008. Transnational corporations and the infrastructure challenge. United Nations Conference on Trade and Development, Geneva [etc]

UNDP (2010) The real wealth of nations: pathways to human development. Human Development Report 2010. United Nations Development Programme, New York

UNEP (2007) Global environment outlook 4 – environment for development. United Nations Environment Programme, Nairobi

UNEP (2012) Global environment outlook 5 – environment for the future we want. United Nations Environment Programme, Nairobi

Unruh J, Cligget L, Hay R (2005) Migrant land rights reception and 'clearing to claim' in Sub-Saharan Africa: a deforestation example from southern Zambia. Nat Resour Forum 29:190–198

Upadhyay TP, Solberg B, Sankhayan PL (2006) Use of models to analyse land-use changes, forest/soil degradation and carbon sequestration with special reference to Himalayan region: a review and analysis. Forest Policy Econ 9(4):349–371

USAID, ARD (2008) Global land tenure master database. Unpublished data. Original graphics published in USAID and ARD, Inc. (2007) Land tenure and property rights tools and regional reports. United States Agency for International Development, Washington, DC; ARD, Burlington

Warner K, Hamza M, Oliver-Smith A, Renaud F, Julca A (2010) Climate change, environmental degradation and migration. Nat Hazards 55(3):689–715

Warren A (2002) Land degradation is contextual. Land Degrad Dev 13(6):449–459

White JS, Speizer IS (2007) Can family planning outreach bridge the urban–rural divide in Zambia? BMC Health Serv Res 7:143. doi:10.1186/1472-6963-7-143

Whitmore TC (2009) An introduction to tropical rain forests, 2nd edn. Oxford University Press, Oxford

Wilmsen B, Webber M, Duan Y (2011) Involuntary rural resettlement: resources, strategies, and outcomes at the three Gorges Dam, China. J Environ Dev 20(4):355–380

Wilson S, Fesenmaier DR, Fesenmaier J, van Es JC (2001) Factors for success in rural tourism development. J Travel Res 40(2):132–138

Wolf S (2009) Water and sanitation for all? Rural versus urban provision. Int J Serv Econ Manag 1 (4):358–370

World Bank (2009) Reshaping economic geography. World Development Report 2009. World Bank, Washington, DC

World Bank (2010) World development indicators. World Bank, Washington, DC. Available online at http://data.worldbank.org/indicator/. Accessed 29 Apr 2011

World Bank (2012a) Logistics performance index. International LPI ranking. Available online at http://info.worldbank.org/etools/tradesurvey/mode1b.asp. Accessed 21 Jan 2012

World Bank (2012b) The worldwide governance indicators, 2012 update. Available online at www.govindicators.org. Accessed 14 Feb 2013

WRI (1992) World resources 1992–1993: a guide to the global environment. World Resources Institute/Oxford University Press, New York/Oxford

WRI (2011) Earth trends: environmental information. World Resource Institute, Washington, DC. Available online at http://earthtrends.wri.org. Accessed 29 Apr 2011

Xu S (2011) Tactics on the development of modern agricultural logistics in central China. Adv Mater Res 219–220:366–369

Ye L, van Ranst E (2009) Production scenarios and the effect of soil degradation on long-term food security in China. Glob Environ Change 19(4):464–481

Zacharias A, Vakulabharanam V (2011) Caste stratification and wealth inequality in India. World Dev 39(10):1820–1833

Zimmer Z, Kaneda T, Tang Z, Fang X (2010) Explaning late life urban vs. rural health discrepancies in Beijing. Soc Forces 88(4):1885–1908

Chapter 4
Land Management Systems at the Interface Between Forestry and Agriculture

Gerald Kapp and David Butler Manning

Abstract Trees cultivated on agricultural land are playing an ever more significant role in rural development while at the same time also providing a range of ecosystem services such as soil protection, biodiversity conservation and greater carbon sequestration. All over the world, at the interface between forestry and agriculture, a wide variety of agroforestry systems are being implemented as a means to provide much needed wood of different dimensions and qualities, other non-timber forest products as well as animal and agricultural crop produce. An overview of some of the main agroforestry approaches making a contribution to rural development globally are presented, including innovative examples of modern agroforestry. Subsequently, another land use system flanking forestry and agriculture that has been growing in importance internationally in recent years is presented, namely short rotation coppice management. Distinct from traditional agroforestry systems in that in most cases wood is the only production goal, short rotation coppice plantations are a highly productive form of agricultural land use providing farmers with a great deal of flexibility, potentially high yields and certain ecosystem services.

Keywords Agroforestry systems • Short rotation coppice • Farm forestry systems • Non-timber forest products

G. Kapp (✉)
Climate Policy & Finance/World Bank Institute – Climate Change, 701 18th Street, MSN J3-300, Washington, DC 20433, USA
e-mail: geraldkapp@worldbank.org

D. Butler Manning
Technische Universität Dresden, Institute of International Forestry and Forest Products, Pienner Straße 19, Tharandt 01737, Germany
e-mail: david.butler-manning@forst.tu-dresden.de

J. Pretzsch et al. (eds.), *Forests and Rural Development*, Tropical Forestry 9, DOI 10.1007/978-3-642-41404-6_4, © Springer-Verlag Berlin Heidelberg 2014

4.1 Introduction

In this chapter tree-based land use systems implemented by farmers and agricultural enterprises are presented from the managerial perspective. It includes a brief overview of the different agroforestry and farm forestry systems backed up by some case studies. The principal focus, however, is on short rotation coppice; an innovative tree-based form of agriculture that has steadily grown in importance over recent years. In contrast to the traditional agroforestry systems, short rotation coppice is often geared solely towards the production of wood – for either energetic or material use. In most cases no other crops or livestock are integrated within the land use system, although incorporation within agroforestry systems is also possible. The physical, labour and financial characteristics, conflicts and interactions of tree-based subsystems within the farm economy are assessed.

In most industrialised countries wood production and agriculture are neatly separated between departments in public administrations and in land use systems on the ground. There are some overarching reasons for this. Modern agriculture generally produces much higher benefits per hectare and year, but it needs fertile sites, predominantly on flat land easily accessible to machines. Over time this has led to the situation that forests are principally confined to mountains, less fertile, dry or wet areas. Yet a glance at many landscapes reveals that trees are not confined to forests and that in fact the agricultural-pastoral landscapes contain quite a lot of trees, scattered in little groves, lines, hedges or plantations, and that in many, often degraded forests, livestock grazes and browses and shifting agriculture is still practised.

Due to the fact that in many temperate countries agroforestry systems were more abundant in the past, some people tend to conclude that agroforestry is an outdated remnant of the past. This is certainly not the case. Spread all over the world, agroforestry systems are constantly adapted to the prevailing demands and the framework conditions of changing societies.

Advances in agroforestry can play a significant part in the achievement of virtually all of the UN's millennium development goals (Garrity 2004a). The means by which agroforestry can contribute to rural development according to the World Agroforestry Centre (ICRAF) are by helping to provide food, generating income and building assets, advancing health and nutrition, conserving biodiversity, protecting watershed services and assisting adaptation to climate change. A continued expansion of agroforestry land use systems globally, and also of short rotation coppice management, will bring with it substantial benefits in terms of the local environment, in turn producing benefits relating to energy security, agriculture, a revival of biodiversity and also greater employment. The role of trees cultivated on agricultural land as a means to mitigate greenhouse gas emissions is also of considerable importance and has been widely underestimated to date (Pachauri 2012). This maintenance of high carbon stocks in terrestrial ecosystems and agri-ecosystems is vital to reducing emissions in both developing and developed countries (Minang et al. 2012).

Smallholder timber production is already an important source of wood in many countries, especially where the forest cover is low. In Kenya, for example, two thirds of the country's woody biomass stems from non-forest areas (Holmgren et al. 1994) and it is estimated that in Bangladesh 90 % of the wood used is produced on agricultural land. In India the figure is approximately 50 % (Garrity 2004b). These values underline the importance of trees in agriculture and support the recent assertion by Minang et al. (2012) that, "tree-based and managed agroforestry systems are beginning to emerge at some scale."

4.2 Some Definitions

4.2.1 Types of Agroforestry and Farm Forestry Systems

Throughout the world a large variety of 'tree'-related systems can be found with a direct connection to agriculture. They are either integrated into the arable or pastoral landscape, where border effects are predominant and no 'forest climate' is present, or they are farm forests with direct links (e.g., labour, finance, use of farm equipment, wood, local climate) to the farm economy. Agroforestry generally includes all land use systems where woody perennials fulfil agricultural as well as forestry functions. More explicitly, agroforestry can be defined as a land use where agricultural crops and/or animals are managed together with woody perennials on the same plot, simultaneously or in succession, to profit from this combination. Farm forestry involves a different focus on land use systems, with forest trees managed within a farmstead. The following figure (Fig. 4.1) provides an overview of common agro- and farm forestry systems found in temperate and tropical countries. It should be noted that 'trees' are often understood in the sense of woody perennials, that is, including shrubs, palms and bamboos. The various systems each usually have quite distinct physical appearances, growth characteristics, management requirements and financial returns. Nevertheless, a limited overlapping of the systems may occur.

A similar, but slightly different approach is taken by the FAO with its concept of 'trees outside forests,' which refers to "trees on land not defined as forest and other wooded land" (thus excluding farm forests). It embraces trees and shrubs on agricultural land, barren land and built-on areas, including agroforestry systems, orchards, small clumps of trees, permanent meadows and pastures, trees growing on farms and in urban and per urban zones, in lines along rivers, canals and roads, and in gardens, parks and towns (see Bellefontaine et al. 2002, presenting many case studies from around the world).

As many traditional agroforestry systems are well documented (e.g., Nair 1993; Huxley 1999; Bellefontaine et al. 2002; Guuroh et al. 2012) the focus of this chapter is on selected examples of general interest, highlighting recent and innovative

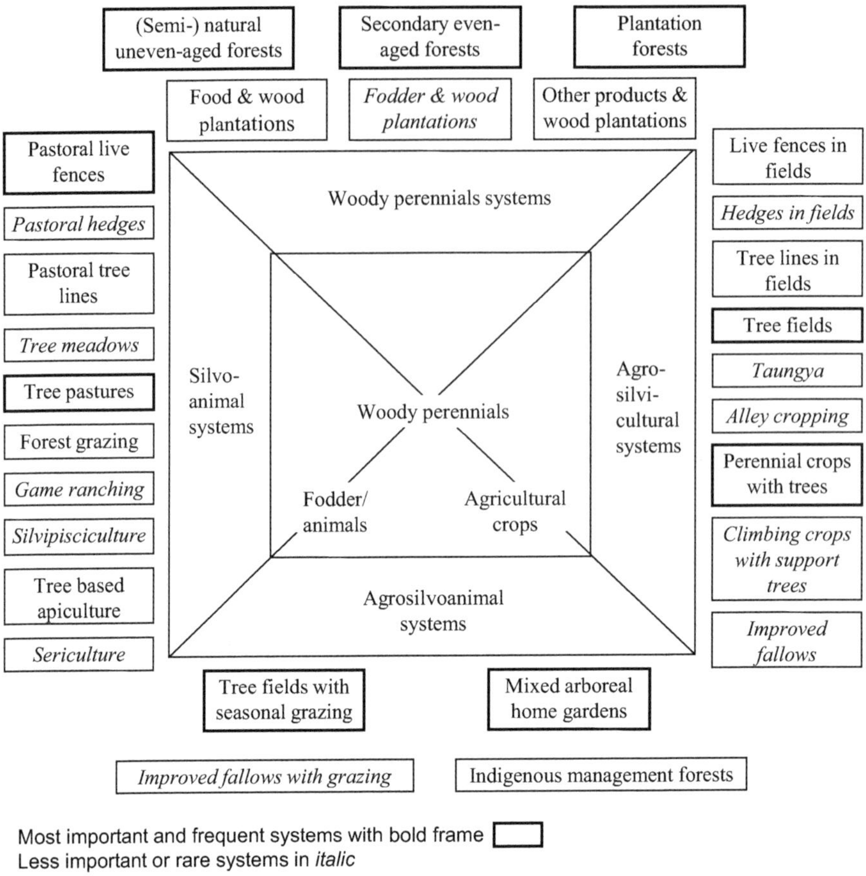

Fig. 4.1 Overview of agroforestry and farm forestry systems in temperate and tropical zones (Source: Kapp (1998), modified)
The term *silvoarable systems* is used synonymously with *agrosilvicultural systems*. All pasture related *silvoanimal systems* are also described as *silvopastoral systems*

systems. Their strengths and weaknesses, and the corresponding opportunities and threats are summarised in SWOT analyses.

4.2.2 Short Rotation Coppice

A short rotation coppice is a plantation of fast growing trees or shrubs, the primary objective of which is to produce high yields of wood over short periods of time. Confusingly, short rotation coppice (sometimes abbreviated as SRC) is one of a number of terms frequently used interchangeably in the literature although in many cases the terms are not synonymous. Various studies refer to short rotation coppice as short rotation crops (also inconveniently abbreviated as SRC), short rotation woody crops, short rotation forestry, arable coppice, short rotation intensive culture, etc. The term short rotation coppice is preferred here for its specificity: coppicing is central to the management system and relates specifically to woody species. Short rotation forestry, for example, does not necessarily involve coppicing and short rotation crops need not necessarily incorporate either coppicing or even woody species.

As part of the management system, the trees are cut back to just above ground level at the end of the first – and after each subsequent – rotation, with the following generations based on sprouts emerging from the stools remaining in the ground (i.e., coppicing). The system is particularly flexible in that plantations may be established in conventional blocks of differing sizes and with widely diverging stocking densities, or they may be established in strips resembling hedgerows or windbreaks. The differing rotation lengths are another source of flexibility. Short rotation coppice plantations are usually harvested in 2–5 year rotations, but may also have longer rotations of between 10 and 20 years, depending on the production goals, the market situation, labour capacities, etc. Plantations established to produce fibres for material use in the pulp and paper industry, for instance, will have longer rotations and lower stocking densities than plantations established to produce wood chips or billets to be used as woodfuel. Generally the second and later generations of trees will not produce wood suitable for use in fibre production. Once established, a plantation can be harvested for 25 years or more, depending on the species and site, before the vitality of the stools begins to decline and a renewal of the plantation is required.

The annual biomass yield attained in short rotation coppice plantations depends on the tree species and growth conditions, and varies between roughly 7–18 t dry bulk (bdt) per hectare and year when grown on suitable sites in temperate regions, and 15–30 bdt/ha/a in the tropics. Although most fast growing tree species are generally less site demanding than traditional bioenergy crops, even these have certain site requirements, which, if they are not met, will result in very low yields or even complete crop failures. The net biomass productivity of short rotation coppice is considerably higher than in natural and in sustainably managed high forests (e.g., Bemmann et al. 2011). In many cases it achieves levels of productivity comparable to traditional agricultural production. It is not only the high levels of productivity that make short rotation coppice attractive, it is also the potential benefits in terms of counteracting damage to watersheds, wind and water erosion, pollution and as a means to enhance biodiversity relative to annual crops (e.g., Blick et al. 2003; Burger 2006; Kroiher et al. 2008).

The establishment of short rotation coppice plantations resembles a land use hybrid nestled somewhere between agriculture, horticulture and forestry. In certain countries there is a strict legal separation between land used for agriculture and land that is designated forest land. Contrasting rights and obligations are associated with each land classification, as well as different opportunities for income generation, and different values are placed on the two land types as a consequence. Arable land allows landowners the greatest freedom to make decisions on crops, to adapt to market conditions in the short term and so to avail of the greatest income generation opportunities. In these countries, therefore, it is important that the agricultural status of the land is maintained when trees are planted for short rotation coppice management. In many countries short rotation coppice is now recognised as an agricultural form of land use rather than as forestry.

4.3 Evolution of Land Uses and Lines of Thought

4.3.1 Shifting Cultivation and Farm Forestry in the Tropics

In the past, and still to this day in many places, farmers established their property in areas of natural forest by clearing trees using both tools and fire so that they could either plant their crops or establish pastures. Often other parts of the property remained under forest. This remnant area of *(semi-)natural uneven-aged forest* has in many cases been exploited. The most valuable trees have usually been harvested by timber companies or local individuals, and the logs or boards sold on to timber merchants. These remnant forests are located further away from the farm house or on land less suitable for tillage or pastoral use due to their steepness, the soil properties or the water regime. At the agricultural frontier in developing countries, farm property sizes are still relatively large, often in the range of 10–100 ha. Due to limitations in terms of family labour force, most farmers cannot clear and manage all of this land in the beginning. However, in many countries unmanaged natural forest land is subject to an unclear property status. Often the clearing and the non-intensive use of the land, for example, with grazing of less than one animal per hectare, is accepted as a demonstration of land ownership.

Such remnant forests are also a continuous source of new fertile land; areas cleared and burnt to provide a substrate to replace other plots where crop growth has decreased as a consequence of weed invasion, soil erosion, compaction or nutrient depletion, if no special techniques and fertilisers are available. This frequently leads to a shifting cultivation cycle as depicted in Fig. 4.2.

This process of temporarily, and sometimes permanently, abandoning tillage or pasture leads to the establishment of *secondary, even-aged forests*. For the first 5–10 years these stands can barely be used, except occasionally for some non-timber forest products. Exploitation usually restarts once individuals of timber tree species within the secondary forest have grown to harvestable dimensions.

Farm forests are a constant source of timber for fencing and housing, of wild fruit, game, medicine, materials for handcrafts and roofing, and preserve ground water and springs. Beyond these material uses, farm forests sometimes also have cultural benefits, as landscape attractions, the sites of graveyards and as places of spiritual value. These less tangible benefits often ensure the continued existence of remnant forests.

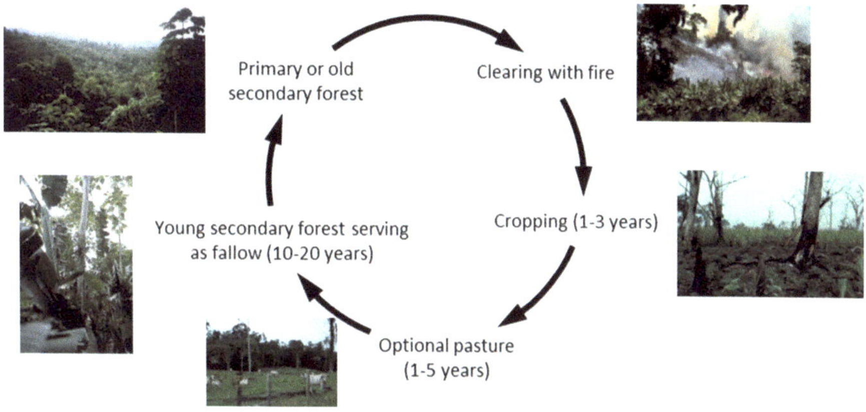

Fig. 4.2 A typical shifting cultivation cycle, a type of sequential agroforestry system. This cycle may or may not include a pasture phase

In addition to (semi-)natural and secondary forests there is a third forest type. Many farmers experiment with the growing of commercial, frequently also exotic timber trees in *plantation forests*. Such plantations are usually small in size.

The frequency and size of forests situated on farms in two neighbouring areas of Costa Rica and Panama were analysed as part of a comprehensive field study (Table 4.1). At the time of study the areas had been colonised 10–30 years.

The results of the survey indicated that in both regions about half of the farms still had substantial forest land that was in acute danger of being gradually converted to agriculture. Equally, the local farmers also recognised that 30–59 % of their forest areas were not suitable for agricultural use. More than half of the farmers in both regions had young secondary forest serving at that time as forest fallow and about 10 % had established plantation forests. These findings are still characteristic of many parts of the developing world.

In view of the important functions of (semi-)natural farm forests as a means to complement livelihoods from farming, for soil conservation, the protection of biodiversity and in terms of carbon sequestration, it is vital that farmers be provided with technical and political incentives to engage in appropriate management of such forests (Kapp 1996).

Some technical and financial information on the feasibility of the management of (semi-)natural farm forest is shown in the following summary of a case study from Panama. The case study highlights that although farm forest management may be less attractive economically in the short term, in the long term it is more feasible than uncontrolled exploitation, both ecologically and economically (Box 4.1).

Similar polycyclic silvicultural models are applied in many countries; for example, in Venezuela (Plonczak 1989) and in a large private enterprise in Paraguay (Sociedad Agrícola Golondrina et al. 2004), as well as in Africa and Southeast Asia.

The main features of *(semi-)natural, uneven-aged forest management* are presented in the form of a SWOT analysis in Table 4.2.

Table 4.1 The use of forest land and land use planning on farms in two neighbouring regions in Costa Rica and Panama

Object	Unit	Talamanca, Costa Rica	Changuinola, Panama
Area surveyed	km^2	600	1,500
Number of smallholders visited and surveyed	–	197	271
Average farm size			
of all farms	ha	20	28
of farms with forests only	ha	33	39
Proportion of farms with forest plantations	%	12	9
Size of forest plantation	ha	3	8
Proportion of farms with forest fallows	%	52	57
Size of forest fallow	ha	5	8
Proportion of farms with (semi-)natural forest	%	40	47
Size of (semi-)natural forest	ha	11	12
Proportion of farms where forest products are used	%	53	57
Proportion of farms that will keep at least part of their forest	%	48	35
Proportion of forest area the farmer deems suitable for pastoral use	%	43	70
Proportion of forest area the farmer deems suitable for cropping	%	41	53

Source: Kapp et al. (1991)

Table 4.2 SWOT analysis of on-farm (semi-)natural, uneven-aged forest management

Strengths	Risk is spread as a consequence of the diversification of farm production
	Higher labour productivity than agriculture
	Environmental co-benefits
	Use of land unsuitable for agriculture
Weaknesses	Lack of silvicultural knowledge
	Market access often problematic, difficult to aggregate sufficiently high timber volumes
	Lower land productivity than agriculture
Opportunities	Additional income from payments for environmental services (PES), carbon credits
	Land titling with forest management plan
	Increasing demand in many countries for wood for construction, paper and energy
Threats	Large proportions of non-commercial tree species
	Restrictive forest laws

Box 4.1 Technical and Financial Feasibility of (Semi-)Natural Farm Forest Management: A Case Study from Panama

The 40 ha making up the moistest area of the Sánchez' 71 ha farm was still covered by natural forest, which had been partly exploited. Some 29 ha of the

(continued)

Box 4.1 (continued)

area provided grazing for 21 cattle. Cocoa and maize were each planted on 1 ha. Traditionally, the farmer dedicated 120 working days to timber exploitation, yielding US $3,810/year, 50 days to cattle management yielding US $600/year, 50 days to the cocoa crop and 17 days to the maize, yielding US $60 and US $150, respectively. The extractive exploitation of the forest led to degradation. The main commercial species were *Carapa guianensis, Pentaclethra macroloba* and *Pterocarpus officinalis*, which contributed 104 m^3/ha to the total standing harvestable stock of 129 m^3/ha. The most abundant species was the palm *Raphia taedigera*, accounting for 45 % of the overall stocking density of 428 trees/ha.

The core of the newly executed on-farm forest management plan, which was introduced in cooperation with a research project coordinated by CATIE (Turrialba, Costa Rica), was a polycyclic silvicultural system oriented towards final crop trees with a timber harvesting cycle of 20 years. With yearly cuttings on 1.55 ha, only 27 % of the 21 m^2 tree basal area is removed, corresponding to a timber volume of 97 m^3 (= 62 m^3/ha). Prior to cutting, 100 potential final crop trees/ha are selected, marked and, where necessary, liberated, with the latter operation to be repeated after 10 years. These potential final crop trees constitute the collective of commercially valuable but still young trees to be cut in future harvest cycles; that is, in 20 or 40 years. Immediately after cutting, the logs are sawed into boards by the farmer using a chainsaw. To transport the sawn timber out of the forest, the farmer installed a cable logging system, an adapted version of the cable system employed by a neighbouring banana company. The gross income from this new system is in the range of US $2,015/year, which is 6 % less than the traditional destructive exploitation system. This provides the farmer with an income of US $38/working day, which is attractive compared to the average local wage of US $8/working day. Forest stocks will be improved under this new form of management, and after 20 years yields are expected to rise to 126 % of the traditional exploitation system, which, had it continued, would have left a degraded forest or cleared land.

Source: Kapp (1998)

4.3.2 *Recent Developments in Agroforestry Systems in Tropical and Temperate Zones*

Agroforestry systems are most prominent in tropical and developing countries, where they have a decisive role in providing the rural and to a lesser degree the urban population with food, fuelwood, timber and other products. There are studies from many countries demonstrating that over half of the total wood supply comes from such trees outside the forest (Bellefontaine et al. 2002).

Agroforestry systems are of increasing importance for a sustained agricultural production in the context of adaptation to climate change. They play a prominent role in 'climate smart agriculture' (The World Bank 2011), which is defined as an agriculture that, "sustainably increases productivity, resilience (adaptation), reduces/removes greenhouse gases (mitigation) while enhancing the achievement of national food security and development goals" (FAO 2012). Trees are also increasingly being integrated into 'conservation agriculture' and form a substantial pillar of 'sustainable land management' (SLM) technologies (Liniger et al. 2011). Trees can have exceptional effects, such as those of *Faidherbia albida*, prominent in large parts of semi-humid and semi-dry Africa. Researchers from GART in Zambia stated that crop yields under the canopy are around 30 % higher (The World Bank Institute, TerrAfrica 2012). This is due to its reversed leaf cycle, nitrogen fixing, deep rooting and nutritious pods, which are also appreciated by livestock.

Agroforestry hedge systems ('alley cropping'), regularly pruned and providing materials for use as mulch or fodder, can successfully stop erosion and increase soil fertility and farm income on sloping lands, as demonstrated in the Sri Lankan–German Upper Mahaweli Watershed Management Project (GTZ 1998) incorporating a variation of this system in conjunction with their sloping agricultural land technology (SALT).

Another prominent agroforestry system throughout the tropics is mixed arboreal home gardens. In a recent study in the Bieha department of Burkina Faso (Guuroh et al. 2012) all of the 80 households surveyed managed home gardens composed of trees (25 species were found), crops and animals. Including their animal component, they contributed 60 % of the farm income.

The most prominent agroforestry research institution, the World Agroforestry Centre (ICRAF), stated that, "it is possible to have higher yields, more carbon in the soil and biomass and greater resilience to droughts and temperature stress if trees are integrated into agricultural production landscapes through locally appropriate forms of agroforestry that modify local climate" (World Agroforestry Centre 2011). Constraints to agroforestry are mainly seen in misperceptions concerning the interactions between trees and crops, and in traditional agricultural policies that often ignore trees on farms and do not provide the necessary incentives.

In recent years agroforestry systems have been undergoing a revival in temperate zones as well, with potentially interesting implications for tropical agroforestry. Whereas forest pastures have largely vanished in temperate Europe, a modified form of tree pasture is a prominent modern land use system in New Zealand (Gordon and Newman 1997). In central Europe, hedgerows have gained new importance as aesthetic landscape elements and as ecologically valuable biotopes with a connective function (Reif and Schmutz 2001), as shelters for year-round pasture for robust sheep and cattle races and as sources of bioenergy.

Another form of modern agroforestry re-emerging in central Europe is the planting of tree species producing valuable timber in single rows in fields and pastures (Bender et al. 2009; Dupraz and Liagre 2008). Only in recent years have such modern forms of agroforestry been promoted by European research projects

Table 4.3 SWOT analysis of on-farm agroforestry systems

Strengths	Risk is spread as a result of the diversification of farm production
	Better use of the space above and below ground leading to increased overall production and higher growth rates of freestanding trees
	Improved protection of agricultural soils against erosion
	Increased soil fertility (nutrient cycles, organic matter, less fertiliser needed)
	Investment and saving functions of valuable timber trees
	Environmental co-benefits
	Sustainable use of lower quality sites for agricultural production
Weaknesses	Competition of trees with crops for space, light, water and nutrients leading to reduced crop yields
	Impediment to mechanisation of plant production
	Management more complex and difficult
Opportunities	Availability of better tree cultivars with improved properties in terms of bole form, growth rates, branch diameters
	Better legal and administrative support for agroforestry land use systems
	Increasing demand in many countries for wood for construction, pulp and paper, and energy
Threats	Insufficient support from agricultural extension services
	Impeding laws and regulations

(e.g., Dupraz et al. 2005; Reeg et al. 2009) and agroforestry associations (e.g., Association Française d'Agroforesterie, www.agroforesterie.fr and EURAF - European Agroforestry Federation, http://euraf.isa.utl.pt/). The basic idea is to achieve higher benefits from the farm production area by integrating a small number (e.g., 20–30) of moderately fast growing valuable tree species (e.g., *Acer pseudoplatanus*, *Juglans spp.*, *Fraxinus excelsior*, *Prunus avium*) that produce highly sought after wood. Under essentially free growing conditions, and with repeated pruning, marketable stem diameters of 40–60 cm can be reached in 30–70 years. The intermedia nut hybrid of *Juglans regia x Juglans nigra*, for example, exhibits very satisfactory growth rates and second generation nut hybrids (*Juglans x intermedia x Juglans mandshurica*) have reached an average stem diameter of 29 cm and an average tree height of 27 m in only 18 years (Mettendorf 2008).

Investments in the on-farm production of hybrid nut can yield an internal rate of return on capital of up to 10 % (Kapp 2009). Out-grower schemes could expect similar returns on investment.

Useful fruit, fodder and wood-producing trees (e.g., *Acer* sp., *Castanea crenata*, *Evodia dainellii*, *Juglans sinensis*, *Robinia pseudoaccacia*) have been planted in contour lines as a means to check erosion in many countries, such as in the case of the North Korean Restoration of Sloping Lands Project. These woody structures also represent good economic prospects (Kapp 2010) and could be adapted in other developing, including tropical countries.

These and other experiences clearly show that agroforestry systems have positive effects but also trade-offs with respect to physiology, labour and cash flow. To be successful, it is essential that these systems be specifically designed to suit the site so that the positive effects are optimised and the potentially negative effects kept to a minimum (Table 4.3).

4.4 Short Rotation Coppice in Industrialised and Developing Countries

In parts of the world where the supply of wood that may be obtained from forests by means of sustainable forest management is abundant, the scope for the establishment of short rotation coppice is limited. However, where the wood supply is short and insufficient to meet existing energy needs, or where the greater use of wood for energy provision is likely to impact negatively on the pulp and paper and wood processing sectors, short rotation coppice plantations are worthy of consideration as a means to enhance the wood supply.

The production of wood in plantations over short rotations is an alternative form of agricultural land use with considerable promise for the future. In addition to providing economic benefits for landowners, short rotation coppice brings with it ecological benefits for society as a whole (Liebhard 2007). Sustainably managed short rotation coppice plantations are a source of renewable energy with very low net carbon dioxide emissions and low levels of nitrogen and sulphur pollutants (Patterson et al. 1994; Matthews and Robertson 2001; Börjesson 2006; BMELV 2007; Bemmann and Große 2011). The reasons for this are that these plantations are a much less management intensive crop than many other plants cultivated for bioenergy and because the root systems, stools and leaves are not removed from the site during the infrequent harvesting operations, so that the impact on soils is also less. Short rotation coppice plantations require a lower chemical input – fertilisers, herbicides and pesticides – than conventional arable crops, and they can serve to increase landscape structure and biodiversity.

There are currently very few areas across the globe where short rotation coppice is practiced to any degree, with two notable exceptions being Brazil and Sweden. As yet, it remains in many ways more a political wish than a land use reality. However, while still a marginal land use, the area of short rotation coppice has expanded steadily over the last decade or two, but to what extent is uncertain. Given that the area is increasing from year to year, the figures quoted in individual studies are often dated. Obtaining clear figures from the literature is also complicated by the aforementioned lack of clarity in the use of alternative terms such as short rotation crops. Moreover, as a relatively new form of land use, short rotation coppice plantations are not separately recorded in most national statistics. On the basis of the data available, Don et al. (2011) estimated a total of somewhere in excess of 30,000 ha short rotation coppice in Europe. For the USA, Wright (2006) calculated an area of short rotation woody crops of around 500 km^2, used for both fuel and fibre, compared to 30,000 km^2 of *Eucalyptus* plantations in Brazil grown solely for woodfuel and between 70,000 and 100,000 km^2 in China. Just what proportion of these woody crops is managed as coppice is not known, however.

In spite of the relatively small area of short rotation coppice cultivated in Europe, the European Union and many of its individual member states rate short rotation coppice highly as a means of biomass production for the purposes of energy generation globally. Indeed, many view short rotation coppice as a welcome alternative to the maize, oilseed rape and other annual energy crops increasingly consuming the arable land area. This latter development is controversial from ecological, nature conservation and landscape protection perspectives. It is also highly questionable in terms of energy and land use efficiency. Short rotation coppice is characterised by numerous advantages over these more traditional energy crops (e.g., Makeschin 1994; Rowe et al. 2010; Glaser and Schmidt 2010), and also by considerably better rates of energy efficiency (Börjesson 2006; Styles and Jones 2007; Bemmann and Große 2011).

For the decision makers at the grass root level, namely farmers, however, there are not only benefits associated with short rotation coppice but also a number of drawbacks and too many uncertainties (Skodawessely and Pretzsch 2009). In the following a general overview of some of the species used for coppice and their characteristics is given, as well as a discussion of the benefits and impediments, and the uses of the wood produced.

4.4.1 Characteristics of Typical Short Rotation Coppice Species

The primary objective of short rotation coppice is the maximisation of wood yield over short rotations. In order to achieve this, the tree species and clones used in short rotation coppice plantations must be characterised by certain traits, whereas other characteristics often valued in tree species used in forestry are irrelevant. Some of the following traits will be of greater or lesser importance in certain parts of the world, but the trees used in short rotation coppice are generally (Liebhard 2007; Schildbach et al. 2009):

- Light demanding species;
- Highly tolerant of both late and early frosts (temperate zones);
- Readily produced in large quantities (e.g., by means of vegetative propagation);
- Relatively easy to establish and characterised by high survival rates;
- Capable of rapid early growth and high biomass increment;
- Able to withstand high levels of competition in closely spaced stands;
- Characterised by high assimilation rates;
- Species with a long vegetation period and characterised by late leaf-fall (tolerant of variable day lengths);
- In possession of a high capacity to coppice or re-sprout from the harvested stool;
- Highly stable;

- Characterised by a small, compact crown with an acute branch angle;
- Not very susceptible to disease;
- Highly tolerant of pests and not prone to browsing; and
- Producers of wood with good thermal characteristics (high calorific value).

Depending on the management objectives and the site, certain other characteristics may also be desirable in short rotation coppice species. In the case of plantations established to grow wood for material use, straight stems may be important. On certain sites species tolerant of either drought or waterlogging may be necessary (Schildbach et al. 2009).

Few of the species potentially used for the purposes of short rotation coppice worldwide have been the subject of breeding programmes to date. Where breeding work has been carried out and different clones and varieties are available, it is important growers choose a variety that is adapted to both the site and the production aims as the differences in biomass production and even survival between the varieties can be considerable.

Box 4.2 Biohof Böhme: A Largely Self-Sufficient Organic Farm Practising Short Rotation Coppice in Eastern Germany

The Böhmes established their first 20 ha of short rotation coppice on their farm in Obercarsdorf, Germany, in 2005. In the years since they have planted another 20 ha. The large yields achieved by the willows and poplars grown on the farm are such – approximately 10 bdt/ha/a – that for every hectare cultivated enough wood chips are produced each year to replace approximately 5,000 l of domestic heating oil. Established for a period of between 25 and 30 years, after which a new tree crop will be planted or the plots restored to conventional agricultural use, the plantations are harvested every 3–5 years, depending on the species and productivity of the individual lots. A 75 kW wood chip heating system uses between 102 and 150 bulk m^3 of the wood chips (17–21 bdt) produced to heat the farm facilities and an adjacent building containing five apartments annually. The heated area amounts to approximately 550 m^2.

As a certified organic farm, no pesticides or herbicides are used on the Böhmes' land. To remove competition from weeds, which can threaten the successful establishment of a crop, the poplar plantations are weeded using Shropshire sheep. The trampling effect of the sheep has the added effect of keeping mice and voles out of the plantation.

The Böhmes were one of the first in their region to participate in short rotation coppice. As a result of this early involvement they are now in a position to market their services as consultants providing advice on the establishment of short rotation coppice and as contractors providing planting services. In addition to using the wood to meet their own heating needs, the Böhmes also produce planting material sold on the regional market. Open to

(continued)

Box 4.2 (continued)

innovation and experimentation, the Böhmes work closely with scientists and engineers from a nearby university in the development of new planting and harvesting equipment and approaches for the tending of plantations. As a result, the enterprise has earned itself a reputation throughout Germany as a pioneer and innovator, and is a model that other enterprises look to for inspiration.

4.4.2 Species Used for Short Rotation Coppice

The principal tree species used in short rotation crops worldwide are few in number, and are limited mainly to eucalypts, poplars, willows, *Robinia* and *Acacia* (Verwijst 2003). Locally other woody species may also be important (refer, for example, to work by Pereira and Pereira 1994; Dalianis et al. 1996; Shackleton 2001; Sims et al. 2001; Rédei 2003; Fang et al. 2004, 2011; Geyer 2006; Buchholz and Volk 2007; Noh et al. 2007; Seebauer 2008; Avohou et al. 2011). The main species used for short rotation coppice in Europe are poplar and willow, and within Europe there are regional differences. Further to the east and south east, as water availability becomes critical, *Robinia* or black locust becomes more prevalent. Short rotation coppice plantations in North America are based primarily on poplar, whereas in Brazil eucalypts predominate. Eucalypts are also commonly used in short rotation coppice in Australia and New Zealand and have considerable potential for plantations in Africa.

4.4.3 Uses of Wood Produced in Short Rotation Coppice

The wood produced from traditional coppice stands and from short rotation coppice plantations can be put to a number of uses. The primary use of wood from modern short rotation coppice, however, is for energy, more particularly for the provision of heat. The generation of electricity from wood is also possible through processes such as gasification and biomass to liquid (BTL), but these are complex and not yet fully developed (see Box 4.3). Heat generation on the other hand represents a simpler and more efficient use of the resource and requires considerably less investment in technology. Wood can be used as the sole raw material or in combination with other fuels.

In many parts of the world, wood-derived fuel still accounts for up to 50 % of the total energy consumed, predominantly in the form of simple woodfuel and charcoal applications (Abell 2005). Indeed, woodfuels account for a greater proportion of energy consumption than all other forms of renewable energy combined (FAO 2010). The advantages of wood produced for fuel in short rotation coppice over other crops such as Jatropha, for example, is that there is no need for a complex refining process in order to produce the fuel. The wood is simply converted to billets or chips and dried. The cultivation and use of the wood can occur locally, with transport over long distances for processing or to reach markets not necessary. Nevertheless, opportunities for export do exist. Certain large global energy providers are currently engaged in transporting woodchips from both Africa and Asia to fuel large power plants in Europe.

Grown under certain conditions, and using appropriate species, the wood produced in short rotation coppice can also be put to material use; for example, in the pulp and paper industry. There is also a growing market for the co-products of short rotation coppice, such as in the oil mallee industry in Australia, where either activated charcoal and leaf oils may be produced. In Brazil charcoal is also produced from eucalypts (Verwijst 2003). Poplar bark may also prove a useful material for use in tanning, whereas the salicylic acid in the bark of willow may have medicinal applications. The wood from short rotation coppice is also being used for gardening purposes in some countries.

Box 4.3 Using Wood to Provide Heat and Electricity

When carefully controlled to ensure complete combustion, modern woodfuel burning systems can be used to provide energy very efficiently and cleanly using automatic, highly efficient boilers and power installations. Approximately 80 % of the energy produced in the burning of wood is released by the combustion of gases formed during heating. The remaining 20 % is produced from the charcoal. Air introduced to the woodfuel aids efficient burning of the gases. Domestic open fires burn at temperatures of between 200°C and 500°C; at around 35 % efficiency. Efficiency is greatly improved in stoves and boilers, up to in excess of 90 %. Introducing hot secondary air above the fuel increases the efficiency of the burn and the gases produced burn at 600–700°C.

Essentially there are three means of converting woodfuel to energy; pyrolysis, gasification and combustion:

Pyrolysis: Pyrolysis involves the heating of wood in the complete absence of oxygen, during which the wood degrades to produce gases, liquids and charcoal. The principle product of the process is a liquid that can be used as a fuel to power a generator. A distinction is made between two forms of pyrolysis: slow pyrolysis and fast or flash pyrolysis. Slow pyrolysis has been used for years to produce charcoal as a solid, stable fuel for industrial and domestic use. Flash pyrolysis, alternatively, is the thermal conversion of biomass at high temperatures over a short time, with reaction times of only a few seconds or less. The aim is to produce gases and liquids for use as biofuel. The gas produced in the fast pyrolysis of wood can be fed back in and burnt to further drive the process. The resulting charcoal can also be

(continued)

Box 4.3 (continued)

burnt. Fast pyrolysis is a relatively new technology, however, and has not been applied in the burning of wood on a commercial scale as yet. As the liquid fuel produced in fast pyrolysis can be stored and readily transported there is considerable interest in further developing the technology. A further form of 'mild' pyrolysis is torrefaction, which is currently also being investigated as a means to improve the transport-worthiness of woody biomass.

Gasification: Combustible gases are given off when the supply of air is restricted as wood is heated. These gases can be cleaned and used to drive an engine, which is in turn used to generate electricity. Gasification is generally more efficient than combustion, especially when generating electricity. The application of the technology to produce electricity from wood is still in the early stages and developments are ongoing. Nevertheless, gasifiers produced by a small number of manufacturers are already available on the market.

Combustion: During combustion wood is completely oxidised. The resulting hot gases are used to produce steam, which drives a steam turbine in order to produce electricity. This is a less efficient means of converting woodfuel to electricity but, unlike pyrolysis and gasification, the technology is well established and the capital costs are relatively low. Almost all woodfuel-powered electricity plants are based on combustion. By using the heat produced by the combustion process as part of a combined heat and power (CHP) scheme, improvements in both efficiency and costs can be achieved.

Source: Ireland et al. (2004, 2006)

4.4.4 *Benefits, Drawbacks and Impediments to Short Rotation Coppice*

As with so many forms of land use, the pros and cons of short rotation coppice are not absolute and are a function of the site and the land use they replace (Cossalter and Pye-Smith 2003; Elliot 2003; Bielefeldt et al. 2008). Where short rotation coppice replaces a portion of a large scale, intensively managed maize crop situated in an open, poorly structured landscape the plantation will serve to increase biodiversity, for example, and provide benefits in terms of soil and wind erosion. If a coppice plantation replaces natural forest, however, the impact in terms of biodiversity and erosion will be negative. Just some of the benefits will be summarised in the following before the factors currently mitigating against the expansion of short rotation coppice are outlined.

4.4.4.1 Benefits

There is a range of benefits associated with short rotation coppice. Importantly, these plantations represent a means to increase the *wood supply* on the market in the short term. The growing demand globally for wood for energy purposes is leading to conflict between those seeking to use wood for material purposes and those requiring wood for energy. The growing demand for wood is placing a greater strain on the Earth's forests, in tropical, boreal and temperate zones, thereby posing a threat to their sustainable management. Even in countries where the concept of sustainable forest management is well established and widely implemented, the pressure to make greater use of forest residues normally left in the stand after harvesting, so that nutrients may be returned to the soil, is increasing.

The advantages of short rotation coppice plantations extend beyond simply relieving the wood market, however. These plantations also provide ecological advantages relative to many traditional annual cultures. As they are not harvested every year and as the roots remain in the ground for many years the pressure on the *soil* is greatly reduced. This is beneficial in terms of soil erosion but also in that the loss of nutrients from the soil is lessened. Plantations have also been found to positively influence soil structure (Kahle et al. 2005; Lamersdorf et al. 2010).

Short rotation coppice plantations are also characterised by greater *biodiversity* than many traditional agricultural crops (Gustafsson 1987; Berg 2002; Weih et al. 2003; Weih 2009; Glaser and Schmidt 2010; Schmidt and Gerold 2010), provide a better habitat and also represent an additional *structure*-giving element in the landscape (Skärbäck and Becht 2005; Rode 2005). Also of considerable importance in terms of ameliorating climate change is the fact that short rotation coppice is a largely *carbon neutral* form of land use because the carbon released upon burning is carbon that was removed from the atmosphere during the standing time of the plantation (BMELV 2007; Styles and Jones 2007; FAO 2010), with carbon bound for varying lengths of time in the wood of the roots, stools, stems and in the humus remaining on site. The low management intensity and the very limited need for fertiliser application mean that the energy inputs required for cultivation are less than for other, annual biofuel crops. Additionally, there is some evidence to suggest that such plantations can play a role in the *phytoremediation* of contaminated soils and in the treatment of wastes (Aronsson and Perttu 2001; Guo et al. 2002; Klang-Westin and Eriksson 2003; Dimitriou and Aronsson 2005; Cavanagh et al. 2011). Short rotation coppice is also being investigated as a means to recultivate sites previously used for surface mining (Grünewald 2005; Rockwood et al. 2006).

Short rotation coppice plantations also provide *management and economic advantages* for landowners. In many parts of the world the price of wood chips is increasing gradually, and is generally more stable than the values of other agricultural crops. Plantations represent a means to diversify production profiles and provide landowners with a buffer against negative market developments for certain produce, or against crop failures caused by weather extremes or pest outbreaks. The introduction of plantations also represents a means for farmers to adapt their production to possible changes to the local climate generally. In temperate regions of the world, the harvesting of plantations occurs in winter, a season that is otherwise relatively quiet for arable farmers, so providing for a better distribution of labour over the entire year.

4.4.4.2 Drawbacks and Impediments

In spite of the many advantages of short rotation coppice and the strategic importance often attributed to this form of land use at scientific and policy level, in many parts of the world the steps necessary to encourage a more widespread implementation of the practice are not being made. As yet there are very few countries or regions with an unambiguous and definitive stance with regard to short rotation coppice at the policy level. Whereas in some cases general targets have been set in terms of land area or wood volume, no clear indication has been given as to how these goals are to be met and few concrete measures have been taken to remove the obstacles to the establishment of woodfuel plantations. This is also reflected in a general lack of, or the often contradictory nature of the information regarding short rotation coppice available at the level of various national, regional and local agricultural advisory bodies and services. As a consequence, many landowners lack the necessary knowledge in relation to the management and the perspectives offered by short rotation coppice (Weih 2009; Bemmann et al. 2010), and also regarding the marketing of wood chips.

The *high capital investment* relative to traditional agricultural crops required in the establishment of short rotation coppice means that a plantation must remain in place for a considerable period of time. How long will vary depending on local economic and environmental factors. In regions where the costs of land and labour are high, a plantation lifetime of 20–30 years with harvests every 3–6 years is necessary. Where natural productivity is higher and labour costs are lower, the investment and plantation lifetime will decline correspondingly. The consequence of this long term binding of the site is a reduction of the landowners' *flexibility* to react to positive market developments for other crops. Accustomed to an annual income for their produce, a return only every 3–5 years is also a hindrance for many landowners.

Farmers' *knowledge* with regard to the establishment, management, harvesting and marketing of short rotation coppice plantations and their produce is still very limited. Whereas traditional agricultural crops have been the subject of breeding programmes, and developments in the corresponding technology, marketing structures and advisory services have evolved over time, this is not yet the case for short rotation coppice. The knowledge required by the landowner relates, among other things, to the choice of tree species and site, site preparation, the selection of varieties and clones, the possible diseases and pests, growth rates, the calculation of returns, harvesting and marketing of wood chips.

Wood chips are a relatively new product and as such in many parts of the world the corresponding market is only just developing. As a result there is no reliable *market price*, which is a cause of concern for both potential producers and consumers of wood chips and a disincentive to participate in short rotation coppice. Three basic marketing approaches exist for farmers, namely private use, sale to local or regional consumers (community facilities, local industry) and sale to large-scale, supraregional consumers (combined heat and power plants) (Bemmann et al. 2010).

The *equipment and technology* used in the establishment and harvesting of short rotation coppice plantations is often machinery that farmers do not use in the cultivation of traditional crops and so do not have ready access to. This includes planting machines, harvesters, chippers and wood chip drying solutions. As short rotation coppice is still a relatively new form of land use, and as yet not very widely practiced, the corresponding technology is in some cases not yet very advanced (Bemmann et al. 2010).

Another more fundamental issue in many parts of the world is farmers' (lack of) *acceptance* of trees on agricultural land. Where the clearance of trees has been associated with progress, the decision to restore trees to hard-earned agricultural land is not one that is made lightly.

In spite of the scope for environmental benefits, short rotation coppice can impact negatively on the *environment* where due consideration is not given to the surroundings. Short rotation coppice should not replace natural forest or be established within areas with a high nature conservation value. Where water availability is a problem, the suitability of short rotation coppice plantations must also be carefully considered as they may impact negatively upon the local hydrology. It is essential that a landscape approach be adopted when planning the establishment of a plantation to ensure that there is no adverse impact on ecosystems, biodiversity or on the hydrological cycle. The social context must also be taken into consideration to ensure that plantations benefit local communities rather than work to their detriment (Cossalter and Pye-Smith 2003).

The main features of short rotation coppice are summarised in the SWOT analysis presented in Table 4.4.

4.5 Outlook

4.5.1 The Outlook for Agroforestry

As highlighted by Nair and Garrity (2012), the fact that nearly a billion hectares of agricultural land currently have a tree cover in excess of 10 % demonstrates the global importance of agroforestry. It is now widely recognised as being a viable land use alternative with a key role to play in banishing hunger and poverty, and in rebuilding resilient natural environments. Increasingly, centuries-old traditional agroforestry approaches are being implemented on the basis of a solid scientific foundation.

Table 4.4 SWOT analysis of short rotation coppice plantations

Strengths	Risk is spread as a consequence of the diversification of farm production
	Environmental co-benefits
	Use of land unsuitable for agriculture
	Greater efficiency of wood as an energy crop
Weaknesses	High capital investment relative to other energy crops
	Lack of knowledge relating to species, plantation management and marketing of produce
	Market access often problematic, difficult to aggregate sufficiently high timber volumes
	Lower land productivity than intensive agriculture
Opportunities	Additional income from payments for environmental services (PES), carbon credits
	Increasing demand in many countries for wood
Threats	Insufficient support from agricultural extension services
	Competing demands for land
	Failure to develop local markets for wood chips

Against a background of ongoing deforestation and forest degradation, a growing demand for additional sources of timber, woodfuel and many other tree products is leading to improved prices for farm-grown wood; a development to which more and more land owners are responding. An increase in the extent of tree cover on farms is, therefore, likely to continue (Nair and Garrity 2012).

Especially for many old tropical soils with unfavourable chemical compositions and physical structures, providing shelter and the incorporation of organic matter by means of woody perennials is key to a sustainable production of crops and fodder, and to the adaptation of agriculture to climate change. This carbon sequestration and storage also provides a valuable climate mitigation option, which can generate some additional carbon income in agricultural projects, as evidenced, for example, in Kenya (Woelcke and Tennigkeit 2009; Friedman and Shames 2012). At the same time, agroforestry systems conserve biodiversity in agricultural landscapes, protect groundwater and relieve the pressure on the remaining forests.

In some parts of the world the removal of legal and policy constraints previously serving to hinder the development of agroforestry are having positive effects (Minang et al. 2012; Ajayi and Place 2012), while elsewhere economic incentives to land managers are beginning to provide an impetus for agroforestry (Kumar et al. 2012).

4.5.2 The Outlook for Short Rotation Coppice

The outlook for the expansion of short rotation coppice globally over the coming years is good but many barriers need to be overcome and uncertainties expunged. The growing demand for wood worldwide and the increasing recognition of the much higher energy efficiency and positive carbon balance associated with short rotation coppice over other energy crops favour this expansion. In various parts of the world studies of the most suitable species and management approaches have been carried out, or are currently underway, and a good basis for coppice systems has been established. Elsewhere, however, there are still considerable knowledge gaps with regard to the appropriate species, cultivation practices, the expected yields, technology and so on. The opportunities for farm diversification and the creation of an additional source of income, as well as the possibility to achieve greater independence from fossil fuels and large energy concerns are just some of the incentives that will potentially motivate small and large landowners to consider participating in short rotation coppice for the purposes of producing wood chips. Production on a scale going beyond mere subsistence use, however, requires the development of markets for this relatively new product. In order to expand, existing and successful wood chip networks beginning with the producer and ending with the consumer, and including all of the various partners in between, need to be highlighted and used as models upon which a wider application can be based.

References

Abell TM (2005) Forestry and biomass production – lessons from the temperate regions and the tropics. Paper presented to the international conference 'issues for sustainable use of biomass resources for energy,' Colombo

Ajayi OC, Place F (2012) Policy support for large-scale adoption of agroforestry practices: experience from Africa and Asia. In: Nair PKR, Garrity DP (2012) (eds) Agroforestry – the future of global land use, vol 9, Adv in Agrofor. Springer, Dordrecht/Heidelberg/New York/ London, pp 175–201

Aronsson P, Perttu K (2001) Willow vegetation filters for wastewater treatment and soil remediation combined with biomass production. Forest Chron 77(2):293–299

Avohou TH, Houehounha R, Glele-Kakai R, Assogbadjo AE, Sinsin B (2011) Firewood yield and profitability of a traditional Daniellia oliveri short-rotation coppice on fallow lands in Benin. Biomass Bioenerg 35:562–571

Bellefontaine R, Petit S, Pain-Orcet M, Deleporte P, Bertault J-G (2002) Trees outside forests. Towards a better awareness. FAO Conservation Guide 35, Rome http://www.fao.org/docrep/ 005/y2328e/y2328e00.htm. Accessed 9 Jan 2013

Bemmann A, Große W (2011) Effiziente Landnutzung – ein Beitrag zur Zukunftssicherung. Vision der Professur für Forst- und Holzwirtschaft Osteuropas. In: Bonn S, Erler J, Herzog S (eds) Tharandt 2011 – 200 Jahre Ideen für die Zukunft. Forstwissenschaftliche Beiträge Tharandt/Contributions to For Science. Beiheft 12:16–37

Bemmann A, Nahm M, Brodbeck F, Sauter UH (2010) Holz aus Kurzumtriebsplantagen: Hemmnisse und Chancen. Forstarchiv 81(6):246–254

Bemmann A, Butler Manning D, Röhle H, Skibbe K (2011) Potenziale, Chancen und Hemmnisse einer Kurzumtriebswirtschaft. AFZ-Der Wald 11:10–12

Bender B, Chalmin A, Reeg T, Konold W, Mastel K, Spiecker H (2009) Moderne Agroforstsysteme mit Werthölzern. Leitfaden für die Praxis. BMBF und Forschungszentrum Jülich, meisterdruck, Reute

Berg Å (2002) Breeding birds in short-rotation coppices on farmland in central Sweden. The importance of salix height and adjacent habitats. Agr Ecosyst Environ 90:265–276

Bielefeldt J, Bolte A, Busch G, Dohrenbusch A, Kroiher F, Lamersdorf N, Schulz U, Stoll B (2008) Energieholzproduktion in der Landwirtschaft. Chancen und Risiken aus Sicht der Natur- und Umweltschutzes. Naturschutzbund Deutschland (NABU) e.V., Berlin, Warlich Druck Meckenheim GmbH, Meckenheim

Blick T, Weiss I, Burger F (2003) Spinnentiere einer neu angelegten Pappel-Kurzumtriebsfläche (Energiewald) und eines Ackers bei Schwarzenau (Lkr. Kitzingen, Unterfranken, Bayern). Arachnologische Mitteilungen 25:1–16

Börjesson P (2006) Livscykelanalys av Salixproduktion. Lunds Tekniska Högskola, Institutionen för teknik och samhälle. Rapport nr 60

Buchholz T, Volk T (2007) Designing short-rotation coppice based bioenergy systems for rural communities in east Africa. Final report: biosyrca project. USAID

Bundesministerium für Ernährung, Landwirtschaft und Verbraucherschutz (BMELV) (2007) Nutzung von Biomasse zur Energiegewinnung – Empfehlungen an die Politik. Studie des wissenschaftlichen Beirates Agrarpolitik des Bundesministeriums für Ernährung, Landwirtschaft und Verbraucherschutz. Berlin

Burger F (2006) Zur Ökologie von Energiewäldern. In: DRL (Deutscher Rat für Landespflege) Die Auswirkungen erneuerbarer Energien auf Natur und Landschaft. Schriftenreihe des Deutschen Rates für Landespflege 79:74–80

Cavanagh A, Gasser MO, Labrecque M (2011) Pig slurry as fertilizer on willow plantation. Biomass Bioenerg 35:4165–4173

Cossalter C, Pye-Smith C (2003) Fast-wood forestry: myths and realities. Centre for International Forestry Research, Jakarta

Dalianis C, Djouras N, Sooter C (1996) Very short rotation and dense eucalypt plantations for energy. In: Chartier P, Ferrero GL, Henius UM, Hultenberg S, Sachau J, Wiinblad M (eds) Biomass for energy and environment. Proceedings of the 9th European bioenergy conference, Copenhagen, Pergamon, pp 725–732, 24–27 June 1996

Dimitriou I, Aronsson P (2005) Willows for energy and phytoremediation in Sweden. Unasylva 221(56):47–50

Don A, Osborne B, Hastings A, Skiba U, Carter MS, Drewer J, Flessa H, Freibauer A, Hyvönen N, Jones MB, Lanigan GJ, Mander Ü, Monti A, Njakou Djomo S, Valentine J, Walter K, Zegada-Lizarazu W, Zenone T (2011) Land-use change to bioenergy production in Europe: implications for the greenhouse gas balance and soil carbon. GCB Bioenerg 4(4):372–391

Dupraz C, Liagre F (2008) Agroforesterie. Des arbres et des cultures. Editions France Agricole, Paris

Dupraz C, Burgess P, Gavaland A, Graves A, Herzog F, Incoll L, Jackson N, Keesman K, Lawson G, Lecomte I, Liagre F, Mantzanas K, Mayus M, Moreno G, Palma J, Papanastasis V, Paris P, Pilbeam D, Reisner Y, Vincent G, van der Werf W (2005) Synthesis of the 'silvoarable agroforestry for Europe' project. INRA-UMR System Editions, Montpellier

Elliot C (2003) WWF vision for planted forests. Report for the UNFF Intersessional Experts' meeting on the role of planted forests in sustainable forest management, Wellington, 24–30th Mar 2003

Fang S, Li G, Fu X (2004) Biomass production and bark yield in the plantations of Pteroceltis tatarinowii. Biomass Bioenerg 26:319–328

Fang S, Liu Z, Cao Y, Liu D, Yu M, Tang L (2011) Sprout development, biomass accumulation and fuelwood characteristics from coppiced plantations of Quercus acutissima. Biomass Bioenerg 35:3104–3114

FAO (2010) What woodfuels can do to mitigate climate change, FAO forestry paper 162. Food and Agriculture Organisation of the United Nations, Rome

FAO (2012) Climate smart agriculture for development. Food and agriculture organisation of the United Nations. http://www.fao.org/climatechange/climatesmart/en/. Accessed 9 Jan 2013

Friedman F, Shames S (2012) Making money from soil carbon in western Kenya. New agriculturist, http://www.new-ag.info/en/focus/focusItem.php?a=2611. Accessed 9 Jan 2013

Garrity DP (2004a) Agroforestry and the achievement of the millennium development goals. Agroforest Syst 61:5–17

Garrity DP (2004b) Agroforestry and the future of global land use. In: Nair PKR, Garrity DP (2012) (eds) Agroforestry – the future of global land use, vol 9, Adv in Agrofor. Springer, Dordrecht/Heidelberg/New York/London, pp 21–7

Geyer WA (2006) Biomass production in the Central Great Plains USA under various coppice regimes. Biomass Bioenerg 30:778–783

Glaser T, Schmidt PA (2010) Auswirkungen von Kurzumtriebsplantagen auf die Phytodiversität. In: Bemmann A, Knust C (eds) Agrowood – Kurzumtriebsplantagen in Deutschland und europäische Perspektiven. Weißensee Verlag, Berlin, pp 153–160

Gordon AM, Newman S (eds) (1997) Temperate agroforestry systems. CAB Internat, Wallingford

Grünewald H (2005) Anbau schnellwachsender Gehölze für die energetische Verwertung in einem Alley-Cropping-System auf Kippsubstraten des Lausitzer Braunkohlereviers. Cottbuser Schriften zu Bodenschutz und Rekultivierung, Band 28

GTZ (1998) Upper Mahaweli Watershed Management Project: second progress review. Kandy, Sri Lanka, Mahaweli Authority of Sri Lanka/GTZ, Eschborn

Guo LB, Sims REH, Horne DJ (2002) Biomass production and nutrient cycling in eucalyptus short rotation forests in New Zealand. I: biomass and nutrient accumulation. Bioresour Technol 85:273–283

Gustafsson L (1987) Plant conservation aspects of energy forestry – a new type of land use in Sweden. Forest Ecol Manag 21:141–161

Guuroh RT, Uibrig H, Acheampong E (2012) Homegardens as a source of income for rural households – a case study of Bieha district, southern Burkina Faso. J Agr Sci Technol B2:798–813, http://uni-bonn.academia.edu/ReginaldTangGuuroh/Papers. Accessed 9 Jan 2013

Holmgren P, Masakha EJ, Sjoholm H (1994) Not all African land is being degraded: a recent survey of trees on farms in Kenya reveals rapidly increasing forest resources. Ambio 23:390–395

Huxley PA (1999) Tropical agroforestry. Blackwell Science, Oxford

Ireland D, Hall A, Jones DH (2004) Woodfuel information pack. Forest Research, Surrey

Ireland D, Claridge J, Pow R (2006) The quest for sustainable energy: woodfuel meets the challenge. Forest Research, Surrey

Kahle P, Baum C, Boelcke B (2005) Effect of afforestation on soil properties and mycorrhizal formation. Pedosphere 15:754–760

Kapp G (1996) Development-oriented research: peasant rainforest management in the lowland tropics of Central America. Plant Res Dev 43:16–30

Kapp G (1998) Bäuerliche Forst- und Agroforstwirtschaft in Zentralamerika. Untersuchungen über forstliche und agroforstliche Produktionssysteme unter besonderer Berücksichtigung des feuchten Tieflands von Costa Rica und Panama. Margraf Verlag, Weikersheim

Kapp G (2009) Waldbauliches Potentiale von Hybridnüssen für Forstinvestitionen. Internal study. GFA ENVEST, Hamburg

Kapp G (2010) Interim evaluation report: improvement of food security and people's livelihood on sloping lands in DPR Korea. Deutsche Welthungerhilfe e.V, Bonn

Kapp GB, Kremkau K, Dixon F (1991) Manejo sostenido de bosquetes en fincas privadas de los trópicos húmedos. Un estudio efectuado en la zonas de Changuinola (Panamá) y Talamanca (Costa Rica). CATIE, Turrialba, C.R, El Chasqui No. 26:5–25.

Klang-Westin E, Eriksson J (2003) Potential of salix as phytoextractor for Cd on moderately contaminated soils. Plant Soil 249(1):127–137

Kroiher F, Bielefeldt J, Bolte A, Schulter M (2008) Die Phytodiversität in Energieholzbeständen. Archiv Forstwesen Landschaftsökologie 42(4):158–165

Kumar BM, Singh AK, Dhyani SK (2012) South Asian agroforestry: traditions, transformations, and prospects. In: Nair PKR, Garrity DP (2012) (eds) Agroforestry – the future of global land use, vol 9, Adv in Agrofor. Springer, Dordrecht/Heidelberg/New York/London, pp 359–89

Lamersdorf N, Petzold R, Schwärzel K, Feger K-H, Köstner B, Moderow U, Bernhofer C, Knust C (2010) Bodenökologische Aspekte von Kurzumtriebsplantagen. In: Bemmann A, Knust C (eds) Agrowood – Kurzumtriebsplantagen in Deutschland und europäische Perspektiven. Weißensee Verlag, Berlin, pp 170–188

Liebhard P (2007) Energieholz im Kurzumtrieb – Rohstoff der Zukunft. Leopold Stocker Verlag, Graz/Stuttgart

Liniger H, Studer RM, Hauert C, Gurtner M (2011) Sustainable land management in practice. Guidelines and best practices for Sub-Saharan Africa. WOCAT, TerrAfrica-NEPAD, FAO, Rome, http://www.wocat.net/fileadmin/user_upload/documents/Books/SLM_in_Practice_E_Final_low.pdf. Accessed 9 Jan 2013

Makeschin F (1994) Effects of energy forestry on soils. Biomass Bioenerg 6:63–79

Matthews R, Robertson K (2001) Answers to ten frequently asked questions about bioenergy, carbon sinks and their role in climate change. IEA Bioenergy Task 38, Joanneum Research, Austria

Mettendorf B (2008) Erfahrungen im Anbau mit Hybridnüssen. In: Mauer WD, Haase B (eds) Die Walnuss (Juglans regia L.). Mitteilungen aus der Forschungsanstalt für Waldökologie und Forstwirtschaft Rheinland-Pfalz Nr. 66/08, Trippstadt, 61–72

Minang PA, van Noordwijk M, Swallow BM (2012) High-carbon-stock rural development pathways in Asia and Africa: improved land management for climate change mitigation. In: Nair PKR, Garrity DP (2012) (eds) Agroforestry – the future of global land use, vol 9, Adv in Agrofor. Springer, Dordrecht/Heidelberg/New York/London, pp 127–44

Nair PKR (1993) An introduction to agroforestry. Kluwer Academic Publishers in cooperation with International Centre for Research in Agroforestry, Dordrecht

Nair PKR, Garrity D (eds) (2012) Agroforestry – the future of global land use, vol 9, Adv in Agrofor. Springer, Dordrecht/Heidelberg/New York/London

Noh NJ, Son Y, Kim RH, Seo KW, Koo JW, Park IH, Lee YJ, Lee KH, Son YM (2007) Biomass accumulations and the distribution of nitrogen and phosphorus within three quercus acutissima stands in central Korea. J Plant Biol 50(4):461–466

Pachauri RK (2012) Climate change and agroforestry. In: Nair PKR, Garrity D (2012) (eds) Agroforestry – the future of global land use, vol 9, Adv in Agrofor. Springer, Dordrecht/Heidelberg/New York/London, pp 13–15

Patterson W, Booth R, Elliot P (1994) Power from plants – the global implications of new technologies for electricity from biomass. Royal Institute of International Affairs, London

Pereira JS, Pereira H (eds) (1994) Eucalyptus for biomass production. Commission of the European Communities. Brussels. Instituto Superior de Agronomia, Lisbon

Plonczak M (1989) Struktur und Entwicklungsdynamik eines Naturwaldes unter Konzessionsbewirtschaftung in den westlichen Llanos Venezuelas. Göttinger Beiträge zur Land- und Forstwirtschaft in den Tropen und Subtropen, Heft 43. Göttingen

Rédei K (ed) (2003) Black locust (Robinia pseudoacacia L.) growing in Hungary, 2nd edn. Publications of the Hungarian Forest Research Institute, Budapest, 19

Reeg T, Bemmann A, Konold W, Murach D, Spiecker H (eds) (2009) Anbau und Nutzung von Bäumen auf landwirtschaftlichen Flächen. Wiley-VCH Verlag GmbH & Co. KGaA, Weinheim

Reif A, Schmutz T (2001) Planting and maintaining hedges in Europe. Institut pour le dévelopement forestier, Paris

Rockwood DL, Carter DR, Langholtz MH, Stricker JA (2006) Eucalyptus and populus short rotation woody crops for phosphate mined lands in Florida, USA. Biomass Bioenerg 30(8–9):728–734

Rode M (2005) Energetische Nutzung von Biomasse und der Naturschutz. Nat Landsch 80:403–412

Rowe R, Hanley M, Goulson D, Clarke D (2010) Potential benefits of commercial willow short rotation coppice (SRC) for farm-scale plant and invertebrate communities in the agri-environment. Biomass Bioenerg 35(1):325–336

Schildbach M, Grünewald H, Wolf H, Schneider B-U (2009) Begründung von Kurzumtriebsplantagen: Baumartenwahl und Anlageverfahren. In: Reeg T, Bemmann A, Konold W,

Murach D, Spiecker H (eds) Anbau und Nutzung von Bäumen auf landwirtschaftlichen Flächen. Wiley-VCH Verlag, Weinheim, pp 57–71

Schmidt PA, Gerold D (2010) Nachhaltig bewirtschaftete Wälder versus Kurzumtriebsplantagen versus Agroforstsysteme. In: Bemmann A, Knust C (eds) Agrowood – Kurzumtriebsplantagen in Deutschland und europäische Perspektiven. Weißensee Verlag, Berlin, pp 208–216

Seebauer M (2008) Silvicultural study on Markhamia Lutea with regard to short rotation energy plantations. EU Re-Impact Project. Unique Forestry Consultants Ltd, Freiburg

Shackleton CM (2001) Managing regrowth of an indigenous savanna tree species (Terminalia sericea) for fuelwood: the influence of stump dimensions and post-harvest coppice pruning. Biomass Bioenerg 20:261–270

Sims REH, Maiava TG, Bullock BT (2001) Short rotation coppice tree species selection for woody biomass production in New Zealand. Biomass Bioenerg 20:329–335

Skärbäck E, Becht P (2005) Landscape perspective on energy forests. Biomass Bioenerg 28:151–159

Skodawessely C, Pretzsch J (2009) Akzeptanz des Energieholzanbaus bei Landwirten. In: Reeg T, Bemmann A, Konold W, Murach D, Spiecker H (eds) Anbau und Nutzung von Bäumen auf landwirtschaftlichen Flächen. Wiley-VCH Verlag GmbH & Co. KGaA, Weinheim, pp 217–226

Sociedad Agrícola Golondrina SA, UNIQUE WOOD Paraguay SA, UNIQUE WOOD Alemania GbR (2004) Plan estratégico y operativo de Forestaría Certificada en el Paraguay (FORCERPA). Caazapa/Freiburg (unpublished report)

Styles D, Jones MB (2007) Energy crops in Ireland: an assessment of their potential contribution to sustainable agriculture, electricity and heat production. Final report of the environmental RTDI programme 2000–2006. Environmental Protection Agency. Wexford

The World Agroforestry Centre ICRAF (2011) Trees in climate-smart agricultural systems. A submission to the UNFCCC subsidiary body for scientific and technological advice on issues related to agriculture. ICRAF, Nairobi http://www.unfccc.int/resource/docs/2012/smsn/igo/72.pdf. Accessed 9 Jan 2013

The World Bank (2011) Climate smart agriculture – a triple win. http://www.youtube.com/watch?v=i0V2xzEw44Y&feature=relmfu. Accessed 9 Jan 2013

The World Bank Institute, TerrAfrica (2012) Conservation agriculture: how has Zambia scaled it up? http://www.youtube.com/watch?v=qRh6FCvx91g&list=UUoeN5nAIVN_qd_rp0I_jJWA &index=1&feature=plcp. Accessed 9 Jan 2013

Verwijst T (2003) Short rotation crops in the world. Report of IEA-Bioenergy Task 30: Short rotation crops

Weih M (2009) Willow short rotation coppice commercially grown on agricultural land in Sweden – possibilities for improvement of biodiversity and landscape design. IEA Bioenergy Task 30 – short rotation crops for bioenergy systems. Technical review no. 4

Weih M, Karacic A, Munkert H, Verwijst T, Diekmann M (2003) Influence of young poplar stands on floristic diversity in agricultural landscapes (Sweden). Basic Appl Ecol 4:149–156

Woelcke J, Tennigkeit T (2009) Harvesting agricultural carbon in Kenya. Rural 21 – 01/2009: 21:29–31

Wright L (2006) Worldwide commercial development of bioenergy with a focus on energy-crop based projects. Biomass Bioenerg 30:706–714

Chapter 5
Organizational Changes in Forest Management

Eckhard Auch, Jürgen Pretzsch, and Holm Uibrig

Abstract This chapter describes the evolution of forest management organizations in the tropics from pre-colonial times up to the present, based on a review of literature and on case studies. In general, organizations are designed to suit the prevailing political and economic frame conditions. Generic models for forest management organizations are identified. The portfolio of these models is structured according to the sources of their production factors, namely labor, capital and land. Some contemporary organization models are outlined, detailing the inputs, the processes and the outputs. The models' functions, specific designs and relevance for rural development are stressed. Steering, monitoring and enforcement of capacity are understood as core issues with respect to further development of forest management organizations.

Keywords Organizations • Co-management • Community forestry • Cooperative • Farm forestry • Forest enterprises • Forest contractor • Forest management unit • Forest organization • Forest user groups • Joint forest management • Outgrower schemes • Participatory forestry

5.1 Introduction

Forest and tree utilization makes up an essential part of human livelihoods in many tropical countries. Today as in the past, forests and forest products, especially wood, play an essential role in the development of civilizations (Bass 1992). At each stage of development, specific forest management systems have been designed to produce the desired products and services from the forests. To implement these

E. Auch (✉) • J. Pretzsch (✉) • H. Uibrig
Technische Universität Dresden, Institute of International Forestry and Forest Products, 1117, Tharandt 01735, Germany
e-mail: eckhard.auch@tu-dresden.de; pretzsch@forst.tu-dresden.de; holm.uibrig@forst.tu-dresden.de

J. Pretzsch et al. (eds.), *Forests and Rural Development*, Tropical Forestry 9,
DOI 10.1007/978-3-642-41404-6_5, © Springer-Verlag Berlin Heidelberg 2014

systems, the users and right holders designed organizations according to their own rules and institutions. The design of forest organizations depends on the 'purpose', the specific products and services the forests are to produce, and on the particular ecological, social, economic and political context (Honadle 1999). Usually forest resources are a matter of public concern. Users and right holders have to bargain to realize their interests. In general, the dominant power structures (e.g., chiefdom, kingdom, religious, colonial, democratic, economic, managerial) have always dominated the governance of forest products and services (Bass 1992) and influenced the development of forest management organizations (Pretzsch 2005).

Based on the paradigms of forestry (Chap. 2), the evolution of the main forest management concepts and the prevailing organization models are presented. Structured in stages (Pre-colonial, Colonial, Independence and capital formation, Internationalization, Polarization and Globalization), the changes in institutional structures and forest organizations are outlined.

Institutions and organizations implementing forest management are subject to the impacts of complex, ongoing global changes (Chap. 3). Success and failure, threats and opportunities play a part in identifying new ideas, recalling local knowledge and testing new organizational models of administration, cooperation and partnership. Within the globalized world, sustainability-oriented forest management, with all its facets, plays an essential role in the design of potential organizational models with an emphasis on rural development.

The chapter begins by specifying certain definitions relevant to the topic of forest management organizations, and continues with a review of the evolution of forest organizations. The combination of the production factors *labor and capital* with *land* are used to structure the models of forest management organizations. The relevant models, characterized by both socio-economic and ecological attributes, are outlined in the final part of the chapter. In practice the organizational archetypes discussed in this chapter may be modified and adapted according to the practical situation.

5.2 Some Definitions

Organizations are understood as social systems comprising a purpose, people, information and a structure (Chrobok and Büchi 1996). A more operational view describes organizations as formal bodies, established to enforce, facilitate or improve institutions, consisting of resources, responsibilities, operating procedures and rules (Bromley 1989; Dhakal and Bhatta 2009). Forest management organizations are like firms, formed to assume the functions simple market mechanisms cannot provide (Coase 1937). They reduce transaction costs and enable long-term planning. The particular ecological, social, economic and political context determines the design, establishment and performance of forest organizations (Honadle 1999), closely related to diversified general policy objectives (Pretzsch 2010).

Forest management is understood as the controlled flow of inputs and outputs of forest ecosystems in order to meet the objectives of the holders of the property rights. Although the objectives of forest management may be multiple, not fully operational and competing, the appropriate means should be employed in order to reach the stated goals.

Forest management system denotes the technical and silvicultural design of forest management, e.g. the selection system, clear cut system, agroforestry, plantation forestry, forest gardens, slash and burn, etc.

A **forest management organization** is an enterprise, business operation model or institutional arrangement established to realize the forest management operations on the ground. Some sources use the term 'forest management unit' or 'FMU' (ITTO 2003).

Models of forest management organizations are generic forms of a forest management organization, derived from the variety of empirical arrangements in place. While some of these archetypes can be quite clearly delineated, other models overlap, with many mixed and transitional forms occurring.

5.3 Evolution of Forest Management Organizations

5.3.1 Organization of Forestry in Pre-colonial Times

In pre-colonial times tropical forests were either controlled by traditional hierarchical structures, such as kings, chiefs or rajas, or by other authorities (Turyahabwe and Banana 2008). The smallest units were based on groups of households and family-based organizations. Strong local institutions, like sacred forests, not only contributed to protect the interests of the powerless people (Agrawal 2005), but also served to prevent forests from overexploitation (Turyahabwe and Banana 2008).

The centralized control of forests was also practiced, as exemplified by elite hunting reserves like the one established by the Persian King Khosrov II in the seventh century BC (Bass 1992). Ancient central organizations were in place to secure actual and future timber supplies, exploit timber for the benefit of the sovereign, and to maintain environmental services (Larson et al. 2010), as well as and for cultural purposes. In India, for example the rulers controlled large forest areas, but also allowed utilization and management by local communities and princes (Guha 1983; Poffenberger and McGean 1996). Similar structures existed in Mayan and Incan organizations in America.

5.3.2 Colonial Forest Organizations

Colonial administration superimposed and substituted complex centralized and small-scale traditional forest management and its institutions. The objective was twofold, to produce commercial goods in the most profitable way, and to control the territories under consideration. Forests were reserved for the production of military and commercially valuable products (Guha 1983; James 1990), but also to a certain extent for protection too (Wiersum 1999). This was achieved, for example, by means of the demarcation of 'crown forests' for exclusive use by the colonial power (Rueda 2008). The forests were managed by a powerful colonial forest administration, which had adopted the Western model of a bureaucratic administration: a vertical hierarchy with a strong top-down command structure and the self-conception of an authoritarian forest police (Blank 2006). Especially under the British colonial regime, control over forest resources was transferred in part to local authorities. In Uganda, for example control was granted to chieftains (Turyahabwe and Banana 2008), who in turn were under the control of the colonial governor. In British India, a district forestry officer was not only entrusted with the maintenance of law and order, he also controlled the forest land in his district, granting or leasing unoccupied forest land in order to extend the cultivated area, at the expense of forests, so as to ensure the generation of higher state revenues (Guha 1983). Forests were exploited by agencies of the colonial states, by local landlords and rulers for strategic purposes, such as ship-building, railroad construction and for fuel, but also for the export of timber and non timber forest products (NTFPs). Furthermore, other forest land was under the control of landlords who usually sought to convert the forests to cash crop plantations with higher returns. Private and company owned forest organizations were organized in the same hierarchal way as the state administrations.

5.3.3 Forestry Organizations After Independence

Accustomed to colonial forest tenure and administrative bodies, the new national governments took over the former colonial forests as state property. The new state forest administrations adopted a bureaucratic approach to manage the respective 'state' forests, whereby ancestral ownership by indigenous people and customary uses were largely ignored (Turyahabwe and Banana 2008; Rueda 2008).

Smallholder forestry and traditional agroforestry practices provided evidence for time-tested adaptive production systems. The organizational unit for forest management was mostly the family or household, adhering to the characteristics of a household economy (Auch 2007). Parallel to this, communities of various kinds took up traditional ways of managing forests. These models did in fact render the possible continuation or revival of forest utilization according to customary claims. This kind of individual forest management had been sub-summarized under the initial Community Forestry after state independence, as defined by FAO (FAO 1978). Furthermore, the absence of organizations responsible for forest management on a local level led to so-called open-access forests, where previously limited access had existed.

Following the prevailing paradigm of the liquidation of forests for the purpose of developing forest industries (Zivnuska 1966) and in order to the contribute to general economic development of these countries, the state forest administrations granted forest concession rights to private enterprises. Logging became a lucrative business for international companies (Rueda 2008). In other countries the governments initiated the establishment of state forest enterprises effectively to exploit valuable commercial timber, as was the case, for example, in Northern Vietnam and Laos at the time (Le 1988; Hartzsch 1983). Some states, which had never colonized, followed the trend and adopted centralized forest management, for example Nepal under the Rana Dynasty (Nagendra and Gokhale 2008; Kanel Raj and Acharya 2008).

5.3.4 The Internationalization of Tropical Forestry

With the decline of bilateral links which had kept on functioning for a long time after state independence, many tropical countries gradually became integrated in to international structures, following the guidance of FAO and the World Bank. An initial afforestation boom was linked to the first energy crisis and focused on the production of firewood especially for the people concentrated in the towns. However, instead of the liquidation of wood, the trend then turned more favorably towards industrial processing which achieved higher selling prices of the raw material over firewood. With the development of mechanized forestry and the production of genetically improved planting material, more private investors and corporate groups engaged in industrial forestry, driven by the growing demand and rising prices for industrial softwood products on the world market (Bass 1992). These companies were usually organized according to conventional industrial organizational structures (e.g. profit centers) with profit maximization as the overall goal. The large-scale reforestation projects implemented were often poorly integrated into existing social structures and had ecologically destructive impacts (see Chap. 3). Forest concessions continued to expand in natural forests. Over the years a concentration process took place, resulting in the establishment of some very large consolidated companies which operated on different continents.

5.3.5 Polarization

During the stage of polarization, there was a diversification towards three objectives of forest organizations. In addition to state administration and private sector business-orientated enterprises, diverse organization types engaged in various forms of social forestry and nature conservation (Uphoff 1993). Social forestry was launched in India in the early 1970s where the government started a program of tree plantings on degraded village land outside the forests to supply the rural population, especially small holders and landless people, with domestic firewood, small timber for construction purposes and minor forest products (Prasad and Bhatnagar 1995). Organized and implemented by government institutions, the villagers were expected to contribute labor only. This model was adopted by many actors, driven by the insight that the industrial forestry model prevalent at the time had failed to contribute to rural development (World Bank 1978; FAO 1978) or to solve pressing ecological issues, such as the large-scale deforestation of fragile land. Consecutively, modified forms of *Social forestry* were designed, and this became one of the umbrella terms for forestry projects targeting basic needs and subsistence use in numerous tropical countries (Nair 1993). The concept was further developed into a variety of similar models that stressed the pursuance of social objectives and the benefits for poor populations, such as the Joint Forest Management model launched by the Indian government during the 1970s (Colchester et al. 2003).

Striving for more efficient governance and management of forests, a new Community Forestry approach was promoted which had originated and been in use in Nepal since 1976 (Nagendra and Gokhale 2008). Under this model, a substantial portion of government forest land was transferred to local communities (panchayats), the forest administrations were decentralized to the same level, and the role of the Department of Forestry changed from supervisor to that of an advisor and facilitator (Kanel Raj and Acharya 2008). Interested villagers were obliged to form forest user groups in order to apply for a piece of the community forest. The model spread to many other areas and was accepted as a recommended model for sustainable resource management (Arnold 2001).

In time, the notion of the model changed, from the provision of a purely domestic household supply of forest products, to market production for cash income generation. Besides livelihoods and the empowerment of the forest users, the supporters' objectives also included protection goals (soil, habitat, biodiversity, water) (Wiersum 2010; Dhakal and Bhatta 2009). In order to fulfill these sophisticated goals, the forest organization had to further upgrade administration and governance. Each country shaped its community forest designs quite individually, to fit in with the given policies, structures and cultures. Essential elements now include the property rights conferred, the commercial value of the forest produce, the size of the forest, and the number and homogeneity of members of the respective communities. To transfer forest use rights, community members must establish or appoint a legally recognized organization to represent the community, the desired area must be delineated (at the consent of other governmental ministries and the neighboring communities) and be formally attached to the established organization with an agreement between the community body and the government. A management plan is usually required. In most cases forest rights are granted for a limited period of time only, but can be renewed. The government maintains the right to revoke the rights to a forest area if the agreed terms of reference are not fulfilled. The community forest organization requires an additional effort on the part of the community members (relatively high transaction costs), but provides a structure for good governance and empowerment at community level.

However, the outcome of social and community forestry programs did not always meet the high expectations: the degradation of forests could not be sufficiently controlled; forest rehabilitation advanced rather slowly and rural people remained poor. The projects were designed within a technocratic planning process and implemented by governmental institutions in a top-down manner, without respect for or participation by local people (Colchester et al. 2003). The different benefits sought by the government (reforestation and a share of timber trees) and by villagers (cultivation of crops, rearing of animals and harvesting of NTFPs) resulted in contradictory management objectives (van Noorwijk and Tomich 1995) and ultimately in limited success.

When collectivizing the variations of community focused and community based forest management models, the technical terms Participatory Forestry (Schreckenberg et al. 2006) and Collaborative Forest Management (Carter and Gronow 2005) emerged as umbrella terms for the various forms and models of the management of public forest land by local communities.

5.3.6 Globalization

Smallholder forestry on farms and small and medium forest enterprises provide evidence for ancient and adaptive production systems. Many new smallholder properties were created as a result of resettlement programs and unregulated migration into tropical forests (Kinsey and Binswanger 1993). Now, the management of the remaining forest patches on these smallholder properties, together with tree plantations, agroforestry systems and the farm land are at the discretion of the individual farm owner. The organizational unit for forest management is mostly the owner's family or household, adhering to the characteristics of a household economy (Auch 2007). The area of privately owned forest has increased, especially in Asia and South America (12 % in China, 20 % in Colombia). This increase is mainly made up of planted forests (FAO 2010). Today, smallholders are being increasingly recognized as important actors in the context of global change, food production and rural development.

The development of outgrower schemes on a large-scale was a reaction to decreasing land resources. Outgrower schemes are contractual arrangements for vertically coordinated production relations, initially formed for the large-scale production of agricultural export crops. The first reported forestry scheme of this kind dates back to the 1970s, when a state-run Philippine paper mill supported smallholders in planting pioneer trees on fallow land around the mill's concessions, in an attempt to protect the land from illegal settlers (Rueda 2008). Outgrower schemes are either market-driven, where companies contract smallholders to grow raw materials on a long-term contract basis, or they are centrally planned development interventions to combine the supply for processing plants with the incomes of surrounding smallholders (Dillon 1992; Little and Watts 1994; Mayers and Vermeulen 2002). In many cases the growers are not able to negotiate beneficial agreements with the company (Race et al. 2009).

Forestry contractor enterprises emerged as a reaction to the outsourcing of many activities by state and private enterprises. These offer forest management services without owning forest land. Often as small but specialized private business enterprises, the 'contractors' (UNECE/FAO 2008) offer planting, logging, extraction and haulage services employing specialized equipment. In the tropics such contractors can be found in areas with a certain proportion of plantation forest and with established markets for timber (refer to Boxes 5.4 and 5.5). The contractor complements the smallholder tree grower and this enables both to benefit from the economies of scale.

Recently, new investment forestry enterprises have been established. Since the financial crisis in 2008, high yielding forest plantations are increasingly being included in diversified investment portfolios, promising ‘clean’, ‘green’ and secure demand markets, as well as high, sustainable returns. New forms of funding have also evolved, for example ‘green investments’ and ‘crowd funding’ (Chap. 10). The organizational units on the ground, managing the forests according to the investors’ standards and provisions, are, however, usually commercial companies or contracted community organizations. The investment model comprises the investors, a central finance institution organizing and selling investment products as well as holding and monitoring the forest enterprise, and the forest enterprise on site which implements the forest management according to the required standards.

5.4 Model Design

5.4.1 Portfolio of Forest Management Organization Models

The types and varieties of forest management organizations and their operational arrangements are manifold. Some generic models are presented in Fig. 5.1. The ownership of the classical economys’ production factors is used for the models’ structure:

1. *Labor* and *capital*: the models are first grouped according to their legal basis, which is either ‘public’ or ‘civil’ law. ‘Public’ entities are owned by all citizens and are usually controlled by the government. They include entities owned by public bodies or territorial authorities, such as municipalities and communal bodies, but also indigenous groups of people. ‘Civil’ entities may be owned by a group of individuals or by a single individual. The latter comprises family or household-owned entities and commercial firms. Usually, the members of these groups have had to contribute financial resources to the organization and they have the right to make decisions concerning both the affiliation of new members and the entity’s contributions to cooperatives or private companies.
2. *Land*: Despite the fact that land ownership is not a binary variable for multiple kinds of use rights that an individual or community can possess (Nagendra and Gokhale 2008, p. 721), ownership of land in the context of this chapter is defined as the right of alienation, i.e. the right to sell or lease the resource (Schlager and Ostrom 1992, p. 251). In its narrow legal sense (*de jure*), ownership comprises two categories at the first level: ‘public’ and ‘privately owned’ forest land. ‘Public’ land is owned by either the central/national government (all citizens) or by public bodies and territorial authorities, such as communities and

Production factor land

Production factors labor and capital		Owned by state (government)	Owned by groups	Owned by individuals
		Under civil law		Under public law
Government	Under public law	▪ State forest administration ▪ State forest enterprise	▪ State forest management service ▪ Trust forest*	▪ -
Groups	Under public law	▪ Community forest ▪ Co-management ▪ Agro-extractive reserve ▪ Joint forest mgmt.	▪ Community forest ▪ Communal forest enterprise	▪ -
Groups	Under civil law	▪ Concession ▪ **Forest user group** ▪ Co-management ▪ PES**	▪ Forest-user-group ▪ PES** ▪ Co-management ▪ Concession	▪ PES** ▪ Co-operative ▪ Producer association
Individual commercial - household	Under civil law	▪ Taungya ▪ **Leasehold** ▪ Conservation contract	▪ Traditional / customary*** ▪ **Outgrower scheme**	▪ **Farm forestry** ▪ Outgrower scheme
Individual commercial - household	Under civil law	▪ Concession ▪ Leasehold ▪ Service contract	▪ Concession ▪ Service contract	▪ Forestry enterprise ▪ Leasehold ▪ **Service contract**

Fig. 5.1 Generic models of forest management organizations in the tropics
Models structured according to the entities providing the production factors *labor and capital* on the y axis and *land* on the x axis. Bold print indicates models presented in Sect. 5
*Forest owned *de jure* by indigenous groups, managed by state forest administration, e.g. Kakamega forest, Kenya; **Payments for environmental services; ***e.g. Ejidos in Mexico

indigenous groups. The 'private' category refers to land, subject to legal land titles, held by groups or individuals as private property in accordance with the provisions of civil law. Typically the rights relating to privately owned land cannot be terminated unilaterally by a government; this requires a formal process and compensation (White and Martin 2002). Generally privately held property is more secure in terms of the withdrawal of land use rights or expropriation by the government. Groups with private rights have a stronger claim to control over and benefits from the forests than groups with customary rights on public land only (ibid.).

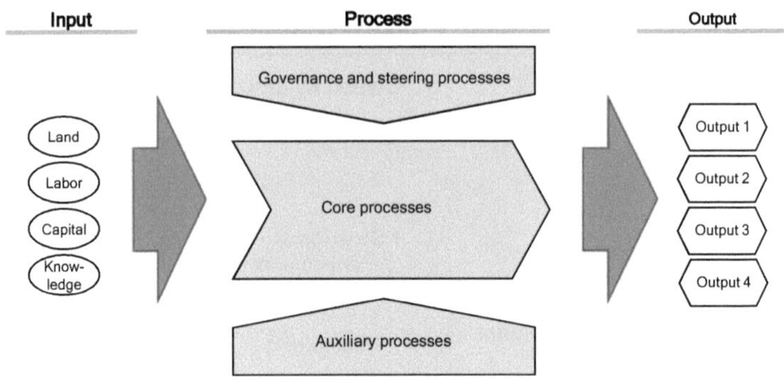

Fig. 5.2 Generic input-process-output scheme for forest management organizations

5.4.2 *Input-Process-Output Scheme*

For a holistic understanding of forest management organizations as social systems, an *input-process-output* scheme is designed (Fig. 5.2). It represents a causal model, applicable to all organizations. The inputs are the available resources, and the outputs result from the transformation processes applied (Andersen 2010; Emery and Trist 1965; Stogdill 1967). Inputs are the classic production factors *labor, capital* and *land* provided by the base organizations. *Knowledge* is specified separately, to allow for its contribution independent of labor and capital (cf. Li et al. 2004). The transforming processes are grouped into *core processes* (production of outcome-relevant products and services), *steering processes* (including strategy, decision-making, coordination, supervision) and *support processes* (enabling the other processes to operate) (Miebach 2009; Gaitanids 2007; Schober 2002). Outputs are the products and services generated with a value (relevance) for the base organizations and their clients.

Some forest management organizations may own all of the production factors, for example a state or a farm forestry enterprise. Others are designed as joint ventures between the owner(s) of the forest land and those tasked with forest management, providing capital and labor.

5.5 Selected Contemporary Models

5.5.1 *Farm Forestry*

Farm systems may integrate agricultural activities with natural forest, planted forest or agroforestry systems as production components. Such mixed farm designs are common in areas where trees dominate the natural ecosystem, as well as in areas with unfavorable conditions for cropping and grazing, such as poor soils or steep slopes.

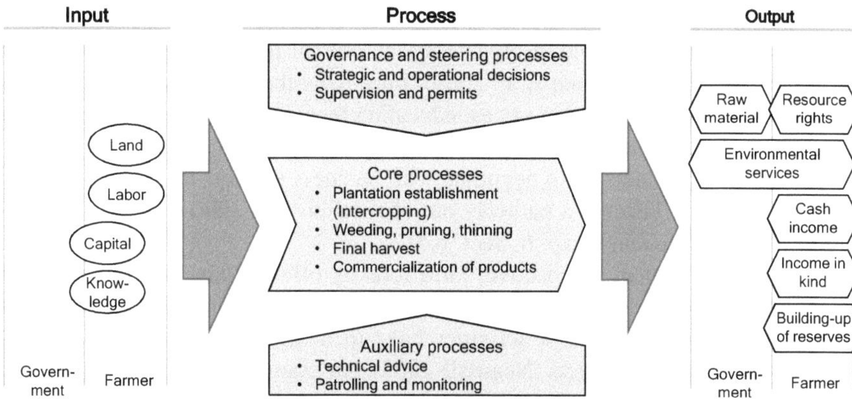

Fig. 5.3 Input, process and output of farm forestry (planted forest)

On-farm trees and forests are typically used to support other farm activities, and are for household consumption and commercial purposes (Pretzsch 2010). They also contribute to environmental services and household resilience. Farm forestry is an option for the use of marginal land and a means to rehabilitate degraded land, diversify income and secure property rights (Nsiah 2010; Darr 2003).

The availability of land, knowledge, labor and equipment are hurdles to negotiate when to entering into forestry activities. Yet, as most farms have at least some of these factors already in place, fixed cost sharing allows mixed farmers to operate farm forestry profitably on a quite small scale; a scale on which purely forestry enterprises could not survive on their own. Figure 5.3 shows the analysis of forest plantations on a farm. Usually the inputs are provided by the farmer, often specialized knowledge is offered by extension services. Due to the long production cycle, insecure or monopolistic output markets hamper farm forestry, as do discontinuous state policies. The establishment of a tree plantation is a high investment in terms of labor (e.g. for land clearing, site preparation and planting) and materials (e.g. seedlings, fertilizer) for the farmer. Given the expected supply of raw materials and environmental services, most governments support tree growing on private land through the provision of technical advice, credit facilities, planting material and marketing support. In addition to the establishment of the forest, the farmer's ongoing activities cover monitoring and fire prevention means, pruning and thinning. Thinning usually brings the first positive returns. Full revenues are realized at the end of the rotation.

With the increasing global demand for agricultural products, the establishment of forest plantations is limited to marginal agricultural and/or degraded land. Agroforestry systems, providing multiple agricultural, animal and forest products, constitute other land-use alternatives (see Chap. 4).

Small scale tree growers often lack experience in the marketing of timber and usually depend on intermediary brokers. Compared to the growers' share, these middlemen often receive a disproportionately large portion of the profits. Due to the poor infrastructure and the farmers' lack of direct market access, middlemen are however often indispensable in organizing the critical links between growers and the processing industry (Race et al. 2009). Improving the supply of information to farmers with respect to market prices, educating them on the functioning of factor and output markets, linking them to competing buyers and increasing their confidence and the skills needed to negotiate with business partners are prerequisites if farm forestry is to become a pathway out of rural poverty (ibid., see also conclusions on value chains in Chap. 6, Sect. 6.5).

From a management perspective, farm forestry offers a land use with reduced labor inputs compared to agriculture, and sometimes with a higher return on labor (see Box 5.1). On-farm forest activities have to fit into the whole-farm activity portfolio and resource budgets (Negussie 2003). An example of farm forestry with a natural forest component is described in Chap. 4, Box 4.1.

To a certain extent, small, family-based farms follow an economic rationale that differs from that of conventional commercial enterprises. Profit maximization in financial terms is not necessarily the prime objective. Often aspects such as overall utility, resilience, risk aversion and household survival are more important (Ellis 1992). Culturally determined norms and group values may also influence management decisions (Zitzmann 1998; Xayvongsa 2001). Small farms in the tropics and subtropics often operate under shaky framework conditions, characterized by a lack of market access, credit facilities and permanent land titles (Pretzsch 2010).

Compared to agriculture, farm forests in the tropics are often lucrative in terms of the return per unit family labor, but not in the absolute turnover per unit of land (Box 5.1). Under competitive market conditions, farm forestry is economically viable in many tropical countries (e.g., Pretzsch 2010; Mayers and Vermeulen 2002), and farm forests are often in a better state than public forests (Royo 2011).

Given their medium to long-term nature, farm forestry investments are difficult or impossible for poor and very small farmers to undertake. Usually, the benefit from forest resources to better-off and middle-income farmers is proportionally higher. Only under stagnant forestry conditions, when better-off farmers look elsewhere for the few opportunities available, is the benefit from forest resources enjoyed by the poor proportionally higher (Pérez et al. 2004, cited in Hobley 2007). Given that farm forestry requires less labor to utilize the land than typical agricultural activities, it particularly suits the livelihood strategies of better educated households, who tend to allocate more family labor to off-farm work (Jolliffe 2004).

Farmers have to decide between marginalization and the exodus of rural populations, or new green revolutions (Otsuka and Larson 2013), or else pursue a new course (third way) which involves combining their skills with elements of the market economy, technology and partnerships. In the rural areas of the future, diversified small farms will provide a basis for the livelihoods of families and their social net. With strong links to value chains, their production and marketing can be organized more efficiently, the value of their products can be increased through grading and processing and their strength can be enhanced by cooperating with other producers.

Box 5.1 Farm Forestry in Ghana

Although less than 10 % of the original forest remains in Ghana, the annual national deforestation rate is still 2 %. A growing timber industry, the demand for firewood and ambitious national development programs are driving the demand for roundwood, which currently lies at about 2.5 times the annual allowable cut. Forest plantations on both state and privately owned degraded land were established to bridge the gap in the supply. This forest plantation program was initiated by the government in the early 1980s, and several thousand hectares of forest plantation have been established since.

In the Offinso District most plantations have been designed as taungya-type agroforestry systems, with teak (*Tectona grandis*) or *Cedrela odorata* in the upper storey, intercropped with food crops for the first 4 years. The average plot size is 2.9 ha and nearly half of all households possess a forest plantation. Farm forestry is generally undertaken by better-off households, characterized by larger land holdings, more secure land titles, more household labor, higher levels of education and a higher per-capita income.

The farmers establish their plantations themselves, with the Forestry Commission providing high quality seedlings. Farmers also use wildlings harvested from existing forest plantations and planting material procured from private nurseries. Weeding and fire protection measures are carried out annually after site preparation and planting, while pruning and thinning start from year five on. Some farmers also establish a fire break around their plantations. The total labor input for a teak plantation over 25 years is 750 man-days/ha (one man-day equals 4 h work and is paid USD 2.20), while intercropping requires almost twice this. Pure cropping necessitates five and a half times as much labor input. The mean annual volume increment of the teak ranges between 12 m^3/ha and 17 m^3/ha.

The food crops produced are sold and/or consumed by the household members. The teak stands are harvested after 10–25 years. Depending on the diameter, logs are used as telephone poles, for furniture or construction, as well as for export to Asia. Farmers are paid for standing trees, so those farmers capable of measuring tree volumes have an advantage over those who cannot. Despite the fact that middlemen they pay less for the trees, most farmers sell to them because they are easier to access than the processing companies.

The buyer applies for a harvesting permit, fells the trees at his own expense, applies for a transport permit and finally hauls the logs after verification by a forest officer. Farmers are paid after felling.

The net present value of a pure teak plantation over a 25 year rotation at a 12.6 % discount rate is USD 53.20/ha. The equivalent for pure maize and plantain cropping is USD 520.50/ha, and USD 962.20/ha for teak with intercropping. Intercropping requires a higher initial investment but provides

(continued)

Box 5.1 (continued)

positive returns in the first 4 years. The profitability per labor unit invested (undiscounted values, costs to finance material and land not considered) is highest for the pure teak stand (USD 17.77/man-day), medium for intercropping (USD 11.62/man-day) and lowest for pure agriculture (USD 2.19/man-day).

Source: Nsiah (2010)

5.5.2 Community-Based Forest Management (CBFM)

Community-based forest management combines sustainable forest management with support for rural livelihoods. It is perceived in the context of devolution as an optional public model to overcome the ineffective management practiced by central government institutions (see Box 5.2). Usually CBFM is organized under public law. Per definition, all legal dwellers in a distinct area, for example, a commune, are automatically members of the community forest and part of the management. Often CBFM is based on a traditional system of community management (Nagendra and Gokhale 2008). The general objectives of CBFM are to: (1) sustain forest resources and environmental services, including the conservation and rehabilitation of forest land; (2) contribute to local livelihoods, including poverty alleviation; (3) protect the property rights of local dwellers; and (4) maintain public property rights and public control via state and government institutions. Community forest management is based on the assumption that only users who benefit will develop a sense of ownership, thus motivating them for active protection of the forest against anthropogenic and natural threats from outside and from the community itself. Box 5.2 illustrates the costs and difficulties associated with the protection of a common pool resource in Vietnam.

Figure 5.4 illustrates the major characteristics of the community forestry model. Essential inputs are: the resource forest land, demarcated with at least the corner points; the formally defined and organized members; a body responsible for management, communication and decision-making; in some cases hired staff and a supervising institution, usually a government body. Often non-governmental organizations support the establishment and management of the forest. The management of the community forest comprises typical tasks of public enterprises, such as maintaining environmental services; and has many other similarities with public forest enterprises, including long-term and annual planning, decision-making controlled by members, budgeting and benefit sharing, transparent financial management and accounting, organization of forest operations and marketing of forest produce, monitoring of the resource, supervision of staff, and communication with members and stakeholders. Operational forest management tasks, such as harvesting, are often implemented by the members on the basis of use regulations for domestic products and permits for high value commercial forest products.

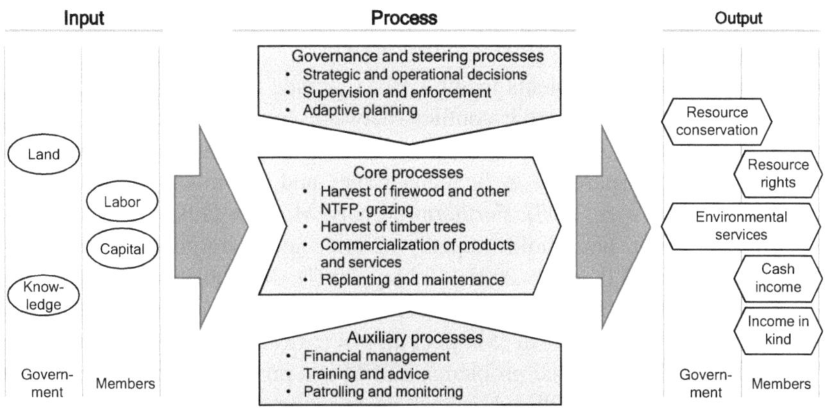

Fig. 5.4 Input, process and output of the CBFM model

Outputs include improvement of the forest resource, secured ecosystem services, higher land tenure security for members, a range of consumable forest products and services both for household use and for sale, and indirect benefits such as increased community resilience through the provision of a safety net in times of emergency, increased employment through the development of downstream processing enterprises, and direct income from salaries and compensation measures. Occasionally cash incomes from community forest are quantified, whereas products and services consumed for subsistence and the increased community resilience are difficult to measure and are therefore often underreported (cf. Larson et al. 2010).

Evidence of the effectiveness of community forestry is mixed. Looking at ecological sustainability and resource rehabilitation, reports from Asian countries typically claim positive results, whereas the findings from Africa and Latin America are less optimistic (Larson et al. 2010; Pokhare and Nurse 2004; Dougill et al. 2001; Nagendra and Gokhale 2008). The CBFM model works where the forest resource provides enough net benefits to fund the organization (see footnote 1: (2) *congruence*). Van Laerhoven (2010) identified the following as being the most prominent features of successful sustainable community forest management: the ability to learn from collective problem-solving, the presence of leaders and the freedom to design governance regimes, while competing groups of members complicate the good governance of community forests. Other limiting factors are weak state control, unmotivated proponents and community forest members lacking information and a capacity to defend their rights (Cronkleton et al. 2012). If community forestry does not curb the absolute demand for forest products, the ecological status of community forests might improve at the expense of neighboring forest areas that are being depleted more rapidly, as observed in Madagascar (personal communication with K. Ackermann). In general, community forestry has had a positive effect on livelihoods, with members mostly experiencing improvements, and with only a few negative impacts on livelihoods (Larson et al. 2010).

Livelihood improvements are largely determined by the amount of harvestable high-value timber available (Iversen et al. 2006).

The most common problems facing CBFM include the allocation to communities of poor-quality forest land; conflicts between communities provoked by the transfer of exclusive forest use rights; programs over-emphasizing collective activities; a lack of transparency in financial issues and too much bureaucracy and paperwork (Kaimowitz 2005). Furthermore, CBFM often fails to benefit the less powerful and poor households equally (Carter and Gronow 2005; Adhikari et al. 2004; Campbell et al. 2001; Hobley 2007). Monitoring and sanctioning have been found to be the most critical factors for the success of sustainable community forest management (Singh et al. 2011; Coleman and Steed 2009).

Despite these fundamental problems, CBFM is a promising option, particularly for forests required to fulfill public welfare functions, such as conservation and environmental services (cf. Colchester et al. 2003; Bakarr 2005). The vital governance principles were identified by Ostrom (1990, p 90)[1] and confirmed by others (cf. Cox et al. 2010; Singh et al. 2011).

An example of CBFM in Vietnam, in the form of 'village forest management' (VFM), is outlined in Box 5.2. The example shows the collective action weaknesses of the CBFM model. To overcome these problems an organizational model with small member groups was developed, the sub-VFM. This sub-VFM is a forest user group (FUG) model. In this case the steps taken led to greater active participation by marginalized groups, conflict resolution, supervision and enforcement by the authorities.

---

[1] Ostrom (1990, p. 90) identified the following principles for enduring common-pool resource institutions: (1) *clearly defined boundaries* (individuals or households with rights to withdraw resource units from the common-pool resource and the boundaries of the common-pool resource itself are clearly defined); (2) *congruence* (A. The distribution of benefits from appropriation rules is roughly proportionate to the costs imposed by provision rules. B. Appropriation rules restricting time, place, technology and/or quantity of resource units are related to local conditions); (3) *collective-choice arrangements* (most individuals affected by operational rules can participate in modifying operational rules); (4) *monitoring* (monitors, who actively audit common-pool resource conditions and user behavior, are accountable to the users and/or are the users themselves); (5) *graduated sanctions* (users who violate operational rules are likely to receive graduated sanctions, depending on the seriousness and context of the offence, from other users, from officials accountable to these users, or from both); (6) *conflict resolution mechanisms* (users and their officials have rapid access to low-cost, local arenas to resolve conflict among users or between users and officials); (7) *minimal recognition of rights* (the rights of users to devise their own institutions to organize are not challenged by external governmental authorities). For common-pool resources that are part of larger systems: (8) *nested enterprises* (appropriation, provision, monitoring, enforcement, conflict resolution and governance activities are organized in multiple layers of nested enterprises).

5.5.3 *Forest User Groups*

A forest user group (FUG) is a formal association of individuals with exclusive rights to the management of a delimitated public forest area. Sometimes this is also labeled 'community forestry', but in contrast to the CBFM model, only a limited number of individuals (or households) from a whole community (village, settlement, hamlet) belong to the FUG. In some cases, as in Nepal, an individual can become a member of more than one FUG, if he or she is interested in specific products from different forests. Generally FUGs are organized under civil law as cooperatives, with a constitution, member register and committee. The FUG, as a legal body, enters into a legal agreement with the state to use the forest land, subject to the defined rights and obligations of both parties.

The inputs, processes and outputs are similar to those of the community forestry model presented in Fig. 5.4. FUGs often receive highly degraded forests or even bare land for reforestation. In some cases, the FUG members use the allotted forest plot jointly; in others they subdivide the area and each member receives a plot of his or her own. FUG members elect a committee to organize and communicate FUG issues, such as plot identification and allocation, technical advice and support, central facilities and services like nurseries, the drafting of a management plan, logging, marketing and financial management. The members implement the practical operations, such as tree planning, weeding and thinning, and take part in group meetings. FUG membership requires active participation in forest management decisions, plus contributions to forest management operations. The committee keeps a part of the revenues from timber sales to meet administrative costs.

As the sub-VFM groups presented in Box 5.2 illustrate, the FUG model overcomes certain weaknesses of the CBFM model, such as unbalanced power structures, unfair benefit sharing, low motivation of occasionally anonymous members and low economic efficiency (Tuan 2006). The FUG approach suffers less from collective action problems, it is market-oriented and it is able to unlock entrepreneurial engagement (Pretzsch 2010). It may also be a favorable way of improving the efficiency of public forestry as long as the poor can be integrated (Bastakoti 2005). However, the individual design of the model creates the risk of a socially unbalanced use of public forests. The poor and weaker citizens may not have been strong enough to enter the FUGs or been able to establish their own FUG's when the public forest resources were distributed. As a consequence, they are thus excluded from the forest resources which were once available to them and are even more marginalized than before. The CBFM model prevents such formalized disadvantages.

Box 5.2 Village Forest Management in Vietnam

In Vietnam 2.4 million hectares, 13 % of the country's forest land, is managed by local communities. These forests are managed collectively, either by local village communities or by sub-groups of forest users.

(continued)

Box 5.2 (continued)

Community forest management in Vietnam has its roots in traditional forest management by villagers. In 2004 a national law provided the option to formally contract forest land to communities. In this case the community forms a 'village forest management' (VFM) to set up their own 'collective choice rules' (rules to make operational rules) for the purposes of forest management and utilization. These include voting regulations and 'operational rules' like access (who can access what product), harvest (firewood, bamboo, grazing, other NTFPs; and where, when, how much), penalty for violations, contributions of the members (monitoring and patrols, firefighting, planting), conflict resolution and rewards for detecting and arresting rule violators. The rules are made individually in each village. The VFM concept has been further developed to allow the formation of forest user groups (sub-VFM). Even where there are sub-VFM groups, the determination of 'collective choice rules' remains at village level. The sub-VFM groups, however, make their own 'operational rules' in accordance with the 'collective choice rules'. Generally, these are more conservative and more specific, with quantitative limits with regard to harvesting periods and quantities.

In contrast to the sub-VFM groups, the VFM group members rarely invest in enrichment planting or in the maintenance of the forest. Under both models each member has to pay an annual fee (a certain quantity of rice) to compensate the forest guard. All members utilize the forests, mostly for firewood, bamboo and NTFPs. VFM members use forest produce mostly for domestic purposes, whereas use by sub-VFM members is mostly commercial.

In the year 2004, the average forest area per household member (hh) was higher for the sub-VFM (0.64 ha compared to 0.33 ha), the margin per hectare was higher for the sub-VFM forest (USD 52.57/ha compared to USD 41.91/ha), but the average return on one man-day household labor was higher for the VFM (USD 3.52/hh compared to the sub-VFM USD 1.22/hh) due to planting, maintenance and much more frequent patrolling. The fair sharing of benefits and obligations was lower for the VFM. While the forest area of the sub-VFM could be maintained, the VFM lost land as a result of the illegal establishment of private agricultural fields. The sub-VFM groups outperformed the VFMs with regard to the quality and richness of their forests.

Community forests governed by small sub-VFM groups are significantly more sustainably and equitably managed than the forests managed by the larger VFM groups. The sub-VFM groups are ready to invest more labor for long-term benefits and for the resource as a whole.

Source: Tuan (2006)

5.5.4 Leasehold Schemes

Under leasehold schemes a certain area of land is given to an individual user (lessee) for a specified period of time for utilization, with agreed returns for the lessor. Forest lease agreements allocate a distinct area of forest land to individuals or households for their exclusive use; the rent to the lessor is often based on the overall improvement of the vegetation cover (Naina 2011, see Box 5.3) or the quantity of major forest products harvested. Leaseholds are civil law contracts between the landowner and the user, based on customary institutions and traditions or defined juridical understanding. Lease arrangements are usually uncomplicated and less commingled with politics than permanent transfers of land use rights. The leasing of land is well-established in agriculture, but so far relatively uncommon for forest land. The 'leasing out' of well stocked and intact forests is usually done as a concession. Cleared forest is leased out for the establishment of forest plantations (e.g. Maansson 2003), and degraded forests for the purpose of forest rehabilitation or restoration.

Private or community forest land is occasionally leased out to private sector companies for commercial tree production (Maansson 2003; Race et al. 2009; Rueda 2008; Jørgensen and Vivekanandan 2003). Lease arrangements with households to use degraded state forest land for the purposes of rehabilitation are common, as for example in Nepal (Box 5.3), Vietnam (Box 5.2) and Indonesia (Naina 2011). The model is illustrated in Fig. 5.5. The state typically provides land, advice and planting material. The lessee mostly contributes labor for the replanting of trees and useful shrubs, and the sowing of grass. Until the first thinning, the lessee mainly benefits from the extraction of NTFPs like fodder and/or thatching material. To provide additional benefits during the initial period, some lease contracts allow taungya intercropping and the establishment of cash crop trees such as coffee (Naina 2011).

Leasehold forestry is popular for pro-poor development interventions. Experience in Nepal shows that it successfully improves the ecological state of degraded forest land, but that improvements to the lessees' livelihoods are less than expected (Bhattarai et al. 2007). The socially balanced allocation of plots is problematic, creating conflicts and new inequalities (Thoms et al. 2006). Compared to intact forests, degraded forests require high initial investments in labor and material. Amongst the different models for the transfer of public forest land to individual users, the leasehold model offers a robust setting for the securing of high investments from the partners.

Box 5.3 Leasehold Forestry in Nepal

In 1993 the 'Hills Leasehold Forestry and Forage Development Project' was launched. Its aim was to alleviate poverty amongst the poorest of the rural households by leasing them a piece of highly degraded forest land for fodder production and reforestation. As the very poor families with little or no cultivated land depend on husbandry, the project strategy was to integrate forestry and livestock by means of providing exclusive land use.

(continued)

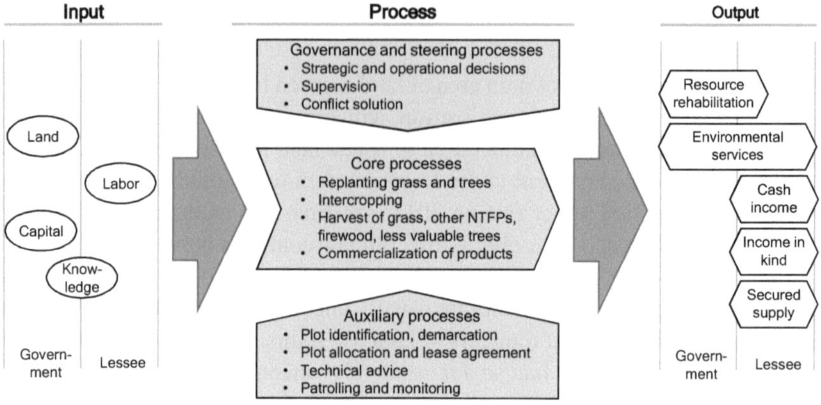

Fig. 5.5 Input, process and output of leasehold forestry in Nepal

Box 5.3 (continued)

The households were selected on the basis of a maximum land ownership of 0.5 ha and a maximum annual income of USD 40 per capita. Demarcated plots of up to 1 ha were leased to the selected families, organized in almost 2,900 small groups of 5–15 households. The plots had to be replanted with grass and different species of trees. Each family was allowed to use the grass, NTFPs and low value timber trees, whilst the right to use valuable timber trees and large trees rested with the state.

The leasing process consisted of three major consecutive steps: submission of application, preparation of operational plan and approval by the District Forestry Office (DFO). Only plans with a positive net benefit were considered. Before the handover of the area to the community group, a 35 day notification of the decision was put in place, and only if there were no objections was the land handed over.

The DFO supported the rehabilitation process by providing planting material and technical assistance. The lessee was responsible for tending the plot and was granted the right to exclusive use in return.

In cases where leasehold forestry was implemented close to nearby community forestry schemes, the leasehold groups frequently experienced encroachment from community forest members, for example illegal grazing or theft of planted seedlings. It was easy for the better-off community members to ignore the rights of the disempowered households from the poorest sections of society. The rehabilitation of the leasehold forests was successful. However, immediate poverty alleviation also required the enforcement of the rights of the poorest people.

Source: Bhattarai et al. (2007); Nagendra and Gokhale (2008); Thoms et al. (2006)

5.5.5 Outgrower Schemes

Forestry outgrower schemes are contractual partnerships, typically between small scale landholders (growers) and commercial companies, for the production of marketable forest products (FAO 2001). Through vertical coordination (Mighell and Jones 1963), outgrower schemes connect rural people with global markets. The core of any such scheme is the outgrower contract, specifying the terms of the cooperation. This includes the deliverables and services provided by the company to the grower (e.g., credits, technical advice, materials); the quantity, quality and timing of produce to be delivered to the company; and finally the price, or references for the price, the company is prepared to pay (Binswanger and Rosenzweig 1986; Little and Watts 1994; Cairns 2000). These contracts also entail the right of the company to reject substandard produce (Glover and Ghee 1992), and often the obligation of the grower to sell exclusively to the contracting company (Watts 1994).

Outgrower schemes can be categorized into two types: arrangements in which growers are responsible for production while the company guarantees purchase of the final produce at a predetermined price; and those in which the company assumes responsibility for production and pays landholders market prices for their produce allocations (Maansson 2003).

While outgrower schemes provide comparative economic advantages for the company in the production of specific commodities, most forestry outgrower schemes are also promoted as a form of development intervention with multiple benefits. As the power of the company is much greater than that of a normal grower, the company takes on a special social responsibility. Good practice schemes only contract landholders with a minimum area of land; enough to secure sufficient production for their own domestic consumption on top of the contract production.

To enter into an agreement on the growing of forest products (on privately owned or designated community land), the plot of land must be suitable for tree growing, including water and conservation restrictions. Usually the company's experts assess the proposed area first. Having entered into the agreement, the grower applies for the necessary reforestation permits, takes on a credit to prepare the land, plants the seedlings and establishes fire breaks – all with the company's technical advice. The grower is responsible for the maintenance and ongoing monitoring of the plantation, and for any other management activities prescribed by the company (Fig. 5.6). While some of the services provided to the grower are free of charge, others, like advance payments and credits, are deducted from the final payment made by the company at the time of harvest. The collateral is the plantation itself. The production risks formally rest with the grower. The harvest is typically organized in coordination with specialized logging and transport service providers. Final payments are based on the amount of timber registered at the mill gate. Outgrower contracts may be agreed for one tree rotation only, or for more. In the latter case, some of the final payment may be used to establish the next crop of trees (cf. Cairns 2000; Howard 2005).

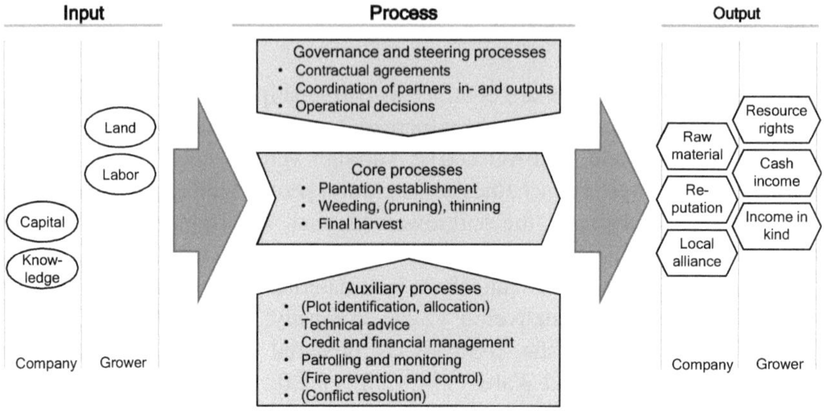

Fig. 5.6 Input, process and output of a forestry outgrower scheme

Motivations for the grower to enter into such an agreement are additional income opportunities, diversified farm production, the option to better use underutilized land, secured sale contracts, technical and financial support, secured land tenure and tree rights. The major advantages of outgrower schemes from the company perspective are access to suitable land in the vicinity of the processing unit, secured future timber supplies, the use of cheap rural labor, a spreading of environmental risks across many small plantations, an improved public image and strengthened relationships with local communities (Maansson 2003; Race et al. 2009; David 1984).

One of the most well-known forestry outgrower schemes is Sappi's 'Project Grow' in South Africa (Box 5.4). Over a period of more than two decades, the project has proved that it is possible to contract rural smallholders for the purposes of industrial pulp production.

Issues of concern from the grower perspective are market uncertainty, the viability of the company partner, environmental risks hampering production, high interest rates on loans, and price and interest rate fluctuations. From the company's viewpoint, there are the risks related to the fact that the growers may choose to sell to other buyers, the pressures to convert the land to other uses, environmental hazards, dispute over prices and contract terms, loan defaults, unstable political framework conditions, conflict with environmental organizations and an unstable local business environment (Maansson 2003).

Transaction costs in outgrower schemes are typically rather high. Direct and indirect cost items, such as program costs or provisions for market risks, reduce the payments to the growers. This limits the attractiveness of outgrower schemes, particularly in times of rising timber prices. At the same time, however, these schemes considerably reduce the risk of collapsing output prices for growers.

Overall, outgrower schemes successfully involve rural people in commercial forestry activities by lowering the barriers to entry into industrial production and by providing access to markets. Cash returns from these schemes usually boost the development of other local businesses. However, their effect in terms of poverty alleviation is limited to an elite (Marcus undated, cited by Cairns 2000).

Box 5.4 Forestry Outgrower Schemes in South Africa: Project Grow

Sappi, one of the big fiber producers in South Africa, started an outgrower scheme for pulpwood plantations with various eucalyptus species in the KwaZulu-Natal province in 1983. Initially designed to provide supplemental income for retired migrant mine workers, it evolved into a viable land use option for community farmers, with access to between 1 and 20 ha of land. In 2010 almost 10,000 growers grew trees on 15,000 ha of community and company land. The project contributes a small but valuable part of the raw material required by the expanding pulp mills (130,000 t of wood fiber in 2010).

The company, through its extension officers, provides technical assistance to the farmers in relation to site selection and site preparation, fertilizer application, planting, weed control and fire management. The company provides improved planting material free of charge, as well as an interest-free loan that can be used to hire labor or to compensate for the grower's own work in planting and monitoring. In addition, the company provides a secure market and competitive prices for the timber at maturity and, at the request of growers, facilitates negotiations between growers and harvesting and transport contractors.

The growers negotiate an agreement with the traditional authority to use community land and water resources for the plantation and contribute their own and/or hired workforce for plantation establishment and management, including monitoring, fire control and final harvesting and transport.

The trees belong to the grower, but the grower is contractually required to sell them to Sappi. Up to 50 % of the final revenues remain with the harvesting and transport service provider, 30 % with the grower as compensation for their own and/or hired labor and 20 % of the final revenues are profit for the grower. In 2005, the total income from a 7-year eucalyptus plantation was USD 3,107/ha, translating into an income of USD 16.86/man-day, and a grower net profit of USD 1,337/ha.

The returns for the grower account for approximately 20 % of the average household income and lift many households above the poverty line. Apart from the reliable financial income, growers also obtain firewood and construction material from their plantations. In addition to growers, non-grower community members also benefit indirectly from increased rural employment opportunities in forest plantations and other business. Several growers have already invested their forestry income in a transport business, forestry service firms or in shops.

(continued)

Box 5.4 (continued)

Challenges facing growers include the theft of timber, competition from other buyers offering slightly higher prices than Sappi, insufficient road infrastructure, bank services that complicate the payout of cash to the growers and relatively high transaction costs for the company.

Source: Howard (2005); Mayers and Vermeulen (2002, 2003); Mncube (2010); Sappi (2006, 2011); Sartorius and Kirsten (2007); Smith (2003)

5.5.6 Forestry Contractors

Forestry contractors are independent, specialized entrepreneurs delivering services for forest management operations (UNECE/FAO 2008; Macqueen 2008). Forestry contractors typically do not grow trees themselves, but offer services required for forest management to other forestry enterprises on a contractual basis. Figure 5.7 shows that the land is with the forest enterprise, nearly all other inputs concerning production and management can be provided by contractors. In the context of increasing division of labor, mechanization and the technological specialization of forest management operations, forestry service contractors horizontally integrate the production processes of various forest enterprises into the forestry commodity chain. Typical services provided are forest inventory, forest management planning, planting, weeding, chemical and fertilizer application, felling, skidding, timber transport and consulting services. Although there are also large commercial forest service enterprises, many forest contractor operations in the tropics are of a small and sometimes even informal character (Osei-Tutu and Nhancale 2010).

The skill sets and assets of forest contractors vary widely. Generally, two archetypal service providers can be distinguished: (1) contractors specializing in services that require capital-intensive equipment, aiming for economies of scale and large turnover; for example, skidder and truck owners; and (2) contractors specializing in services that are infrequent but require a high input of largely unskilled labor, such as tree planting or the peeling of bark (Box 5.5). The diversity of forest management operations means that skilled and unskilled workers alike, the poor and those better-off can find employment as forestry contractors, as demonstrated in the example from Vietnam presented in Box 5.5. Nevertheless, the labor productivity of these people and the share of the added value vary significantly, as do their working conditions.

Forest contractors are of considerable importance where strong and competitive markets for forest products exist. With increasing specialization and investment, contractors can potentially benefit from economies of scale and build their capacity to compete for orders. The required skill sets, level of training and the level of sophistication of the equipment increase simultaneously. In order to cover high fixed costs, capital-intensive equipment and highly specialized employees, contractors need to operate at near maximum levels. Activities outside the core business are typically abandoned to increase efficiency.

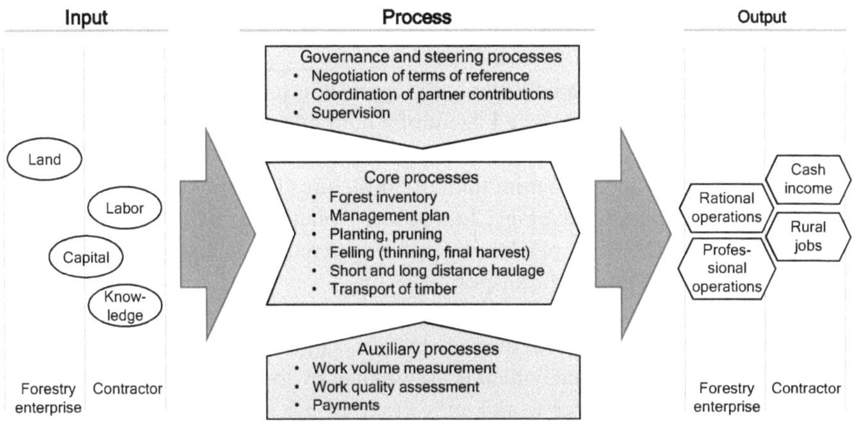

Fig. 5.7 Input, process and output of forestry service providers in Vietnam

For a seamless operation and the accrual of mutual benefits, the partners involved in forest management operations have to establish mutual trust, for example with regard to the quantity of timber delivered, the quality of the produce and the financial transactions. A high level of mutual trust results in significantly lower transaction costs. For instance in the example from Vietnam (Box 5.5), the pulp mill's load weight records are accepted as the basis for the contractor's payments, which saves the stakeholders the effort of making multiple volume/weight measurements. The grower is not part of this alliance, and 'pays' a high price for omission.

Overall, service providers represent a flexible, market-oriented model for forest management through independent and specialized economic units. The biggest advantage of contractors is that they make capital intensive equipment available to small scale producers, thereby allowing them to produce at a cost which is competitive on commercial markets and to cover labor peaks they would not normally be able to manage themselves. Furthermore, forestry contractors potentially create income, jobs and economic demand in rural areas.

Box 5.5 Forest Contractors in Smallholder Tree Plantations in Vietnam

In line with the policy of the Vietnamese government to massively promote tree plantation establishment, villagers from the Thuan Phong village (Phu Cat district), obtained property rights for land set aside for tree plantations in 2005. Since then, approximately 250 households have established more than 800 ha of eucalyptus plantations in rotations of 4–6 years. Planting and tending are typically carried out by the growers themselves. When a stand is ready for harvest, farmers usually sell to local middlemen, obtaining a gross margin of USD 686/ha to USD 897/ha in the process, depending on credit and the costs of hired labor.

(continued)

Box 5.5 (continued)

Five full-time and a small number of part-time local middlemen buy the standing wood for an authorized agent of a pulp mill company. The collector examines the stand and offers the owner a price. If the deal is agreed, the buyer pays an advance of a minimum of 50 %, hires loggers, bark peelers and a truck for transport. On average, 1 ha of forest plantation provides 15 days of employment in the form of planting and maintenance annually and another 11 days for harvesting and transport.

In the village about 20 strong and skilled men work as self-employed loggers, felling the trees with their chainsaws. They are paid according to the amount of wood harvested, which is ascertained by the load weights recorded at the pulp mill. A logger works an average of 3.5 days and receives USD 82.80/ha, leaving him with a net income of USD 10.23/day for his labor. Bark peelers work in teams of 12–17 people, comprising both men and women. One team peels more or less the same amount of timber cut by one feller, approximately 2,000 t per year. No special skills are required to take on the job. Like the others, peelers are paid per ton of peeled timber and earn an income of approximately USD 2.96/day. They also load the truck, earning USD 1.14 per load. Transport is offered by individuals owning a truck. Of the USD 140/ha paid, USD 48.40 remain for labor after the cost of operations, translating into a daily income of USD 16.13. The agent from the pulp mill is paid a provision of USD 30/ha.

Compared to the other stakeholders, the local collectors receive a dis-proportionately large share of the profit; their income amounts to an average USD 310/ha. As most growers are not able to correctly determine the biomass of their trees, collectors usually underestimate the standing biomass deliberately. Up to two-thirds of their income stem from the difference between the true amount and the amount accepted by the tree grower.

Amounts calculated for a 5-year *Eucalyptus urophylla* rotation with a yield of 35 t/ha and an exchange rate of USD 1 $\approx$ 17,500 VND

Source: Kien (2009); Nguyen Quang Tan (2011)

5.6 Outlook

5.6.1 Conclusion

The high competition for resources and the increasing pressure to produce short-term benefits are challenges facing today's forest organizations. This manifests itself in the liquidation of forest resources for the sake of survival, as well as for the maximization of financial profits. The competition for forest land jeopardizes sustainable forest management systems from both ends: by the conversion of forests to more intensive land use systems and by taking forests out of production for purely conservation purposes. In contrast to such segregated, single objective approaches, integrated forestry systems with multiple functions may be more resilient and tend to be more easily reconcilable with natural and social systems. However, these systems require more sophisticated management, are more prone to conflict and incur higher transaction costs. As each challenge offers opportunities, the members of forest organizations are encouraged to identify the opportunities available to them and to take them on. There are many actors in the arena, each with their own individual perception of and agenda for rural development. Ultimately the rural people must obtain an overview and understand the situation, identify the right partners to support their livelihoods in a sustainable way, and band together to negotiate with opponents in order to achieve fair compromises. Their skills and competences, in other words their human and social capital, will determine the future of their organizations and of their forests. There are many technical solutions available. The capacity to take up the challenge, to design adaptive and intelligent solutions, and to implement these, will ultimately determine the performance of forest organizations and their contribution to rural development as well as to other global challenges. The recognition of smallholders as important actors in adapting to global change underlines their prominent position in the future.

The following aspects are deemed essential for the further development of forest organizations towards 'working' and 'sustainable' arrangements for forest management:

5.6.2 Steering, Monitoring and Enforcement

Steering, monitoring and enforcement are based on personal capacity, attitude and motivation. The following factors are elemental in developing these capacities: education, empowerment and responsibility.

5.6.2.1 Education

Traditional knowledge is recognized as a valuable resource for sustainable forest management that complements the available scientific knowledge on forest management. However, under the terms of current and future global competition, forest organizations must also acquire professional business and organizational skills, the capacity to negotiate, to build partnership networks and effective regulating institutions, and they must compete and coexist with other players on the factor markets. The education and training required to develop such managerial skills will become a necessity for the managers and leaders of future forest organizations, if they are to achieve integrated and adapted management characterized by mutual benefits and sustainability (cf. Macqueen 2012; Jolliffe 2004).

5.6.2.2 Empowerment

Often there is a high level of interdependence between forests and poor or disadvantaged groups. Empowering and improving the living conditions of these groups can strengthen sustainable forest management. Group-based forest organizations are one way of attaining this goal.

To maintain established forestry policies and programs beneficial to the poor, these must be reviewed and amended continuously (Bhattarai et al. 2007). While Regmi et al. (2009) focus on capacity building support as a complement to required institutional reforms Hobley (2007) recommends integrating government, private sector and state-incorporated civil society.

5.6.2.3 Responsibility

The struggle for a fair distribution of the benefits derived from forests is at the center of forest governance. To sustain the flow of benefits, the inputs have to be delivered by the various parties:

Coordinated by forest organizations, activities on the ground have to be realized by their members. Monitoring in particular has been highlighted as being crucial to success (Singh et al. 2011; van Laerhoven 2010).

It is the responsibility of governments to stipulate frame conditions which are supportive to sustainable forest management. As superior institutions, they must secure efficient supervision, enforcement and facilitation. The latter includes the coordination of forest organizations and other relevant actors, the dissemination of information, communication, cooperation and negotiation (Carter and Gronow 2005; Aggarwal et al. 2010).

5.6.3 Fair Partnerships

Fair partnerships are crucial for sustainable development (Vermeulen et al. 2003) and for the future of forest organizations. Partnerships can be established through various types of formal and informal arrangement. They are a means to share risks (Mayers and Vermeulen 2002) and to lower transaction costs. Strategic partnerships involve the partner's core business activities. They are widely promoted for civil society-business collaboration (Ashman 2001).

5.6.4 Policies

Policies enable the forest organizations to achieve sustainable forest management. Essential in this respect are the fair allocation of property rights, including land tenure (Hobley 2007; Adhikari et al. 2004); the facilitation of sustainable value chains, connecting rural producers with national and global markets (Mayers and Vermeulen 2002); and the securing of organizational agreements such as outgrower schemes to the mutual benefit of all of the partners.

5.6.5 Forest Organizations in a Diversified, Intelligent Pro-nature Economy

Both the consumption patterns of individuals and the growing population result in a rising demand for an ever larger number of services and products from forest ecosystems. At the same time, (natural) forests tend to be increasingly degraded and converted. The challenge for forest organizations is to use the available resources to 'grow' more market and non-market products and to provide services. This can be attained by higher efficiency of processes within the forest organizations, which also fosters the resilience of the organizations.

5.6.6 Recommendation

In an increasingly complex forest management setting, the plain copy of a generic model for a particular place will not necessarily succeed. First, location specific, essential success factors have to be identified and understood. Then the design of the intended organizational model needs to be reviewed and adapted to serve under the specific natural and social conditions. The badly needed adaptive and learning skills challenge the people assigned with the management of forests. Designing efficient organizational structures and processes for a given set of factors is the 'fine art' of management and leadership.

References

Adhikari B, Di Falco S, Lovett JC (2004) Household characteristics and forest dependency: evidence from common property forest management in Nepal. Ecol Econ 48(2):245–257. doi:10.1016/j.ecolecon.2003.08.008

Aggarwal PK, Baethegan WE, Cooper P, Gommes R, Lee B, Meinke H, Rathore LS, Sivakumar MVK (2010) Managing climatic risks to combat land degradation and enhance food security: key information needs. World Climate Conference – 3. Procedia Environ Sci 1:305–312. doi:10.1016/j.proenv.2010.09.019

Agrawal A (2005) Environmentality: technologies of government and political subjects. Oxford University Press, New Delhi

Andersen ES (2010) The X model – a tool for describing and assessing individual projects. Int J Manag Proj Bus 3(3):369–386. doi:10.1108/17538371011056039

Arnold J (2001) Forests and people: 25 years of community forestry. http://www.treesforlife.info/fao/Docs/P/25y.pdf. Accessed 11 Apr 2012

Ashman D (2001) Civil society collaboration with business: bringing empowerment back in. World Dev 29(7):1097–1113. doi:10.1016/S0305-750×(01)00027-4

Auch E (2007) Überlebensstrategien waldnutzender Familienwirtschaften im Apuseni-Gebirge, Rumänien – sustainable livelihoods analyse und Handlungsempfehlungen. Schriften aus dem Institut für Forstökonomie der Universität Freiburg, vol 27. Verlag des Instituts für Forstökonomie der Universität Freiburg, Freiburg im Breisgau

Bakarr MI (2005) Forested landscapes and community development: science, scale, and sustainability. In: CIFOR (ed) Towards a shared vision and action frame for community forestry in Liberia. Proceedings of the first international workshop on community forestry in Liberia. CIFOR, Bogor Barat, pp 12–16

Bass SM (1992) Building from the past: forest plantations in history. In: Sargent C, Bass SC (eds) Plantation politics. Forest plantations in development. Earthscan, London, pp 41–75

Bastakoti RR (2005) Evaluation of collective action in forest users' groups (Case study from Dhading District, Nepal). Masterarbeit, Technische Universität Dresden

Bhattarai B, Dhungana SP, Kafley GP (2007) Poor-focused common forest management: lessons from Leasehold Forestry in Nepal. J Forest Livelihood 6(2):20–29

Binswanger HP, Rosenzweig MR (1986) Behavioural and material determinants of production relations in agriculture. J Dev Stud 22(3):503–539. doi:10.1080/00220388608421994

Blank W (2006) Die kolonialwirtschaftliche Forstnutzung der ehemaligen Kolonie Deutsch-Ostafrika. Masterarbeit, Technische Universität Dresden

Bromley DW (1989) Economic interests and institutions. The conceptual foundations of public policy. Blackwell, Oxford

Cairns RI (2000) Outgrower timber schemes in Kwazulu–Natal. Do they build sustainable rural livelihoods and what interventions should be made? Instruments for sustainable private sector forestry, South Africa Series, Pretoria

Campbell B, Mandondo A, Nemarundwe N, Sithole B, de Jong W, Luckert M, Matose F (2001) Challenges to proponents of common property resource systems: despairing voices from the social forests of Zimbabwe. World Dev 29(4):589–600. doi:10.1016/S0305-750×(00)00114-5

Carter J, Gronow J (2005) Recent experience in collaborative forest management. A review paper, vol 43, CIFOR occasional paper. CIFOR, Jakarta

Chrobok R, Büchi R (1996) Grundbegriffe der organisation – GBO, vol 1, 2nd edn, GfürO-Reihe organisation. Schäffer-Poeschel, Stuttgart

Coase RH (1937) The nature of the firm. Economia New Ser 4(16):386–405

Colchester M, Apte T, Laforge M, Madondo A, Pathak N (2003) Bridging the gap: communities, forests and international networks. Synthesis report of the project 'Learning Lessons from International Community Forestry Networks'. http://www.cifor.org/publications/pdf_files/OccPapers/OP-41.pdf. Accessed 31 July 2012

Coleman EA, Steed BC (2009) Monitoring and sanctioning in the commons: an application to forestry. Ecol Econ 68(7):2106–2113. doi:10.1016/j.ecolecon.2009.02.006

Cox M, Arnold G, Tomás SV (2010) A review of design principles for community-based natural resource management. Ecol Soc 15(4):38

Cronkleton P, Guariguata MR, Albornoz MA (2012) Multiple use forestry planning: Timber and Brazil nut management in the community forests of Northern Bolivia. Multiple use of tropical forests: from concept to reality. Forest Ecol Manag 268:49–56. doi:10.1016/j.foreco.2011.04.035

Darr D (2003) Farm forestry in semi-subsistence and monetary economies and its interdependency with the land tenure system – case studies from central Laos. M.Sc. thesis, Technische Universität Dresden

David JG (1984) Contract farming and smallholder outgrower schemes in less-developed countries. World Dev 12(11–12):1143–1157. doi:10.1016/0305-750×(84)90008-1

Dhakal B, Bhatta B (2009) An institutional model to explain utilization problems of community forest products. Int J Soc Forest 2(2):23–48

Dillon HS (1992) Contract farming in Indonesia. The dairy and rubber industries. In: Glover D, Ghee LT (eds) Contract farming in Southeast Asia. Three country studies. Institute for Advanced Studies, University of Malaya, Kuala Lumpur, pp 119–156
Dougill AJ, Soussan JG, Kiff E, Springate-Baginski O, Yadav NP, Dev O, Hurford AP (2001) Impacts of community forestry on farming system sustainability in the Middle Hills of Nepal. Land Degrad Dev 12(3):261–276. doi:10.1002/ldr.438
Ellis F (1992) Peasant economics. Farm households and agrarian development, 2nd edn, Wye studies in agricultural and rural development. Cambridge University Press, Cambridge
Emery FE, Trist E (1965) The causal texture of organizational environments. Hum Relat 18 (1):21–32. doi:10.1177/001872676501800103
FAO (1978) Forestry for local community development, vol 7, Forestry paper. Food and Agriculture Organization of the United Nations, Rome
FAO (2001) Forestry out-grower schemes: a global overview. Report based on the work of D. Race and H. Desmond. Forest plantation thematic papers, working paper. FAO, Rome
FAO (2010) Global forest resources assessment 2010. Main report, vol 163, FAO forestry paper. Food and Agriculture Organization of the United Nations, Rome
Gaitanids M (2007) Prozessorganisation. Entwicklung, Ansätze und Programme des Managements von Geschäftsprozessen, 2nd edn. Vahlens Handbücher der Wirtschafts- und Sozialwissenschaften. Vahlen, München
Glover D, Ghee LT (eds) (1992) Contract farming in Southeast Asia. Three country studies, vol 5, Monograph series. Institute for Advanced Studies, University of Malaya, Kuala Lumpur
Guha R (1983) Forestry in British and post-British India: a historical analysis. Econ Polit Wkly 18 (44):1882–1896
Hartzsch H (1983) Wälder und Forstwirtschaft in der VDR Laos – ein Beitrag zum Stand und zu den Entwicklungstendenzen der laotischen Forstwirtschaft und der forstlichen Ausbildung. Dissertation, Technische Universität Dresden
Hobley M (2007) Where in the world is there pro-poor forest policy and tenure reform? http://www.rightsandresources.org/documents/files/doc_658.pdf. Accessed 25 Jan 2012
Honadle G (1999) How context matters. Linking environmental policy to people and place. Kumarian Press, West Hartford, Connecticut, USA
Howard MMPMDBFLFI (2005) Small-scale timber production in South Africa: what role in reducing poverty? Discussion paper. International Institute for Environment and Development (IIED), London
ITTO (2003) Restoring the forests. Introducing the ITTO guidelines for the restoration, management and rehabilitation of degraded and secondary tropical forests. http://www.itto.int/direct/topics/topics_pdf_download/topics_id=3670000&no=1&disp=inline. Accessed 01 May 2012
Iversen V, Chhetry B, Francis P, Gurung M, Kafle G, Pain A, Seeley J (2006) High value forests, hidden economies and elite capture: evidence from forest user groups in Nepal's Terai. Ecol Econ 58(1):93–107. doi:10.1016/j.ecolecon.2005.05.021
James NDG (1990) A history of English forestry, vol 2. Basil Blackwell, Oxford
Jolliffe D (2004) The impact of education in rural Ghana: examining household labor allocation and returns on and off the farm. J Dev Econ 73(1):287–314. doi:10.1016/j.jdeveco.2003.02.002
Jørgensen I, Vivekanandan K (2003) Private forestry based on Paulownia in Sri Lanka. An appraisal of the outgrower scheme by Paulownia Plantations Ltd. Noragric Report, Ås
Kaimowitz D (2005) The international experience with community forestry. In: CIFOR (ed) Towards a shared vision and action frame for community forestry in Liberia. Proceedings of the first international workshop on community forestry in Liberia. CIFOR, Bogor Barat, pp 17–19
Kanel Raj K, Acharya DP (2008) Re-inventing forestry agencies: institutional innovation to support community forestry in Nepal. In: Durst P et al (eds) Re-inventing forestry agencies. Experiences of institutional restructuring in Asia and the Pacific, vol 3. FAO Regional Office for Asia and the Pacific, Bangkok, pp 132–160

Kien ND (2009) Improvement of Eucalyptus plantations grown for pulp production. Ph.D., Sveringes Lantbruksuniversitet

Kinsey BH, Binswanger HP (1993) Characteristics and performance of resettlement programs: a review. World Dev 21(9):1477–1494. doi:10.1016/0305-750×(93)90128-V

Larson AM, Barry D, Dahal GR (2010) New rights for forest-based communities? Understanding processes of forest tenure reform. Int For Rev 12(1):78–96. doi:10.1505/ifor.12.1.78

Le Hung Cuong (1988) Zur Organisation der forstlichen Produktion in den Forstwirtschaftsbetrieben der SR Vietnam. Dissertation, Technische Universität Dresden

Li B, Li Nana, Liu X (eds) (2004) Transformation and countermeasures on business operation models in uncertain environment. The fourth international conference on electronic business (ICEB2004), Beijing

Little PD, Watts M (eds) (1994) Living under contract. Contract farming and agrarian transformation in Sub-Saharan Africa. University of Wisconsin Press, Madison

Maansson M (2003) A review of principles, criteria and indicators of best practice in outgrower schemes. In: FAO (ed) Towards equitable partnerships between corporative and smallholder partners. Workshop synthesis. Food and Agriculture Organization of the United Nations, Rome, pp 27–32

Macqueen D (2008) Supporting small forest enterprises: a cross-sectoral review of best practice, vol 23, IIED small and medium forestry enterprise series. IIED, London

Macqueen D (2012) Enabling conditions for successful community forest enterprises. Small-Scale Forest 1–19. doi: 10.1007/s11842-011-9193-8

Marcus T (undated) KwaZulu-Natal provincial synthesis report. Land and Agriculture Policy Centre, Land Reform Research Programme

Mayers J, Vermeulen S (2002) Company-community forestry partnerships: from raw deals to mutual gains? An international review with proposals for improving forests, enterprise and livelihoods, Instruments for sustainable private sector forestry series. Earthprint, London

Mayers J, Vermeulen S (2003) Company-community forestry partnerships: from raw deals to mutual gains? In: FAO (ed) Towards equitable partnerships between corporative and smallholder partners. Workshop synthesis. Food and Agriculture Organization of the United Nations, Rome, pp 133–144

Miebach B (2009) Prozesstheorie. Analyse, organisation und system. VS Verlag für Sozialwissenschaften/GWV Fachverlage GmbH Wiesbaden, Wiesbaden

Mighell RL, Jones LA (1963) Vertical coordination in agriculture. Farm economics division, economic research service, U.S. Dept. of Agriculture, Washington

Mncube D (2010) Sappi's project grow – presentation by Dinga Mncube. Executive insight. Mark Roux, Deloitte South Africa, Deloitte Eminent Speaker Event

Nagendra H, Gokhale Y (2008) Management regimes, property rights, and forest biodiversity in Nepal and India. Environ Manag 41:719–733. doi:10.1007/s00267-008-9073-y

Naina HA (2011) Impact of taungya implementation in the collaborative forest management (CBFM) program towards the livelihoods improvement of the forest village community in West Java, Indonesia. Masterarbeit, TU Dresden

Nair PKR (1993) An introduction to agroforestry. Kluwer Academic Publishers in cooperation with International Centre for Research in Agroforestry, Dordrecht

Negussie AD (2003) Farm forestry decision-making strategies of the Guraghe households, Southern-Central Highlands of Ethiopia. Dissertation, TU Dresden

Nguyen Quang Tan (2011) Chopping for chips. An analysis of wood lows from smallholder plantations in Vietnam. Working paper, vol 65. CIFOR, Bogor

Nsiah B (2010) Contribution of smallholder farm forest plantation management to livelihood strategies of farm households in the high forest zone of Ghana. Doctoral, TU Dresden

Osei-Tutu P, Nhancale B (2010) Hidden forestry revealed. Characteristics, constraints and opportunities for small and medium forest enterprises in Ghana. International Institute for Environment and Development (IIED), London

Ostrom E (1990) Governing the commons. The evolution of institutions for collective action, vol 1. Cambridge University Press, Cambridge

Otsuka K, Larson DF (2013) Towards a green revolution in Sub-Saharan Africa. In: Otsuka K, Larson DF (eds) An African green revolution. Finding ways to boost productivity on small farms. Springer, Dordrecht, pp 281–300

Pérez MR, Belcher B, Fu M, Yang X (2004) Looking through the bamboo curtain: an analysis of the changing role of forest and farm income in rural livelihoods in China. Int For Rev 6 (3–4):306–316. doi:10.1505/ifor.6.3.306.59968

Poffenberger M, McGean B (eds) (1996) Village voices, forest choices: joint forest management in India. Oxford University Press, New Delhi

Pokhare BK, Nurse M (2004) Forests and people's livelihoods: benefiting the poor from community forestry. http://communityforestry-nscfp.org/files/listingfiles/MicrosoftWord-Propoorframe_DP303mike071103.pdf. Accessed 24 Jan 2012

Prasad R, Bhatnagar P (1995) Social forestry experiences over a decade. International Book Distributors, Dehra Dun

Pretzsch J (2005) Forest related rural livelihood strategies in national and global development. Forest Trees Livelihood 15:115–127

Pretzsch J (2010) Forest organisations in change. Examples from the tropics and subtropics. In: Spathelf P (ed) Sustainable forest management in a changing world: a European perspective. Springer, Heidelberg, pp 191–215

Race D, Bisjoe AR, Hakim R, Hayati N, Julmansyah N, Kadir A, Kurniawan A, Kusumedi P, Nawir AA, Nurhaedah AA, Perbatasari D, Purwanti R, Rohadi D, Stewart H, Sumirat B, Suwarno A (2009) Partnerships for involving small-scale growers in commercial forestry: lessons from Australia and Indonesia. Int Forest Rev 11(1):88–97

Regmi BR, Albano A, Kumar C, Sharma GB (2009) Improving the livelihoods of the poor and marginalized in Nepal through leasehold forestry and livestock program: a review of institutional constraints and opportunities. Banko 18(2):10–17. doi:10.3126/banko.v18i2.2168

Royo N (2011) Why tenure reform is needed in Asia. ITTO Trop Forest Update 20(4):9–10

de Rueda RA (2008) Re-inventing the forestry agencies of the Department of Environment and Natural Resources, Philippines: a case study. In: Durst P et al (eds) Re-inventing forestry agencies. Experiences of institutional restructuring in Asia and the Pacific, vol 2. FAO Regional Office for Asia and the Pacific, Bangkok, pp 79–102

Sappi (2006) Project grow report 28 June 2006. http://environment.yale.edu/tfd/uploads/SAPPI_Project_Grow_Report.pdf. Accessed 19 Feb 2012

Sappi (2011) Southern Africa sustainability report 2011. http://www.sappi.com/regions/sa/Sustainability/Documents/2011 Sappi Southern Africa Sustainability Report.pdf. Accessed 26 July 2012

Sartorius K, Kirsten J (2007) A framework to facilitate institutional arrangements for smallholder supply in developing countries: an agribusiness perspective. Food Pol 32(5–6):640–655. doi:10.1016/j.foodpol.2007.03.001

Schlager E, Ostrom E (1992) Property-rights regimes and natural resources. A conceptual analysis. Land Econ 68(3):249–262

Schober H (2002) Prozessorganisation. Theoretische Grundlagen und Gestaltungsoptionen. Gabler Edition Wissenschaft. Dt. Univ.-Verl, Wiesbaden

Schreckenberg K, Luttrell C, Moss C (2006) Participatory forest management: an overview – background paper. http://www.odi.org.uk/resources/docs/3781.pdf. Accessed 16 Apr 2012

Singh VS, Pandey DN, Prakash NP (2011) What determines the success of joint forest management? Science-based lessons on sustainable governance of forests in India. Resour Conserv Recycl 56(1):126–133. doi:10.1016/j.resconrec.2011.09.015

Smith D (2003) Sappi forests: project grow. In: FAO (ed) Towards equitable partnerships between corporative and smallholder partners. Workshop synthesis. Food and Agriculture Organization of the United Nations, Rome, pp 169–173

Stogdill RM (1967) Basic concepts for a theory of organization. Manage Sci 13(10): B-666. doi:10.1287/mnsc.13.10.B666

Thoms C, Karna BK, Karmacharya MB (2006) Limitations of leasehold forestry for poverty alleviation in Nepal. Soc Nat Resour 19:951–958. doi:10.1080/08941920600902179
Tuan DA (2006) Governance structure and performance. A comparative institutional analysis of community forest management. Dissertation, TU Dresden
Turyahabwe N, Banana A (2008) An overview of history and development of forest policy and legislation in Uganda. Int Forest Rev 10(4):641–656
UNECE/FAO (2008) Guide to good practice in contract labour. Report of the UNECE/FAO team of specialists on best practices in forest contracting. FAO, Rome
Uphoff N (1993) Grassroots organizations and NGOs in rural development: opportunities with diminishing states and expanding markets. World Dev 21(4):607–622. doi:10.1016/0305-750X(93)90113-N
van Laerhoven F (2010) Governing community forests and the challenge of solving two-level collective action dilemmas—a large-N perspective. Governance, complexity and resilience. Glob Environ Chang 20(3):539–546. doi:10.1016/j.gloenvcha.2010.04.005
van Noorwijk M, Tomich TP (1995) Agroforestry technologies for social forestry. Tree-crop interactions and forestry-farmer conflicts. In: Hartadi SH, Sumardi SS, Iswantoro H (eds) Social forestry and sustainable forest management. Perum Perhutani, Jakarta, pp 168–193
Vermeulen S, Nawir AA, Mayers J (2003) Better livelihoods through partnership? A review of the impacts of deals between communities and forestry companies on local development. Paper presented at the international conference on rural livelihoods, forests and biodiversity, Bonn, 19–23 May 2003
Watts M (1994) Life under contract: contract farming, agrarian restructuring, and flexible accumulation. In: Little PD, Watts M (eds) Living under contract. Contract farming and agrarian transformation in Sub-Saharan Africa. University of Wisconsin Press, Madison, pp 21–77
White A, Martin A (2002) Who owns the world's forests? Forest tenure and public forests in transition. Forest Trends, Washington
Wiersum KF (1999) Social forestry. Changing perspectives in forestry science or practice? Ph.D. thesis, Wageningen Agricultural University, Wageningen
Wiersum KF (2010) Community forestry between local autonomy and global encapsulation. http://www.forestrynepal.org/images/02-Presented Papers and Powerpoints/Papers/05-Freerk Wiersum Paper.pdf. Accessed 11 Apr 2012
World Bank (1978) Forestry. Sector policy paper. http://documents.worldbank.org/curated/en/1978/02/440307/forestry. Accessed 07 July 2012
Xayvongsa L (2001) Beurteilung bäuerlichen Teakanbaus mittels Betriebssystemanalyse in Luang Prabang, Laos. Assessment of Teak related Farming Systems in Luang Prabang, Laos. Ph.D. thesis, Technische Universität Dresden, Dresden
Zitzmann G (1998) Multiple use and livelihood strategies in mopane woodland. The case of Ditladi, North-East District, Botswana. M.Sc. thesis, Technische Universität Dresden, Dresden
Zivnuska J (1966) The integration of forest development plans and national development plans. How to make the forestry case at the national level. Proceedings of the 6th world forest congress, Madrid

Chapter 6
Creating Rural Employment and Generating Income in Forest-Based Value Chains

Peter Poschen, Merten Sievers, and Asmamaw Alemu Abtew

Abstract In this chapter the authors present the potential contribution that forest products can make to rural development. The production and marketing of forest products is an important source of rural employment and income. Benefits often extend beyond the forest area itself, as the products move from production to end use through different stages of processing and marketing. The basic concepts of value chain analysis and social accounting matrices are introduced in the chapter, and their relevance and application in understanding and harnessing the employment and income potential of forest product value chains for rural development discussed.

Keywords Forest product • Value chain • Social accounting matrix • Employment • rural development

6.1 Introduction

According to the Food and Agriculture Organization, in the year 2006 the value chains centered on wood products and pulp and paper provided formal employment for 13.7 million people globally, which is equivalent to 0.4 % of the total labor force (FAO 2011). A significant number of the four million jobs in the formal furniture industry are also based on forest products such as wood, rattan and bamboo.

P. Poschen (✉) • M. Sievers
Job Creation and Enterprise Development Department, International Labour Organization, 4 route des Morillons, Genève CH-1211, Switzerland
e-mail: poschen@ilo.org; sievers@ilo.org

A.A. Abtew
Institute of International Forestry and Forest Products, Technische Universität Dresden, Postfach 1117, 01735 Tharandt, Germany
e-mail: asmamawalemu@yahoo.com

J. Pretzsch et al. (eds.), *Forests and Rural Development*, Tropical Forestry 9,
DOI 10.1007/978-3-642-41404-6_6,

The International Labour Organization (ILO) estimated that there are three times as many informal jobs in forestry and in micro and small forest-based enterprises (ILO 2010; ILO 2001).[1]

Timber and non-timber forest products constitute an important income and employment generating factor in rural communities. This is often critical for the conservation and sustainable management of forests as it can offer a viable alternative to either the destructive use of forests or their outright conversion to other land-uses. As a consequence, forest and environmental protection agencies, rural development workers and environmentalists often pin high hopes on forest products obtained from sustainable management.

However, incomes derived from forests are often not high enough to lift people out of poverty and in many cases merely constitute a supplementary income. The challenge in terms of creating employment that is gainful while at the same time being sustainable is enormous due to a wide range of constraining factors in rural areas. These include the geographical distance to markets, little information and knowledge about the end markets for rural products, frequently poor infrastructure and low levels of human capital.

In view of these constraints, the question arises as to what can be done to increase rural employment linked to forestry, and how can it be made more productive and gainful. A possible answer lies in developing a better understanding of the opportunities and constraints faced by different sectors of the rural economy. More insights into the functioning of rural markets and the markets for rural products can help provide an understanding of what channels might provide real opportunities for future income generation and job creation. In this chapter the authors set out to explore how best to approach market development for forestry products in rural areas.

In the past, not enough was done to understand the market opportunities capable of driving more rural job creation. Government and donor programs too often adopted existing rural product supply, and constraints on this supply, as a starting point for rural development. However, while consideration of the resources and constraints on the ground is important, it is not sufficient as a means to establish development opportunities. What is needed is a different perspective not starting from the supply end of rural products, but putting the focus instead on the demand from local, national and international markets. Those behind rural development initiatives need to understand how this demand for rural products is evolving and what it takes to meet this demand. This new perspective cannot solve the problem of the marginalization of certain territories, but it can help to clarify whether, and when, opportunities exist and how these may be seized. Finding new market niches

[1] These are counted as 'full time equivalent jobs' by converting the estimated number of workdays into work years to make them comparable with data for formal employment. Most of the work is actually part-time and the number of people employed and deriving a substantial part of their income from forests is, therefore, much larger. The World Bank (2004) put the number of indigenous people depending mostly on primary forests at 60 million and that of communities near forest areas at 350 million.

and expanding the market channels of existing products needs to inform policy decisions made for the purposes of promoting economies.

In this chapter the market potential of different forestry products is briefly explored and two relevant analytical approaches are then introduced, namely value chain and market analysis for the design of development interventions; and social accounting matrices as a tool to help understand the employment and income dimensions of forest product value chains and to place them in the wider context of the rural economy. The application of a SAM analysis is demonstrated for forest products in the Amazon region of Brazil. This is followed by an examination of two key groups of actors in many of these value chains: rural collectors and middlemen. Some of the key factors are then reviewed for the cases of non-timber forest products value chains in the drylands of Sudan and rattan in central Vietnam. The chapter ends with lessons for the use of value chain development approaches to create jobs in the forestry sector.

6.2 Definitions of Basic Terms

Provided in this section are definitions of the basic terms value chain, value chain development and social analysis matrix, terms used throughout the chapter.

A **value chain** "describes the full range of activities that are required to bring a product or service from conception, through the intermediary phases of production, delivery to final consumers, and final disposal after use" (Kaplinsky and Morris 2001). This includes activities such as design, production, marketing, distribution and support services up to the final consumer (and often beyond, when recycling is taken into account). Unlike 'supply chain', the term 'value chain' refers to the fact that value is added to primary products through combination with other resources (for example, tools, manpower, knowledge and skills, other raw materials or preliminary products). As the product passes through several stages of the value chain, the value of the product increases. Value chains are part of market systems. At the center of the market system are the value chains that bring products and services to the market. The immediate environment is formed by supporting functions (such as business development services and finance) and rules and regulations relevant to the chain. The broader environment around this affects the immediate environment as well as setting its own conditions.

Value chain development describes the process by which the value chain is enhanced through a variety of interventions.

The **social accounting matrix** (SAM) is an analytical and descriptive tool depicting the full flow in an economy, from production and the adding of value to income and demand prompting the next round of production. It provides a snapshot of the state of a national, regional or local economy for a given point in time, consolidating data from different sources (national accounts, sectoral data, household surveys, etc.) into a single, consistent data framework of a square matrix in which the columns show all expenditure flows and the rows reflect all incomes received.[2]

6.3 Forest Products and Value Chain Development

6.3.1 General Scope

Two broad groups of forest products must first be distinguished: timber and non-timber forest products (NTFPs). Both product groups may subsequently be subdivided into a wide range of different products with different end markets characterized by widely differing degrees of profitability. Timber is the most valuable commercial product in most forests (Sunderlin et al. 2005), and provides employment for millions of people. This includes the harvesting and extraction of timber from forests but, more importantly, its processing for the construction and furniture sectors. For example, in Central Java the timber furniture sector employs over two million people (Ewasechko 2005). However, despite the commercial potential of timber, its exploitation often has only limited welfare and development effects for the poor living in rural areas. This is due to a lack of ownership of and/or access to land, the skills and capital required to enter the business, and a lack of the necessary financial means to take high risks and wait for long-term returns (Sunderlin et al. 2005). Much of the processing of timber products will usually take place in urban settings, where the necessary infrastructure and support services are more readily available.

Besides timber, forests provide raw materials for a variety of NTFPs. The majority of NTFPs are used for subsistence and consumed by the collectors themselves (Sunderlin et al. 2005), but some portion is also produced for sale. Pimentel et al. (1997) estimated that about 90 billion USD in non-timber products are harvested globally each year. NTFPs are very important for subsistence, but their value and influence on people's lives are less significant than that of timber. In most cases, NTFPs provide households with only a supplementary income, which is used in combination with revenue from other economic activities. For example, Belcher et al. (2005), in their study of NTFP activities, found that these contributed to only a (small) portion of household income in all cases analyzed. These studies confirmed that activities surrounding NTFPs serve more as an important safety-net during critical times than as a primary source of income.

[2] For a good introduction to SAMs and their application refer to Round (2003) and Thorbecke (2000).

6.3.2 Value Chain Analysis: Understanding Markets for Products from Rural Areas

In order to better understand the actual contribution of forests to rural development, additional questions must also be clarified, for example: 'Which markets are likely to provide opportunities for forest products?' and 'How can new opportunities be identified?' Two elements are crucial to answering these questions:

1. Gaining a better understanding of the local, national and international end markets for forest products;
2. Comprehending the process of how forest products reach these markets and who benefits from this process along the way.

Both issues need to be grasped in order to answer the question of how more value can be created and captured for greater rural employment and income generation. One way of approaching these questions is by means of value chain analysis and development.

The value chain concept originates from Michael Porter's (1985) book on how to create competitive advantage. Porter uses the concept of value chains related to activities *inside* a firm, referring to a chain in which each processing step adds value, moving from inbound logistics through production, outbound logistics, marketing and sales to after-sales service. The concept of a value chain was later broadened and adapted to depict a production chain that includes a range of firms involved in the production of a specific good. It became popular in the 1990s for application in the study of global value chains and their governance systems (Gereffi et al. 1994; Kaplinsky and Morris 2001). During the same period, the concept began to be adapted to development interventions that looked at how agricultural products moved from input suppliers through agri-processing and export to end markets.

Nowadays, value chain development is increasingly seen as a strategy for private sector development, not only in rural areas but also for industrial products (Altenburg 2007; Humphrey and Navas-Alemán 2010). Large multinational companies use similar approaches when analyzing the cost effectiveness of their supply chains and in working towards their improvement. For example, IKEA works actively on the upgrading of its suppliers of home furniture in South East Asia (Ivarsson and Alvstam 2010).

Many organizations working on value chain development today go beyond merely analyzing the relationship between different players in the value chain, taking into consideration the 'market system' also (Springfield Center 2008; Hakemulder and Sievers 2010). As depicted in Fig. 6.1, a market system extends beyond the value chain; it includes (1) the rules and regulations that influence the functioning of the chain and (2) the supporting functions such as the business and financial services that are supplied to the chain. Both aspects can be crucial to the functioning of the chain. In many cases the wider environment, including underlying cultural and social norms, for example gender roles, are also a core aspect of the functioning of a market system.

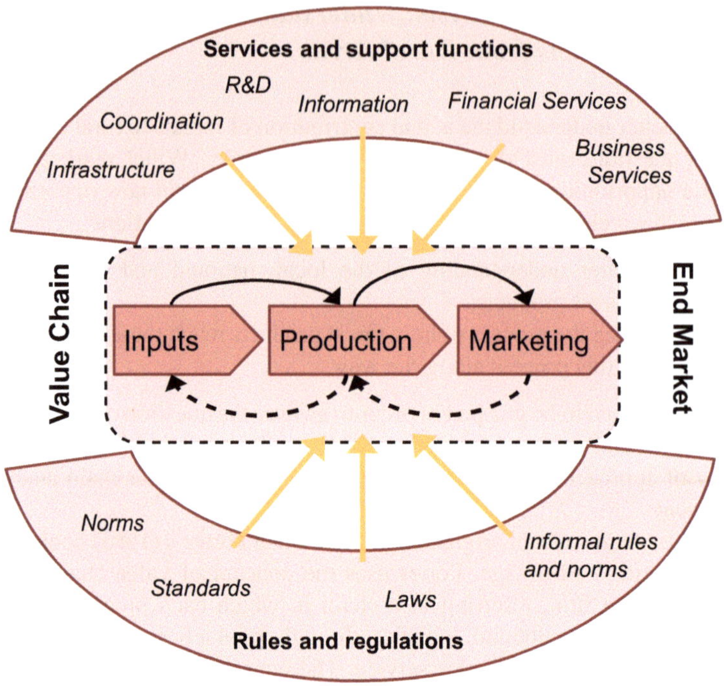

Fig. 6.1 Schematic presentation of a value chain embedded in a market system (Source: © 2009 International Labour Organization. Adapted from DFID/SDC (2008): The M4P operational guide

Value chain development is done in order to increase the competitiveness of the chain as a whole (as it is competing against value chains that produce similar products), to create more employment or to make the supply of inputs to a lead firm more cost-effective. The value chain interventions can be started by the companies within the value chain, actors that are connected to the chain (like financial institutions, business service providers or government) or external actors (like donors, NGOs). Value chain analysis is used in order to identify the issues that need to be addressed to improve the functioning of the chain and possible intervention points. The process of value chain analysis and development described below is based on the ILO's guide to 'Value chain development for decent work' (Herr and Muzira 2009). It proceeds from the selection of the sector to value chain research, mapping and analysis, to the design of measures for value chain development

6.3.2.1 Sector Selection

A key entry point for value chain development is the selection of the product chain. The selection needs to be based on the goals of the intervention (like poverty reduction) and the real potential of the sector to grow and produce positive

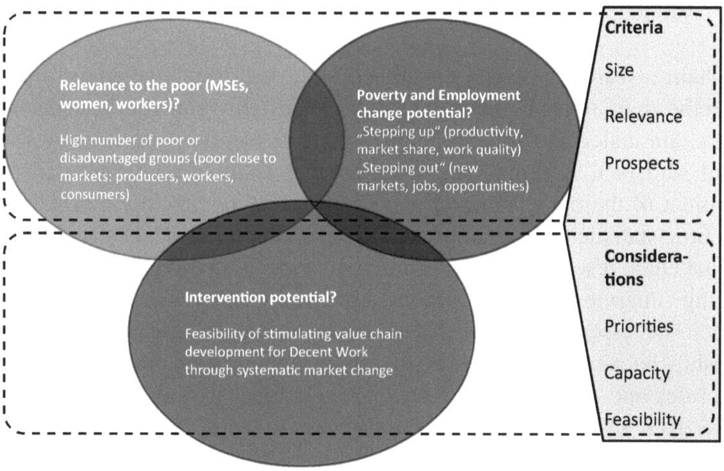

Fig. 6.2 ILO sector selection criteria for value chain development (Source: © 2009 International Labour Organization. Adapted from The Springfield Center (2008): The M4P operational guide for DFID and SDC, Durham, UK

outcomes for those involved. The ILO bases its sector selection criteria on three core factors (see Fig. 6.2). The first is whether the value chain is relevant to the poor or to a specific target group. Another decisive factor is the overall size of the value chain, the number of people and the economic importance of the market system. The size question also relates to the second criterion on the potential to achieve positive change in the chain. Are there new market opportunities that could be tapped into, or is there a possibility to increase productivity and thereby increase competitiveness? What are the future economic prospects of the sector?

The third factor then revolves around the question of the feasibility of the intervention. Does the intervening agent have the capacity to facilitate processes in the market system without disrupting the functioning of the system and distorting the market? Are the interventions feasible within a reasonable timeframe?

Sector selection can also be further informed by macroeconomic analysis, as discussed in the section introducing social accounting matrices below. It has, however, proven to be difficult to select sectors only on the basis of existing data, because a decisive criterion is future growth prospects, which are often difficult to predict. To partially overcome this constraint, one must avail of qualitative information based on material obtained from the value chain stakeholders themselves, gathered in interviews and through practical research as described below under the heading 'Value chain research'. Selecting a sector can be divided into four steps: (1) defining objectives and target groups, (2) deciding on selection criteria, (3) assessing existing and potential sectors, and (4) applying selection criteria in a consultative meeting with stakeholders.

6.3.2.2 Value Chain Research

Value chain research is mainly concerned with describing and analyzing the way actors in the chain relate to one another, and identifying bottlenecks and constraints causing chain under-performance. Value chain research is informed by secondary data and statistical information. In most value chains of relevance in terms of employment of the poor, however, data availability is typically a problem. There is often little statistical information, or the information available is outdated. That is why value chain research relies on empirical research involving value chain actors, combining interviews, focus group discussions, observations and secondary research. The essential purpose of this research is to attempt to establish the relationship between the actors in the chain, the service providers to the chain and the rules and regulations influencing the manner in which the market system functions. Research focuses on looking for ways to realize market opportunities. As in sector selection, research must be dedicated to the end markets and their evolution.

6.3.2.3 Value Chain Mapping

The results provided by the value chain research are often captured in a value chain or market system map clarifying the relationship between the different value chain stakeholders, the market channels of a product or subsector, and the services and business regulations that are crucial to the functioning of the market system. A value chain map is a tool used to illustrate and simplify the complexities of sectors and their value chains. It represents visually, and simply, the connections between businesses in value chains as well as other market players. It may be a simple flow diagram or a more sophisticated grid chart. A value chain map is not an objective in itself (such as understanding the value chain through research and being able to improve it), but a means of realizing these objectives. A value chain map is especially helpful when communicating the results of the research to a wider group of value chain stakeholders (value chain maps are presented in Figs. 6.5 and 6.6).

6.3.2.4 Value Chain Analysis

Findings from the value chain research, as depicted on the value chain map, must be analyzed and understood before moving into value chain development. Value chain analysis is the evaluation of these findings and seeks to identify the root causes of value chain under-performance. The analysis can be divided into three steps: (1) identifying binding constraints and their underlying systemic causes. For example, if actors in the chain experience difficulties accessing finance, the cause may be a paucity of financial management skills and little collateral in the enterprises, or that financial institutions simply do not possess the right lending instruments. (2) Identifying the incentives of market players and agents of change. Identifying incentives is key as it forms the basis for a sustainable change in the market system. If the underlying cause of the lack of access to finance is the absence of adequate lending products, it becomes necessary to explore whether micro-finance institutions or banks have sufficient incentive to develop suitable new instruments. (3) Formulating a vision and strategy for sustainable systemic change. Again, this relates to the incentives and to the institutions and/or enterprises willing to lead the change.

6.3.2.5 Value Chain Development

Value chain development takes the underlying constraints identified during the analysis and produces interventions to address existing constraints. If finance is the missing ingredient, then helping banks to develop new lending products might be helpful in ensuring a better supply of finance to the chain in the long term. Value chain development is a realization of the defined targets. The value chain development process must be effectively monitored and evaluated so as to demonstrate impact. As value chain analysis can lead to a variety of underlying constraints, interventions to address the causes of under-performance may lie in widely diverging technical fields. For example, interventions may be needed in the development of new financial instruments, but also of technical or entrepreneurial skills, or the alteration of an administrative procedure hindering the export of the product to international markets. Herein lies one of the complexities associated with facilitating value chain development, as value chain developers may find it difficult to access the necessary tools and approaches. A value chain developer must be entrepreneurial and inventive. Also, not all changes to the market system may be feasible in the typical timeframe of a value chain development intervention. For example, if fundamental legislative changes must take place before the market system can work, this might exceed the lifetime of a project or program.

6.3.3 Social Accounting Matrices: Putting Employment and Incomes in Forest-Based Value Chains into Context

As has been stressed, meaningful value chain analysis and development requires a good understanding of the broader economic and rural development context, in particular regarding the potential of forest-based value chains with respect to the creation of employment, income generation and poverty reduction. One of the tools that can help to put forest-based value chains into perspective and to get an overview of how they are linked into local economic activity and livelihoods are social accounting matrices (SAMs).

A SAM reflects productive activities and their inputs in an input–output table showing how a final product or service is composed of the contributions from different sectors. An input–output table is in essence an aggregation of sectoral value chains. In addition to the input–output relations, a SAM contains the added value generated and its distribution to different types of households as income from capital and labor. Primary income is redistributed via taxation and transfers for education, health or remittances, for example. Finally, the income after secondary redistribution is spent generating new demand, in accordance with the budgets and expenditure patterns of different types of household. SAMs include savings and finances as well as exports, imports and transfers with other economies ('rest of the world'). SAMs are a basis for other economic modeling and are, therefore, widely available, including in developing countries. They may not be current, the input–output table in particular tends to be updated only at longer intervals, but the fundamental relationships and underlying production structures change slowly, and useful insights can be gained in many cases.

Moreover, SAMs are a flexible instrument. Products and productive activities (sectors) can be disaggregated in line with the study objective and so can workers and institutions, in particular households, as will be seen below. Workers can be grouped by skill or by wage level, for example. Households can be classified into types depending on the issue being analyzed: rural/urban, by income class, male/female-headed, indigenous/non-indigenous, etc., each with their specific income and its origin, and expenditure patterns. Another advantage of using the SAM approach is that it facilitates an integration of different data sources that can go some way towards filling the data gaps alluded to above, as will be seen in the example for the Brazilian Amazon in the following subsection.

6.3.3.1 A SAM for the Brazilian Forest Sector and for the Amazon

Brazil has the greatest area of tropical forests in the world, but also suffers from one of the highest rates of deforestation, which in the last several decades has been concentrated in the Amazon region. The main driver behind the high deforestation rate has been the conversion of forest to land for cattle ranching and farming. Destructive logging has also played a part, and to some extent infrastructure development. For the last decade, the government has been making major efforts to halt deforestation, especially in the Amazon region. The environmental challenge is further complicated by the negative effects that restrictions on agriculture and on logging can have on the employment and income of local communities.

To obtain a better understanding of these interactions, the Brazilian Forest Service requested the ILO conduct a detailed analysis of economic activity, employment and income derived from forests in the Amazon. To this end, a disaggregated social accounting matrix was developed, separating out 12 activities within the forest industries for Brazil and for the Amazon, from among the 30 activities for the entire economy (Bento and Fachinello 2010).

Forest-based activities are most highly disaggregated in the case of those taking place in the forest areas, including timber harvesting; charcoal making; harvesting (tapping) of natural rubber; harvesting (collection) of natural Brazil nuts; harvesting of açai (edible pulp from palm fruits); harvesting of other NTFPs; silviculture for timber production; silviculture for charcoal production; silviculture for other products and fishing in rivers in forest areas. Processing occurs in conjunction with three activities, namely the processing of Brazil nuts; the processing of Açai and in the wood industries.

The SAM provides quantitative estimates of the value added, employment, income and its distribution. This information is not available from other sources and sometimes contrasts with data frequently quoted in the literature. This includes the total number of people employed in forest value chains in Brazil, which at over 1.2 million is significantly higher than reported in other sources.

For rural areas in the Amazon region it shows that, of an estimated six million people living there, about 790,000, or 8.7 %, are directly involved in forest-based activities such as logging, forest products and fishing. This is much higher than for Brazil as a whole, but less than is the case for both agriculture and animal husbandry. The most important employers by far are wood industries, with over 400,000 workers in the Amazon, and timber harvesting, a distant second with some 120,000 workers. This makes wood-based value chains the sixth largest employer in the Amazon region. NTFPs, in contrast to the political and scientific interest they have attracted, only provide about 60,000 jobs between harvesting and processing.

At only 5.3 %, the contribution of forest-based activities to regional income is much smaller than that of regular employment. Forest-based activities are relatively labor-intensive, but associated with low earnings, as can be seen in Figs. 6.3 and 6.4. Two thirds of workers in the forestry and fishing sectors are in the lowest income bracket. Improvements to incomes through higher productivity will be important therefore, but they can also be supported by social protection transfers for the poorest households. A constraint with respect to the improvement of productivity is the low skills levels in forestry compared to other sectors.

One of the most important benefits of a SAM analysis is the insights gained into linkages between sectors of the economy and with the outside. This is derived from multipliers, i.e. coefficients, that indicate how an increase in demand in the given sector would trigger increases in the production of other sectors, in incomes across the economy and thereby in additional employment. Backward multiplier linkages (i.e. the volume of economic activity that is created upstream through inputs for earlier stages of the value chain) of around 2.0 show that development of most forest-based activities, in particular timber harvesting, is at least as beneficial in terms of economic activity as promoting alternatives such as agriculture. It is somewhat better (multiplier 1.3) than agriculture (1.1) in terms of income generation, but less effective for employment creation (151 jobs per million Brazilian Reals[3] additional demand for agriculture compared to 122 for timber harvesting).

[3] 1 Brazilian Reals = approx. 0.5 USD.

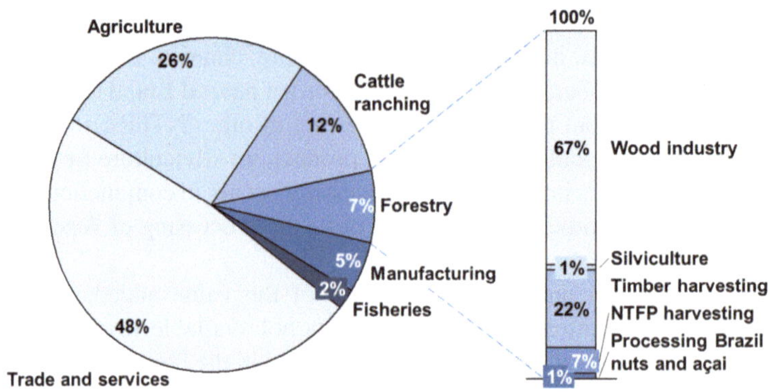

Fig. 6.3 Distribution of employment in the economic sectors and subsectors of Brazil (Source: © 2012 International Labour Organization. Adapted from Ferreira Filho and Fachinello (2010))

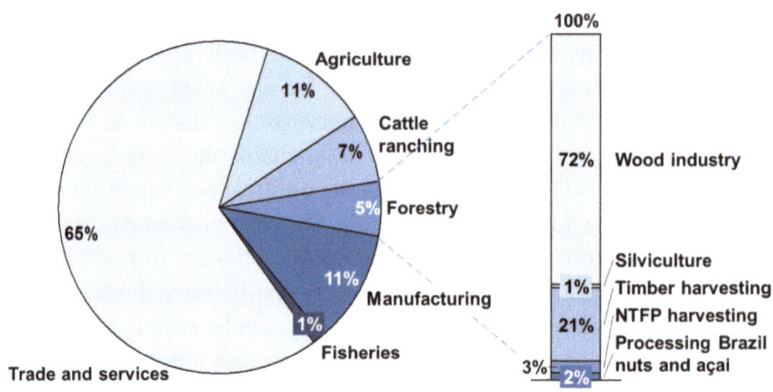

Fig. 6.4 Distribution of income in the economic sectors and subsectors of Brazil (Source: © 2012 International Labour Organization. Adapted from Ferreira Filho and Fachinello (2010))

Forest-based activities that stand out for above-average income and high employment multipliers are rubber tapping and Brazil nut collection and processing, and river fishing.

An increase in forest-based activity would also have a good distributional impact, comparing income shares by categories of household with those of other potential sectors. Linkages to other economic sectors, however, are weak. Due to its relatively small initial size, growth in the forest-based sector also has a limited effect on the regional economy as a whole. Of greater relevance is the low level of integration of forest-based value chains with other economic sectors. The development of higher forward linkages through further processing and value adding in the region are critically important in terms of increasing employment and income benefits to the local population, thereby improving their livelihoods and reducing poverty, but also incentivizing the sustainable use and management of forests (Stoian 2005). The problem of short and poorly integrated value chains exists in most sectors in the Amazon region, but is even more acute for forest-based products.

The insights gained for the selection of products for value chain development are the following:

- Although economically smaller than sectors such as agriculture and cattle rearing, forest-based activities employ more workers and generate higher incomes relative to competitors;
- Like many predominantly rural areas, the Amazon as a whole has short and poorly interlinked value chains. This is worse for forest-based chains than in other sectors;
- Wood products are the backbone of the forest-based economy in the Amazon. The development of wood-based value chains could make a significant contribution to additional growth, employment and income in the region;
- River fishing should be considered for value chain development, because its expansion would lead to high employment and income benefits in both relative and absolute terms;
- Of the NTFPs, rubber, Brazil nuts and açai have high employment and income multipliers, but the relatively small volume will probably limit their overall contribution;
- Low levels of education and skills among workers in forest-based value chains are likely to be a constraint.

These preliminary conclusions are the basis for a value chain development project for sustainable forest use and social inclusion currently under way with the Forest Service of Brazil, the Ministries of Environment and Rural Development, the private sector and the ILO.

6.4 The Forest Products Value Chain in Rural Employment and Economic Development

6.4.1 Rural Employment: The Role of Collectors and Middlemen in Forestry Value Chains

Rural employment in forestry is often linked to activities relating to the collection and transport of forest products. Value chain analysis can contribute to generating an understanding of where most value is accrued in the process of collection and transport. For example, in Senegal, villagers (or village chiefs) control direct forest access but still reap only small profits compared to traders and wholesalers. Thus, controlling forests or trees does not necessarily provide benefits; profits are made by controlling markets (Ribot 1998). Different actors in the chain typically have access to different forms of market information and different marketing channels. In many wood and NTFP value chains, collectors are poorly positioned, mainly because of their poor access to markets and market information. Local harvesters or collectors seldom have access to the end markets for their products and rely on middlemen to provide market access and information. So middlemen play a crucial role at the lower end of the value chain, where a large part of rural employment is based and where poverty is often prevalent.

In the past, the literature tended to stress the exploitative role of middlemen in value chains (Neumann and Hirich 2000), but increasingly the importance of middlemen is being recognized and analyzed in a different way. Belcher (1998) argued that it is instructive to look at the role these middlemen play and the services they provide before accusing them of exploitation. Middlemen typically provide transport and market connections and often bear all or a good part of the financial risks in a range of value chain transactions. All of these functions are essential to making chains work, but rural collectors and forestry workers typically have neither the capital nor the skills to perform them. Velde et al. (2006), in a study of the role of intermediary buyers in NTFP value chains, found that at the level of local collectors, "market contacts and market information were considered the most important barriers to households selling NTFPs." They concluded that the role of middlemen in these value chains is crucial, as they often bring in the information and contacts necessary to make chains work. While this carries with it the risk of intermediary buyers holding a part of the value chains 'captive', the studies find that more often the collectors and producers at the lower end of the chain are likely to benefit from these relationships. These findings were also confirmed by Belcher and Schreckenberg (2007) and Jensen (2009) in other studies.

Whether or not intermediaries exercise their power in a way that benefits small producers and collectors is very much dependent on the local situation. For example, in the organic cocoa value chain in Bolivia, insufficient information at the community level has led to a situation where the intermediary company profits from the growing organic cocoa niche market, but the community receives no price premiums (Marshall et al. 2006). On the other hand, intermediary buyers have helped small producers set up an association that provides additional social benefits to its members in Bolivia (Velde et al. 2006). Often large companies also have a strong influence on their supply chains and can potentially play a positive role further upstream. For example, IKEA helps its suppliers to diversify their marketing channels in order to prevent them from becoming too dependent upon IKEA (Ivarson and Alvstam 2010).

So, while large enterprises and middlemen can play a positive role, there can also be cases where they occupy a monopolistic position in certain market segments or geographical areas. What these recent studies confirm, and what is currently considered best practice in the promotion of value chains,[4] is that the role of each of these actors needs to be carefully studied and taken into account when trying to influence the way value chains function. Instead of attempting to eliminate middlemen, it is more important that their role be understood and that the means to design interventions improving the function and providing a more equitable

[4] See, for example, Miehlbradt and McVay (2005), Hakemulder and Sievers (2010) and the DCED website: www.enterprise-development.org

distribution of benefits without disrupting the chain are developed. Belcher and Schreckenberg (2007), for example, suggested improving the bargaining power of producers without necessarily replacing the middleman.

Value chain analysis can also shed light on gender equality in access to and use of forest products. There are some encouraging signs of progress with greater gender equality in the governance, use and distribution of benefits from forest products and services, as a recent report by CIFOR has shown (Mwangi and Mai 2011). Many of the long-standing issues remain, however; from the male dominance of forestry value chains to the lack of land rights for women, as well as low literacy rates among rural women as an impediment to effective participation (Colfer 2011). The analysis of internationally traded NTFPs in Africa indicates that women perform a variety of functions along the value chains (Abtew et al. 2012; Shackleton et al. 2011). However, their role in the success of these value chains is usually inadequately documented and receives only marginal attention, as they often operate in the informal sector. As in other sectors, a greener and more sustainable forest industry is not automatically more inclusive of women. It will take a range of measures, from organization to support for informal markets, to strengthen women's economic bargaining power (Shackleton et al. 2011) and to increase representation of women in consultations on REDD + (Brown 2011).

6.4.2 Rattan Value Chain in Quang Nam, Vietnam

A participatory value chain assessment was conducted in relation to a rattan value chain in Quang Nam, Vietnam, and constraints were found in both the local supply and in the market demand situation. As depicted in the schematic value chain map below (Fig. 6.5), the availability of rattan raw material in general, especially that of a high quality, is declining due to predatory harvesting, extremely low rates of harvest in rattan plantations and a lack of forest regeneration. Where there is a shortage of the necessary raw materials, local enterprises are pushed out of business. Moreover, the nature of employment is changing; opportunities in natural rattan harvesting are declining and emerging in plantations.

Middlemen play a crucial role in the rattan and bamboo value chain in Quang Nam. Local harvesters sell to collectors with middlemen playing a key part in bringing in large volumes of material and assuring the quality of the input material. Middlemen request rattan stems of a specific size and quality. In some areas, harvesters must venture very far into the forests in order to find rattan meeting these requirements. Some harvesters get money from middlemen in advance so that they can go deeper into the forests and stay longer to better exploit rattan areas (Ha et al. 2005). Paying in advance is precarious and requires capital. Additionally, middlemen are forced to take considerable risks, as they buy rattan on spot markets, but sell it to larger companies which usually delay payment. Access to working

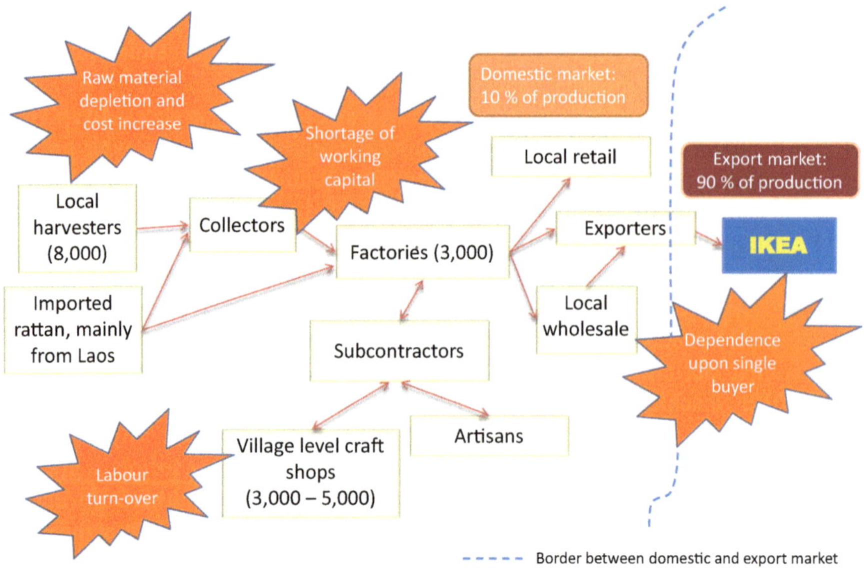

Fig. 6.5 Rattan value chain in Quang Nam, Vietnam

capital is, therefore, a key issue for these middlemen. In some areas harvesters are highly dependent on the middlemen, as there is no processing unit in the area, and harvesters have no other option but to sell their rattan to the middlemen.

The rattan factories rely on a large number of subcontractors in villages that produce the inputs for the factories. The raw material supply and initial processing steps are handled by village-level craft shops. As pay is low and working conditions are often poor, labor turnover is high. This in turn means that new craft shop employees have to be trained constantly, and productive time is lost as a consequence. Both subcontractors and employees in the craft shops perceive this negatively.

The market demand situation is such that increased competition is putting pressure on firms to produce more cheaply and with shorter delivery times. Producers operate under tight margins, have little bargaining power and are highly dependent on one major buyer: IKEA. Firms do not know their competitors and do little benchmarking. As a consequence, they are not able to position themselves on the global market. Some firms are exposed to market requirements and are being supported in meeting them by IKEA. Both local factories and IKEA as the buyer are concerned by the high dependence of rattan factories on IKEA and are interested in diversifying their market channels.

The ILO project working to improve the competitiveness of the rattan sector prioritized interventions in the chain on the basis of the potential effect on competitiveness and working conditions (ILO 2012) through:

- Increasing the raw material supply and access by planting and growing rattan. The project supported the training of 300 farmers in rattan nursery techniques, sustainable rattan harvesting and post-harvest processing of materials. Pilot nurseries were set up. A project to set up a large-scale rattan nursery was developed in collaboration with the provincial rattan association.
- Decreasing post-harvest losses of rattan by introducing in-situ rattan boiling from Indonesia.
- Increasing the appeal of working in village-level craft shops by improving working conditions. A 'Guidebook for homeworkers' was developed in collaboration with IKEA. The book contains suggestions for the improvement of working conditions and productivity in village-level craft shops using simple and low-cost methods.
- Setting up local showrooms and shops in the provincial capital and enabling producers to participate in national fairs facilitating new business linkages in order to decrease dependence upon a single buyer. To achieve this, the project collaborated with the Quang Nam rattan producer association.
- Finally, in collaboration with the provincial Department of Trade and Industry, a sustainability plan for the rattan value chain was formulated, matching private sector objectives with government strategies.

6.4.3 Gum Resin Value Chain in the Drylands of Sudan

The drylands of sub-Saharan Africa are endowed with diverse tree and shrub species, which are sources of economically valuable NTFPs including gum Arabic, frankincense and myrrh. These products perhaps represent the oldest internationally traded NTFPs. The gum and resin subsector in Sudan plays an important socio-economic and ecological role. It constitutes an important source of employment and income for rural communities involved in the production and marketing of the products. The subsector contributes up to 10 % of the non-oil export income of the country.

More than 50 % of the country's frankincense production is used domestically as a traditional medicine and for religious and cultural rituals. In the year 2007 the country earned more than 172,000 USD from the export of frankincense. This figure is very small when compared to the potential production that could be achieved through the sustainable utilization of the resource base. A case study conducted in the Nuba mountains region, South Kordofan, identifies a range of stakeholders engaged in the frankincense value chain. The main actors in the value chain include hired tappers, producers and licensees, village traders and urban merchants, as well as processors, exporters, domestic market wholesalers and retailers, tribal leaders and the government (Fig. 6.6).

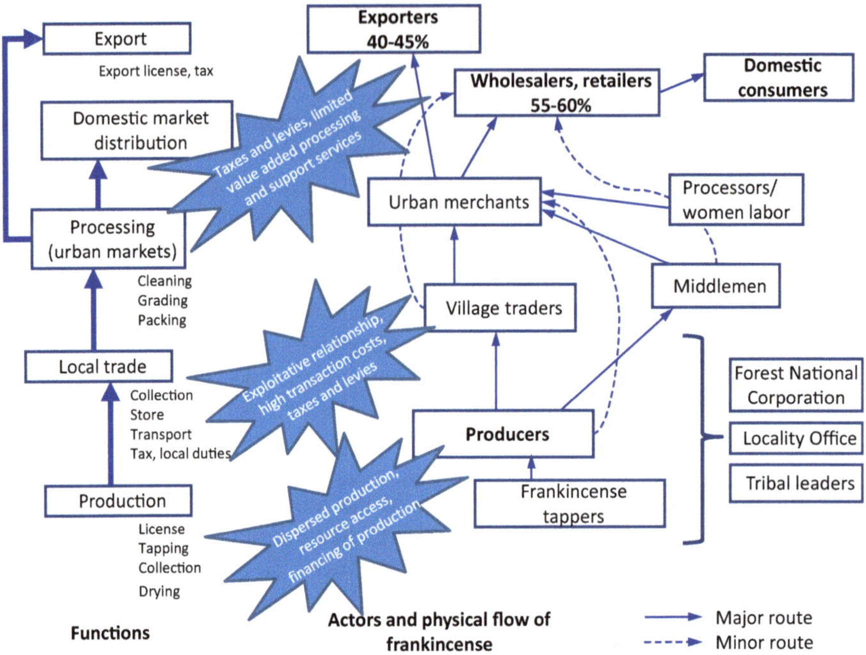

Fig. 6.6 The frankincense value chain in South Kordofan, Sudan (Source: Abtew et al. (2012))

The tapping and collection of the product is mainly done by men. This is due to physical and technical constraints associated with the harsh climatic conditions and the long-distance travel to production areas along with the technical skills required for tapping. The involvement of women in the value chain appears at the processing level and at retail in urban areas. Primary processing involving cleaning, sorting and grading is done solely by women. It provides part-time jobs for many women in urban areas. About 25–30 women work daily in one processing firm and earn 4–6 USD/day. However, the role of women in the value chain is not well recognized, requiring more attention to ensure gender equity.

The production, processing, transportation and marketing of frankincense is labor intensive and provide employment opportunities for local and migrant laborers as well as urban dwellers. In the Nuba mountains region, more than 30 % of households (about 820 households at the local level) were fully engaged in frankincense production during the dry season. However, an upward skewed distribution of benefits is evident. The average annual net incomes of tappers, producers, village traders and urban merchants were estimated to be 74, 740, 1,300 and 11,230 USD, respectively. Local actors are in a weak position to add and capture more value from their products. This is attributed to the lack of market information, credit access and poor infrastructure. For example, the frankincense tappers depend on external actors (village traders, urban merchants and producers/licensees), due to their limited access to credit, and thus capital, to obtain permits for resource access. Private entrepreneurs play an important role in organizing production activities at the local level and linking the producers to domestic and export markets. Generally, the potential added value from the products, including employment creation, is not fully realized. Possible interventions that could enhance the competitiveness of the chain include:

- Organizing production and marketing through collective action;
- Improving the quality of the product through proper post-harvest handling;
- Empowering local actors through training and provision of access to credit;
- Introducing added value processing and grading systems consistent with the requirements of end users.

As frankincense is of importance for the pharmaceutical and cosmetic industries, interventions that would lead to better quality control, certification and further processing, such as the extraction of essential oils, could also generate employment opportunities and promote the development of small and medium-sized enterprises.

6.5 Outlook

The starting point of many development projects in the past were the poor in rural areas, the nature of the production in which they engaged or the wealth of species and potential products in the forests surrounding them. A value chain perspective alters the viewpoint and places a stronger emphasis on real market opportunities and on the market system around the poor producers. Social accounting matrices help to compare different value chains and put them in the context of a rural economy in terms of the value added, employment and income levels and distribution, as well as linkages with the rest of the rural economy. This provides an initial assessment of the relative weight of different value chains, the employment and income potential associated with their expansion and the social groups most likely to benefit.

A key contribution of value chain analysis is to simplify market transactions and relationships in a model (the value chain map) that can be used to identify bottlenecks and shortcomings in the functioning of a value chain. Based on an understanding of how these markets influence the chain, value chain development has a greater chance of coming up with lasting solutions. The value chain perspective places an emphasis on how products get to the market, on the people in rural areas who can potentially benefit from the higher added value or greater competitiveness and on how to create more benefits at the bottom end of the chain.

Value chain analysis can inform policy decisions by making bottlenecks visible and by working towards incentive-based solutions. If this is done using participatory methods, interventions can be driven by value chain stakeholders themselves. A buy-in by value chain actors is crucial for success.

Challenges remain, however. A key difficulty is identifying those value chains and sectors that have the potential for future growth. Estimating future growth potential is, and will remain, problematic. A second challenge relates to the capacity needed to do a thorough value chain analysis and the ability to respond to the findings of value chains analysis with the right policy measures. This response capacity is often limited, especially in rural areas. Policy-makers should, however, resist the temptation to try to resolve issues in value chains by means of heavy-handed direct interventions that might run the risk of distorting the market system rather than creating a model for sustainable change.

References

Abtew AA, Pretzsch J, Mohamoud TE, Adam YO (2012) Commodity chain of frankincense from the dry woodlands of Nuba Mountains, South Kordofan State, Sudan. Small-Scale Forest 11 (3):365–388

Altenburg T (2007) Donor approaches to supporting pro-poor value chains: report prepared for the Donor Committee for Enterprise Development. Working group on linkages and value chains. German Development Institute, Bonn

Belcher B (1998) A production-to-consumption systems approach: lessons from the bamboo and rattan sectors in Asia. In: Wollenberg E, Ingles A (eds) Incomes from the forest. Methods for the development and conservation of forest products for local communities. Center for International Forestry Research, Bogor

Belcher B, Schreckenberg K (2007) Commercialization of non-timber forest products: a reality check. Dev Policy Rev 25:355–377

Belcher BM, Ruiz-Perez M, Achdiawan R (2005) Global patterns and trends in the use and management of commercial NTFPs: implications for livelihoods and conservation. World Dev 33(9):1435–1452

Brown PHC (2011) Gender, climate change and REDD + in the Congo Basin forests of Central Africa. Int Forest Rev 13(2):163–176

Colfer C (2011) Forests and women: some encouraging signs. http://www.cifor.org/online-library/polex-cifors-blog-for-and-by-forest-policyexperts/spanish/detalle/article/1230/forests-and-women-some-encouraging-signs.html. Accessed 05 May 2012

DIFID/SDC (2008) The operational guide for the Making Markets Work for the Poor (M4P) approach. Department for International Development (DFID); Swiss Agency for Development and Cooperation (SDC), London, Berne

Ewasechko AN (2005) Upgrading the Central Java wood furniture industry: a value chain approach. International Labour Office (ILO), Geneva

FAO (2011) State of the world forest. Food and Agriculture Organization, Rome

Filho F, Fachinello AL (2010) A geração de emprego e renda na Amazônia brasileira [The generation of employment and income in the Brazilian Amazon]. International Labour Office, Brasilia

Gereffi G, Korzeniewicz M, Korzeniewicz RP (1994) Introduction: global commodity chains. In: Gereffi G, Korzeniewicz M (eds) Commodity chains and global capitalism. Praeger, Westport, pp 1–14

Ha TT, Ban NK, An LT, Chung DH (2005) Comprehensive report from fact-finding mission in Quang Nam province in rattan value chain in Quang Nam. GTZ-Office in Vietnam on behalf of the Project "Promotion of Small and Medium Size Enterprises", Hanoi

Hakemulder R, Sievers M (2010) Practical guidelines for a more systemic approach to sustainable enterprise development. International Labour Office (ILO), Geneva

Herr ML, Muzira TJ (2009) Value chain development for decent work: a guide for development practitioners, government and private sector initiatives. International Labour Office (ILO), Geneva
Humphrey J, Navas-Alemán L (2010) Value chains, donor interventions and poverty reduction: a review of donor practice. IDS research report 2010(63). Institute of Development Studies at the University of Sussex, Brighton
Kaplinsky R, Morris M (2001) A handbook for value chain research. Working paper prepared for International Development Research Center, Brighton
ILO (2001) Globalization and sustainability: the forestry and wood industries on the move, report for discussion at the Tripartite Meeting on the Social and Labor Dimensions of the Forestry and Wood Industries on the Move. International Labour Office (ILO), Geneva
ILO (2009) Value Chain Development for Decent Work. A guide for development practitioners, government and private sector initiatives. International Labour Organization (ILO), Geneva
ILO (2010) Report on the participatory value chain assessment for bamboo, rattan, home accessories and furniture sector in Quang Nam, Vietnam. International Labour Organization (ILO), Hanoi
ILO (2012) Case study: strengthening the rattan value chain in Quang Nam. Project: "Youth Employment through Local Economic Development in Quang Nam", International Labour Organization (ILO), Hanoi
ILO (2012) Working towards sustainable development. Opportunities for decent work and social inclusion in a green economy. International Labour Organization (ILO), Geneva
Ivarsson I, Alvstam CG (2010) Supplier upgrading in the home-furnishing value chain: an empirical study of IKEAs sourcing in China and South East Asia. World Dev 38:1575–1587
Jensen A (2009) Valuation of non-timber forest products value chains. Forest Policy Econ 11: 34–41
Marshall E, Schreckenberg KJ, Newton AC (eds) (2006) Commercialization of non-timber forest products: factors influencing success: lessons learned from Mexico and Bolivia and policy implications for decision-makers. UNEP World Conservation Monitoring Centre, Cambridge, UK
Miehlbradt AO, McVay M (2005) From BDS to making markets work for the poor: annual BDS seminar. International Training Centre of the International Labour Organization (ILO), Turin
Mwangi E, Mai YH (2011) Introduction to the special issue on forests and gender. Int Forest Rev 13(2):119–122
Neumann R, Hirch E (2000) Commercialization of non-timber forest products: review and analysis of research. Center for International Forestry Research (CIFOR), Bogor
Pimentel D, NcNair M, Buck L, Pimentel M, Kamil J (1997) The value of forests to world food security. Hum Ecol 25(1):91–120
Porter ME (1985) Competitive advantage: creating and sustaining superior performance. Free press, London
Ribot J (1998) Theorizing access: forest profits along Senegal's charcoal commodity chain. Dev Chang 29:307–341
Round J (2003) Social accounting matrices and SAM-based multiplier analysis. In: Bourguignon F, Pereira da Silva LA (eds) Evaluating the poverty and distributional impact of economic policies (techniques and tools). World Bank, Washington, DC
Shackleton S, Paumgarten F, Kassa H, Hausselman M, Zida M (2011) Opportunities for enhancing poor women's socioeconomic empowerment in the value chains of three African non-timber forest products (NTFPs). Int Forest Rev 13(2):136–151
Stoian D (2005) Making the best of two worlds: rural and peri-urban livelihood options sustained by non-timber forest products from the Bolivian amazon. World Dev 33(9):1473–1490
Springfield Centre (2008) A synthesis of the making markets for the poor (M4P) approach. http://www.springfieldcentre.com/wp-content/uploads/2012/10/sp0804.pdf. Accessed 13 May 2013
Springfield Center (2008) The M4P operational guide for DFID and SDC, Durham, UK
Sunderlin WD, Angelsen A, Belcher B, Burgers P, Nasi R, Santoso RM, Wunder S (2005) Livelihoods, forests and conservation in developing countries: an overview. World Dev 33: 1383–1402

Thorbecke E (2000) The use of social accounting matrices in modelling. Paper prepared for the 26th general conference of the international association for research in income and wealth Cracow, Poland, 27 August to 2 September 2000

Velde DW, Rushton J, Schreckenberg K, Marshal E, Edouard F, Newton A, Arancibia E (2006) Entrepreneurship in value-chains of non-timber forest products. Forest Econ 8:725–741

World Bank (2004) Sustaining forests: a development strategy. World Bank, Washington, DC

Chapter 7
At the Interface of Culture, Development, and Forests: Insights from Bolivia and Kenya

Stephan Rist, Barbara Darr, and Patrick Bottazzi

Abstract The first part summarises the origins, definitions and debates around the general notions of development, culture and associated more specific concepts such as identity, tradition, exogenous and endogenous knowledge, institutions, governance or territoriality. A second part highlights how culture and development got related to the debates around sustainable governance of natural resources and forests. The third part illustrates on the basis of a case study from Kenya and Bolivia how culture as a transversal element of forest governance is expressed in empirical terms. Moreover it is shown how the cultural dimension affects positively or negatively the outcomes of culturally shaped forest governance outcomes and the role these effects play in shaping the sustainability of the socio-ecological systems of forests in Africa and South America.

Keywords Sustainable forest governance • Culture and sustainability • Bolivia • Kenya

7.1 Introduction

From a phenomenological point of view, separating "culture" from "development" is simply impossible. If we understand "development" as any kind of real or imagined form of personal and/or social change, and "culture" as any kind of

S. Rist (✉) • P. Bottazzi
Centre for Development and Environment (CDE), University of Bern, Hallerstrasse 10, CH-3012 Bern, Switzerland
e-mail: Stephan.rist@cde.unibe.ch; patrick.bottazzi@cde.unibe.ch

B. Darr
Institute of International Forestry and Forest Products, Technische Universität Dresden, Postfach 1117, 01735 Tharandt, Germany
e-mail: barbaradarr@gmx.de

J. Pretzsch et al. (eds.), *Forests and Rural Development*, Tropical Forestry 9,
DOI 10.1007/978-3-642-41404-6_7,

presupposition about what reality is (ontological assumptions), what we can know about it (epistemological assumptions), and which basic values are related to them (axiological foundations), it becomes clear that both these concepts – "development" as well as "culture" – are meaningless unless they are understood in relation to one another.

It seems astonishing, therefore, that until the 1980s the concept of "culture" remained widely unacknowledged in debates about development and cooperation. Three factors may have contributed to this situation:

1. The institutional specialisation and ideological orientation of academic disciplines engaged in studying "development": For decades, "development" was studied mainly within disciplinary fields that focused on either liberal or Marxist theories, such as economy and sociology. It is only since the 1990s that the emerging field of development studies has grown into the interspace between the economic sciences, sociology, and the classical disciplines dealing with cultural issues, that is, cultural studies, the humanities, and anthropology.
2. The reluctance of international organisations dealing with "development" to reflect on the fact that at the global level, "cultures" constitute a highly diverse, ambiguous, and moving epistemic "subsoil" that often questions the apparent consensus about the universality of the "codes of conduct" based on Western culture. The latter are often expressed in the concepts of modernisation, progress, democracy, free markets, human rights, or in a specific way of appropriating "nature" and human labour through capitalist relationships.
3. The effects of past and present foreign and domestic colonisation of the approximately 6,000 ethnolinguistic groups worldwide. Colonisation means that a dominant culture suppresses and marginalises cultural identities that challenge that dominant culture, and keeps them from growing stronger. This makes it difficult for colonised cultures to express their identity in culturally independent forms of social, economic, technological, and spiritual organisation. Thus, as a consequence of widespread colonisation, cultural diversity today appears to be progressively fading away, further reinforcing the false impression of a global convergence towards values that correspond to the Western or other dominant cultures e.g. related to "modern" health traditions or the expansion of the use of uniform modern technologies. Until recently, these issues were missing from the global policy agenda. This changed in 2005, when all UN member states – with the exception of the United States and Israel – adopted the *Convention on the Protection and Promotion of the Diversity of Cultural Expressions* (UNESCO 2005). The convention provides a global framework that is generally favourable to cultural diversity; however, it has yet to be effectively implemented.

Against this background, it becomes clear that putting "culture" in relation with "development" challenges established political and academic structures. It also challenges the ways in which international cooperation deals with cultural diversity, and the ways in which non-Western cultures have internalised the supposed universality of the Western understanding of "development". However, questioning the universality of the Western concepts of "culture" and "development" is not equal to rejecting them by default. We argue that becoming aware of the ethnocentricity of "cultures" and their related styles of "development" must be the *intra*cultural starting point of an *inter*cultural dialogue between Western and non-Western cultures or between dominant Non-western cultures that opress other non-western cultures, e.g. Hindu culture and various tribal cultural groups in India.

While such dialogues remain open in terms of their contents, the procedural conditions under which they must evolve are predefined: in order to ensure equal participation by all partners in dialogue, the organisation of intra- and intercultural relationships should be guided by the ideal of the non-authoritarian situation of "free speech" in the sense of Habermas (1970).

What is new in establishing the intra- and intercultural dialogue is that it cannot advance on the basis of unreflected power relations within and between the cultural groups involved. Accordingly, the potentials, limitations, challenges, and specificities of one culture compared to another are no longer defined by cultural, socio-economic, military, or religious hierarchies. Instead, establishing such a dialogue involves a collective attempt by each cultural group to inquire how its own cultural assumptions influence the establishment of intercultural relationships to others, and how these assumptions can be reshaped and freed from intended or unintended tendencies towards colonisation or domination.

In the following sections we highlight the importance of adopting "culture" as both a core element and a context in the search for more sustainable forms of governing and managing forests and agroforestry systems. In a first step, we discuss the basic concepts of "development" and "culture" and show how they relate to identity, traditions, endogenous or local knowledge, governance, territoriality, and transdisciplinarity (Sect. 7.2). In a second step, we present how sustainable forms of development are linked with culture. This paves the way for a conceptual model of sustainable development that takes account of culture and can guide research and action towards integrating the cultural dimension in efforts to understand, and achieve more sustainable forms of, governance of forest and agroforestry areas (Sect. 7.3).

In a third step, we present a case study from Kenya in which this conceptual model was applied, summarising results and drawing conclusions from the experience. Based on a second case study from Bolivia, we illustrate how transdisciplinary research and action helps to enhance co-production of knowledge and the development of norms for more sustainable forest governance (Sect. 7.4).

The last section of this chapter summarises the main lessons learned and outlines challenges that have to be addressed when linking culture to development in the context of forest governance and management.

7.2 Development, Culture, and Related Basic Concepts

7.2.1 Development

The term **development** is so broadly used that it is often understood in a rather undifferentiated way, meaning everything and nothing all at once. The term has multiple roots. On the one hand, it is commonly used in cellular biology to refer to the evolution of microscopic life, and from there was transferred to the societal sphere to express the idea that non-Western countries should follow the same unilinear pattern of economic and technological change that characterized the evolution of Western capitalist countries (Rist 2001). On the other hand, the concept of development also emerged as part of an analysis of the negative consequences of "capitalist modernization" in European societies. As a consequence of such – largely Marxist – analysis, dating back to the nineteenth century, intellectuals, political parties, and states started to "ameliorate the disordered faults of progress" by means of "development", understood as "one means to construct the positive alternative to the disorder and underdevelopment of capitalism" (Cowen and Shenton 1996). By relating to the notion of "progress", the concept of development clearly connects to the basic ideas of the eighteenth-century period of "Enlightenment". It represents a set of interrelated ideas, values, and principles that shape the "enlightened" understanding of the social and the natural worlds and the relationship between them. It is placing the human capacity of reason at the centre of social evolution. Although the political and technological expressions of the basic assumptions of the "Enlightenment" have changed over time, they are still important building blocks for better understanding current concepts of "development". Hamilton (2003) summarised the main idea of "Enlightenment" as based in: *rational thought* producing ideas independent of experiences, but tempered by experience and experiment; *empirical evidence* produced by *sciences*; *universalism*; *progress* based on the application of reason and scientific knowledge; *individualism*, that is, viewing the individual as the basic cell of social organisation; *secularisation* and *tolerance* as fundamental societal attitudes; all of these sustaining *freedom* as a basic feature of modernity.

Beginning in the 1950s, "development" became part of a political agenda designed by "developed" Northern countries for the "underdeveloped" countries of the South. This "new era of development" represented a next stage in the history of the "development" concept. It gained highest political and ideological importance when Harry Truman stated in 1949 that half of the world's population suffered from hunger, poverty, curable diseases, and other problems, and that the West had developed knowledge, technologies, and ways to overcome these problems:

> What we envisage is a program of development based on the concepts of democratic fair dealing ... Greater production is the key to prosperity and peace, and the key to greater production is a wider and more vigorous application of modern scientific and technological knowledge. (cited in Escobar 1995, p. 3)

This essentially technocratic and inherently hegemonic understanding of development based on economic growth experienced its first severe crisis in 1972, when the Club of Rome presented their highly influential report on *The Limits to Growth* (Meadows 1972). In this report, the authors demonstrated two things: that the material resources fuelling economic growth are limited, and that their exploitation is benefitting only about 20 % of the world population (Meadows 2005). Around the same time, the environmental externalities of this type of development – air and water pollution and declining food quality, among others – became increasingly visible to vast parts of the populations benefitting from this development model. Strong social environmental movements emerged, mainly in industrialised countries. They criticised exclusively growth-based development for its externalities, and called for revising the then current understanding of development. An important alternative appeared with the concept of sustainable development, which was first proposed at the global level by the World Commission on Environment and Development in its famous "Brundtland report" (WCED 1987).

The concept of sustainable development was globally endorsed by the 1992 United Nations Conference on Environment and Development in Rio de Janeiro (UNCED 1992). However, 20 years after this declaration to implement sustainable development, the record of achievements is rather disappointing: although scientific knowledge about global change has increased, degradation of the main global resources (land, forests, biodiversity, oceans and the atmosphere) has continued, and political action to revert these degradation processes has clearly been insufficient (Schreurs 2012). As a consequence, a growing number of voices are calling for a more systematic revision of the relationships between the policy and the politics of sustainable development (Sneddon et al. 2006) and how they link up to the dominant capitalist system (Magdoff and Bellamy Foster 2010). Hence any inquiry into future development pathways must seek new answers to the older question of how to achieve a convergence of social and environmental justice beyond the historically existing forms of capitalism and socialism (Martínez-Alier 2012). Suggestions to build on and expand emancipatory modernity – an idea summarised under the label of "ecosocialism" (McLaren and Houston 2004; Löwy 2005) – are increasingly being discussed. Whether this solution is adequate remains to be further analysed.

7.2.2 *Culture*

The English term **culture** originates from the Latin word *cultura*, which denotes a cleared and cultivated piece of land. In addition, *cultura* also denotes the act of taking care of something that is biologically growing, such as children, animals, grains, or young trees. In contemporary social and human studies, the term "culture" carries a number of different meanings. Scientific disciplines like anthropology, sociology, philosophy, psychology, and ecology use the contested term or "culture" in order to explain cognitive processes, attitudes, and behaviour (e.g. see Andrade et al. 1993; Brechbühl et al. 1995; Ember et al. 2004; Harris 1989;

Haviland 1993; Ingold 2000; Seeland 1997). Accordingly, in the context of the perception and valuation of natural resources, the meaning of "culture" embraces all kinds of behaviour, values and attitudes towards natural resources, as well as knowledge about them, which have accumulated over generations and are expressed in the specific way in which each particular society defines and uses natural resources.

For a long time, researchers and policy makers interested in the relationship between culture and natural resource management used the term "culture" according to how it had been defined in anthropology. Representing an important branch of the social sciences, the discipline of anthropology had to position itself within the overall structure of scientific disciplines. It did so by claiming to be in charge of studying "culture". In order to set anthropology clearly apart from other disciplines occupied with the study of certain aspects of culture – such as, for example, sociology, law, education, history, and psychology – anthropologists carved out their main object of study by defining "culture" as a "discrete, bounded entity, consisting of particular sets or structures of social relations, practices, and symbolic systems which forge a cohesive unity for the group, whether as society, nation, community, or class" (Schech and Haggis 2000).

This anthropological understanding of "culture", however, turned out to be poorly suited for dealing with the permanent transformations and internal contradictions that characterise "cultures" as well. As a result, the use of the anthropological definition was increasingly challenged. The growing field of "cultural studies", instead of perceiving "culture" as a discrete, bounded entity, began to understand it as a "network of representations – texts, images, talk, codes of behaviour, and the narrative structures organising these – *which shapes every aspect of social life*" (Frow and Morris 1993). This understanding of "culture" appears much more adequate for relating "culture" to concrete activities, such as, for example, forestry projects or the management of natural resources. In this relationship, "culture" is nor apart from, nor a discrete – and hard-to-find – space of social life; rather, "culture" is perceived as part and parcel of the production, reproduction, and transformation of social life in all its dimensions. Social groups construct a dynamic set of values, principles, and social practices directed towards creating and maintaining a shared identity; this identity is generally circumscribed around basic categories such as class, gender, age, sex, languages, and nation, and is often additionally supported by references to rural or urban, regional, or religious contexts.

When it comes to understanding the dynamics of cultural processes and related changes in the institutional, political, economic, and cognitive spheres, the concept of **acculturation** is crucial. It refers to the processes through which members of one cultural group acquire the knowledge, skills, attitudes, values, beliefs, and behaviours that enable them to become functioning members of another cultural group. The concept of acculturation is particularly important in an environmental context, as it is often directly related to varying environmental concerns: the higher the degree of acculturation of non-Western cultural groups to the Western culture, the lower their interest in forest sustainability (Caro and Ewert 1995). In fact, in this case, *acculturation* has turned into **assimilation**, which denotes the almost complete abandonment of previous cultural patterns. However, looking at the factors leading to acculturation, it is important to point out that acculturation is most often induced by external environmental or political change, such as, for example, by restriction of the movement of indigenous migrant groups, land reforms that simply ignore such groups, or attempts to convert these groups into (sedentary) citizens of a "modern nation-state" (Kingsbury 2001).

Although acculturation is frequently used to describe the process of a minority group adopting the habits and language patterns of a dominant group, acculturation can also be reciprocal; in this case, the dominant group also adopts patterns typical of the minority group. Assimilation of one cultural group into another may be evidenced by changes in language preference, adoption of common attitudes and values, membership in common social groups and institutions, as well as loss of a separate political or ethnic identity.

In more concrete terms, *acculturation* often refers to what happens at the interface of Western and non-Western or endogenous knowledge, and describes the relative importance of interaction between the two cognitive systems. When Western scientific knowledge adopts the principles of sustainability as a main reference, as in the case of sustainability science, endogenous forms of knowledge often appear highly relevant. This phenomenon has fuelled a neo-traditionalist movement among many Non-western communities. Through this they revaluate certain aspects of their own corpus of endogenous knowledge based on how this knowledge corresponds with sustainability science. Neo-traditionalist tendencies often first become apparent in the context of knowledge related to the perception and management of natural resources (Haller et al. 2008). The more neo-traditional elements are recognised by both the endogenous and the scientific communities involved, the more neo-traditionalist movements advance towards reconstruction of the normative, epistemic, and ontological foundations of practices related to the management of natural resources.

Reconstruction of the cultural foundations underlying social practices related to natural resource management thus becomes the centre from which neo-traditionalism evolves. Once these fundamental cultural principles receive high levels of attention, the process of reconstruction becomes multidimensional. It leads to the modification of individual and collective identities and generates forms of social, economic, and political organisation that place emphasis on the reconstruction of socioecological territories, as a basis for renegotiating their insertion into the wider society. This multidimensional process is accompanied by changes in how rural development is understood and promoted. Understanding these issues requires a conceptual shift from the paradigm of natural resource *management* and knowledge and technology transfer to the new paradigm of natural resource *governance* (Rist et al. 2007). Some fundamental features of a culturally sensitive type of forest governance are laid out by Göhler et al. (2013) in Chap. 12 of this book.

7.2.3 Concepts Related to "Culture" and "Development"

7.2.3.1 Identity

Identity is a concept derived from the Latin word "idem", which means "the same" (Hoffmann 2009). In relation to humans, the term "identity" denotes the sum of a human being's characteristics and traits that make him or her similar to or distinct from other people, in the extended socio-psychological sense. These characteristics and properties allow identifying a person or a group, ideal-typically without ambiguity. In sociological thinking, identity is seen as a distinctively human capacity which is rooted in the "self", that is, the capacity of humans to reflect on their own nature, the societies they are building, and the wider environment to which they belong. Given that identity is thus bound to the self – an inner psychological phenomenon – identity also implies drawing a line between an inner and an outer world (Mead and Morris 2000).

Communication, language socialisation, and education play a key role in shaping individual and collective identities. Accordingly, identity plays an important role in defining how humans perceive the environment, how they position it in normative terms, and what attitude they adopt towards it. Identity cannot fully be explained by individual characteristics and properties: alongside the personal identity that refers to the self, a person also has a social identity relating them to their social environment (Werth and Mayer 2008, p. 164). The latter is based on group membership (Barnard and Spencer 2003, p. 561). The theory of social identity assumes that every person longs to achieve and maintain a certain level of self-esteem. They succeed in this endeavour, among other things, by belonging to groups (Werth and Mayer 2008, p. 563). The link between personal identity and group membership is highly important in the context of natural resource governance. It shapes group members' collective commitments to act according to their joint interests and institutions (norms and regulations). Such collective decisions tend to be more efficient than individual ones. This is the reason why management of forests or pastures as common-pool resources has a high potential for leading to resource governance that is in line with the principles of sustainable development (Agrawal 2001). However, the group identities also become apparent in conflicts, where they can have negative consequences. As pointed out in Opotow and Brook (2003), it is precisely in situations of conflict that groups come to differentiate and strengthen themselves based on their group values. Accordingly, environmental conflicts cause group identities to become distinct; this, in turn, may even sharpen the conflict. A possible approach to solving this problem is to create a more inclusive identity that comprises all groups involved, without, however, abandoning the more specific group identities.

The management strategies of nature reserves with which the local population identify themselves in spiritual and cultural terms increasingly take account of these people's cultural identity, as it influences nature conservation objectives in a supportive manner (Harmon 2004, pp. 13–14). Arora (2006) underlines that sacred groves in India are not only indicative of the relevant cultural group's perception of and attitude towards the environment, but also strengthen the group's identity. Based on these considerations, a growing body of literature suggests that the conservation of cultural and biological diversity could and should go hand in hand (Maffi 2005; Mathez-Stiefel et al. 2007; Bottazzi 2008).

7.2.3.2 Traditions

Traditions are patterns of action which, unlike instincts, are socially constructed. The Latin verb "tradere" means "to hand down" or "to pass on" (Hoffmann 2009) and is the origin of the English word "tradition". In this context, "to pass on" means that whatever is passed on is handed over to the new proprietor together with all related "rights and duties". Adorno (1977) speaks of relationships between generations. Besides representational objects, action schemas – such as regulations, customs, proverbs, and stories – are also handed down and absorbed from generation to generation (Dittmann/no year). Within the scope of cultural development, traditions influence the comprehensive religious-ethical, political, academic, and economic systems. They are adjusted and passed on in more or less complex systems of socialisation. In this process, a certain portion of information gets lost, making room for innovations; these can make reference to the "modern" or contemporary world, but they can also consist of "re-invented" traditions.

Regarding the role of traditions in processes of colonisation and, more recently, in attempts of decolonisation, "culture" is seen as part and parcel of the establishment and transformation of current states and societies in the countries of the global South. It is viewed as connected with, rather than separate from, the economic and political system. This implies that traditions need not necessarily be conceived of as fixed and essential features of a nation that are confronted with a rapidly changing and rootless modernity, as often proposed by Western colonisers, modernisation theorists, and some of the elites in the South (Schech and Haggis 2000).

Another aspect of this idea has been dealt with by Hobsbawm and Ranger (1983, p. 1ff.), who coined the concept of "invented traditions". These are understood as a

> set of practices, normally governed by overtly or tacitly accepted rules and of a ritual or symbolic nature, which seem to inculcate certain values and norms of behaviour by repetition, which automatically implies continuity with the past [...] However, in so far as there is such reference to a historic past, the peculiarity of 'invented' traditions is that the continuity with it is largely factitious. In short, they are responses to novel situations which take the form of reference to old situations, or which establish their own past by quasi-obligatory repetition.

The above authors further differentiate between (more or less invented) "traditions", which they consider to be fixed, and "customs" of which they consider to be open towards certain degrees of change. The concept of "invented traditions" is an adequate response, especially in times of rapid change due to globalisation. Traditions came under increasing pressure because they were either simply swept aside in the course of change, or explicitly were rejected as "obstacles" to progress or domination; both scenarios can lead to the invention of new traditions. As other authors have pointed out, although these "new traditions" draw on the past,

> they tend to be less specific and binding than their 'old' predecessors. They are important in public life, but occupy a much smaller place in the private lives of people than old traditions did. (Schech and Haggis 2000)

Hobsbawm and Ranger (1983, p. 8) distinguished three overlapping types of invented traditions in recent historic times:

- Traditions that establish and legitimise institutions, status, or authority;
- Traditions that establish or symbolise social cohesion of communities;
- Traditions that serve to socialise, or inculcate, beliefs, values, and conventions of behaviour.

This more dynamic concept of traditions provides a better entry point for understanding the making and unmaking of "traditions", for example with respect to "customary" rights of indigenous or native communities to their territories, forests, and natural resources.

The phenomenon of invented traditions plays an important role in understanding how ontological, epistemological, and axiological foundations shape social practice, as expressed, for example, in indigenous institutions related to property rights or the norms regulating access to forests or the distribution of benefits derived from them. As shown in the case studies from Kenya and Bolivia below (Sect. 7.4), the forced or voluntary incorporation of new cognitive and spiritual elements through colonisation or other forms of exposure to Western cultures has triggered a constant process of reconstruction of traditions; this process, however, does not necessarily compromise the worldviews of the respective native people from South America or Africa.

7.2.3.3 Ontologies

Invented traditions often play an important role in the transformation of indigenous *ontologies*. They are sets of basic assumptions that cultural groups make about the constitution of the "reality". These assumptions also constitute the basic elements of a world view. Research on indigenous worldviews in Bolivia (Rist and Dahdouh-Guebas 2006) showed that indigenous people make an important difference regarding the "ontological quality" of mind and matter in comparison with materialist worldview. The materialist view suggests a "monism of matter", while the indigenous position is essentially a "monism of mind". For this reason the establishment of an intercultural dialogue between groups adhering to these two positions translates into a dialogue between different "ontological communities". Such dialogues often explore the practical and theoretical aspects of the specific mind–matter relationship characterising the worldview of each participating group. This means that a

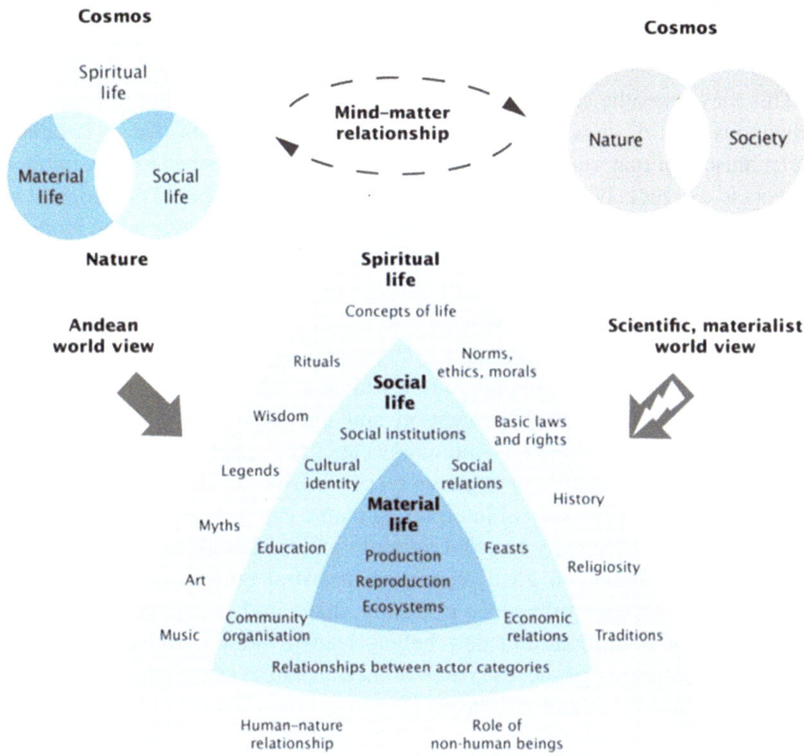

Fig. 7.1 Ethnoscientific comparison of key features of an indigenous (in this case, Aymara) and a materialist Western (natural-) scientific ontology (Source: Rist et al. (2011))

> dialogue on any aspect related to the material sphere of life (e.g. agriculture, forestry, ecology), social sphere of life (e.g. economic relations, social organisation) or spiritual sphere of life (e.g. concept of life, religiosity) must be carried out considering the close relationship – always more or less explicit – between the practical, moral, epistemological and ontological dimensions of knowledge. (Rist and Dahdouh-Guebas 2006, p. 481)

Figure 7.1 illustrates that indigenous people from Bolivia perceive and interpret every phenomenon as a result of interaction between the social, material, and spiritual spheres of life to which human action is directly related (Rist et al. 1999). Viewed from this angle, the mainstream dualist-materialist ontology that underlies the Western sciences is simply one specific situation within a broader and more comprehensive ontological perspective. Therefore, indigenous people can easily accommodate scientific concepts in their own worldview. For example, they might interpret forest degradation as a moral failure of humans for maintaining intact the environments not only for the living human beings, but also for the ancestors or non-human (spiritual) beings of 'Nature' or Mother Earth. Negative socio-economic, ecological or cultural effects of deforestation are thus seen as a reaction or "punishment" of the spiritual beings co-existing in nature with human beings.

This view puts the scientific concept of forest degradation into a context of an indigenous world view enabling people to think beyond the a secularised, Christianised, or materialist worldview having as a consequence to give indigenous answers to exogeously defined problems. This configures a broad ontological space in which indigenous people are constantly reconstructing – or, to use Hobsbawm and Ranger's term, reinventing their "traditions". These reinvented "traditions" express the manifold ways in which ontological principles can be combined within the range of options existing between the ideal-typical expressions of the native Andean worldview, on the one hand, and Western worldviews, on the other.

Similarly to the Andean indigenous worldview, African traditional worldviews are also based on a holistic perception and interpretation of the environment. Everything living and dead is interconnected, and each activity of the living generation elicits a positive or negative response from the ancestral spirits. This – sometimes ambivalent – way of interpreting nature can lead to ecologically responsible as well as ecologically harmful attitudes and traditions; often, these attitudes or traditions are based on great respect of ancestral punishment (Taringa 2006, p. 213). Contemporary influences gradually shape and transform traditions. The combination of traditional and new beliefs leads to new traditions, such as, for example, Christian use of sacred groves or the adaptation of the period of traditional forest ceremonies to school holidays.

7.2.3.4 Endogenous Knowledge

Endogenous knowledge is knowledge that originates within a given ontological community and is rooted in the local expressions of culture and the local environment (Haverkort and Rist 2007). Endogenous knowledge is the knowledge that a community has generated from within, as opposed to knowledge imposed from outside, for example by scientists, researchers, extensionists, and dominant political actors. Such knowledge from outside is referred to as "exogenous" knowledge (see next section below).

Endogenous knowledge is referred to by many different terms, including "indigenous knowledge", "local knowledge", "native expertise", "traditional knowledge", and "cultural cognition" (Antweiler and Mersmann 1996, p. 8). We prefer the term "endogenous knowledge" because it does not restrict these types of knowledge to indigenous people ("indigenous knowledge"), the local level ("local knowledge"), or communities that value their traditions highly ("traditional knowledge"). Devisch and Crossman (2002, p. 108) propose that all forms of non-Western knowledge could be defined as

> ... endogenous knowledge as being a community-, site- and role-specific epistemology governing the structures and development of the cognitive life, values and practices shared by a particular community (often demarcated by its language) and its members, in relation to a specific life-world.

Endogenous knowledge has become increasingly important in efforts to achieve sustainable development. It is deeply interrelated with farming practices and is both cause and effect of specific farmer strategies that are based on the co-evolution (or co-production) of nature and society (Ploeg van and Dijk van 1995). This is especially true in the case of indigenous communities, as many of them perceive humans as a part of nature and understand sustainability based on this holistic perception.

Finally, although endogenous knowledge and associated forms of endogenous development seem to be more important in the global South than in the global North, there is a rather long line of attempts to conceptualise and tap endogenous potentials in the context of European rural development (Dax 2001; Muhlinghaus and Walty 2001). In this context Ray (1999a, b) presents an interesting effort aiming to generalise empirical characteristics of European endogenous rural development initiatives. He suggests that such a meta-framework of endogenous development should revolve around three main components: (1) Ensuring that local actors can access a related territory. This will allow them to become local territorial actors with access to a repertoire of resources; this, in turn, will enable local collective agency, which is required for the pursuit of local objectives. (2) Promoting critical thinking and the development of political strategies within local contexts. This could help to renew representative democracy by adding elements of deliberative or direct democracy based on increased local agency. (3) Defining *collective rights* operating at the interface of local territorial actors and extra-local actors and levels. This would help to protect and enhance locally driven endogenous development initiatives. This type of conceptualization of empirical endogenous knowledge towards a meta-framework of endogenous development is not yet much advanced by researchers working on endogenous knowledge. Exchange between ontological communities that are still using endogenous knowledge in the North and South should be further enhanced in order to learn from each other and explore new ways of cooperation across the local context in which these ontological communities generally operate.

7.2.3.5 Exogenous Knowledge

Scientific is the most disseminated exogenous knowledge in contexts where endogenous prevails. It represents a systematic, objective, and rational type of knowledge. Its production follows generalized epistemological and methodological protocols. Knowledge production in the *natural sciences*, for example, follows a procedure of inquiry and investigation in which researchers try to create a maximum distance between themselves as the observers and the objects they are observing, based on the assumption that this will lead to a maximum of objectivity. This distance is achieved mainly through measurement and quantification, coupled with a positivist perspective. The *social sciences* largely define themselves in contrast to the natural sciences: while the natural sciences study the natural world, the social sciences are occupied with the social world. The social world is essentially constituted by subjects and subjectivities, which reach common understandings through intersubjective validation of propositions, assumptions, and claims. From an epistemological perspective, it is seen as impossible to know to what degree the results of intersubjective validation form part of a social reality that is "objective" in the natural-scientific, positivist sense of the term. Social science, therefore, cannot build exclusively on a positivist epistemology. Instead, social scientists have adopted a diversity of epistemological positions ranging from rationalist to constructivist and idealist approaches (Lele and Norgaard 2005). As a result, unlike the natural sciences, the social sciences are not characterised by one dominant epistemology but by a variety of epistemological approaches whose representatives engage in a constant dialogue. Each of these approaches has its strengths and weaknesses and builds on certain philosophical underpinnings. A common epistemological feature of modern social science is that instead of "measuring", the main methodology for producing scientific knowledge is the "reconstruction" of actor-specific systems of meanings, followed by intersubjective validation among peers belonging to the different thought collectives that constitute the overall community of the social sciences.

In an intercultural context it is often unclear what type of science is meant by "Western" or "scientific" knowledge. In a majority of cases, however, these terms refer to a largely positivist understanding of science as being rooted in a more or less dualist epistemology that draws a clear distinction between the scientist/observer and the object/observed phenomenon and by implication between the natural and social world (studied by the corresponding branches of the academia). This leads to the conclusion that non-Western cultures perceive Western mainstream sciences as investigating the workings of nature and society from an outside perspective, viewing nature and humankind as a dichotomy (Ingold 2000).

According to Descola (2005), different societies around the world combine these two components in more complex ways. A main principle of how people differentiate between human and non-human beings is by comparing their own mind and body – or "interiority" and "physicality", in Descola's terms – with those of the elements surrounding them, and finding resemblances or differences. The combination of perceived resemblance or difference of interiorities (or spiritual aspects) and physicalities (or material aspects) – in other words, whether or not humans attribute to non-human entities an interiority and a physicality that is similar to their own – can vary from society to society. According to the four possible combinations, Descola distinguishes four types of ontology. In the case of *animism*, humans perceive non-humans as having a similar interiority but a different physicality. This means that animist societies, for example many Amazonian societies, consider humans to differ from non-human entities only in their physical aspects, whereas their moral, social, and spiritual aspects are similar. In the extreme opposite case of *naturalism*, for example in modern Western societies, humans consider themselves to be physically similar to non-human beings but socially, morally, and spiritually different. *Analogism*, found in some Meso-American societies, describes a situation where each individual element is seen as being unique in terms of both its interiority and its physicality, but nonetheless closely interconnected within the cosmos.

Totemism, finally, is the tendency to attribute both physical as well as moral characteristics to groups of humans and non-humans based on an experienced characteristic intrinsic relation between physical body and spirit.

7.2.3.6 Traditional Ecological Knowledge

Traditional ecological knowledge has been used by some authors to denote a comprehensive understanding of traditional knowledge as a dynamic "knowledge-–practice–belief complex" (Berkes 1999). A dynamic concept of traditional ecological knowledge helps to avoid the pitfall of essentialism, according to which traditional societies are considered to be static and closed off from other societies. Several authors have warned that the isolation of traditional communities and the untouched character of their environment has often been exaggerated (Boillat 2007). These observations stress the need to consider the importance of change among traditional societies and adopt a dynamic approach to assessing the components of traditional ecological knowledge, taking account also of historical aspects.

7.2.3.7 Institutions

Institutions and governance are closely related. Institutions are the social norms and regulations that shape the actions of a specific person, group, country, or state in such a way that, ideally, it becomes predictable for other actors. Institutions can be formalised, for example as a code of conduct, laws, or a constitution; or they can remain informal, known and respected only by actors involved in certain interactions, such as in the case, for example, of taboos, customary rights, and rules about how to engage in illegal logging activities or whom and how to bribe in a given situation. Social institutions are thus a centrepiece in establishing, maintaining, or transforming social order in general. In the specific context of the management of forests and other natural resources, common-property institutions are considered to have a high potential for enabling more sustainable forms of management (Ostrom and Nagendra 2006; Berkes 2007). Many studies of common-property regimes demonstrated that users of common-property natural resources are capable of creating institutional arrangements and management regimes that allow them to allocate the benefits derived from resource use in a fairly sustainable way e.g. meaning to share equitably, over long time periods, and with only limited efficiency losses (Agrawal 2001).

7.2.3.8 Governance

Governance, according to Rist et al. (2007, pp. 23–24), can be understood as the set of formal and informal "norms and rules of interaction between actor groups involved in natural resource use, and the resulting power relationships between these groups". Norms and rules of interaction are the same as institutions. More details on how institutions set up governance regimes in natural resource management and the relations they have for sustainable forest governance are given in Chap. 12 of this book.

7.2.3.9 Territoriality

Territoriality is another important concept that allows better understanding how culture, institutions, and governance regimes shape specific links between society and nature. Territoriality is defined as the multiple ways in which groups of humans seek to achieve or legitimise control over a geographical area and its natural and/or human resources. According to Sack (1987), this definition implies consideration of three interdependent relationships:

- First, territoriality entails some form of definition and classification of the geographical area in question. This is often achieved by establishing political boundaries, as in the case of municipalities, provinces, districts, protected areas, extraction concessions, and similar types of area. The delineation of boundaries is often directly influenced by a classification of the area – for example, as containing minerals, oil, gas, water, or timber, along with biodiversity (in the cases of extraction concessions or protected areas), or as the place where a certain group of people live (in the case of political units). Due to the importance of these resources, the definition of boundaries, property rights, and use priorities for an area is closely linked with material, economic, and political interests in this area. Accordingly, control over the definition of a territory and the distribution of benefits derived from it is an object of constant struggle and negotiation. In the case of forests, such struggles and negotiations often revolve around the different claims made on forests by different user groups such as indigenous people, logging companies, wildlife conservationists, energy or mineral companies, and others.
- Second, territoriality requires specific forms of communication. The appropriate type of communication depends on the actors involved and the specific interests they have in controlling the territory in question. When governments, national companies, or transnational corporations claim control over areas where poor, indigenous, or native people have been living for a long time, communication, on the one hand, often involves a combination of legal, administrative, and development-related aspects. On the other hand, indigenous, native, or local people often express their historical, cultural, and social rootedness in the area. The purpose is often to resist territorial claims by outsiders and defend their own "historical right" to territorial self-determination.
- Third, territoriality implies the establishment and implementation of legitimacy regarding the specific forms of controlling a territory. This is done by creating political entities that have the capacity for managing incentives and sanctions and defining levels of inclusion and exclusion of actors from within and outside the area (Bottazzi 2008).

Territoriality is a highly important aspect in the quest for biodiversity conservation. Today, around 12 % of the global land surface has been placed under different forms of nature conservation (natural parks, national parks, biosphere reserves, etc.). Accordingly, territorial claims emphasising biodiversity conservation have become a major issue in terms of conflicts of interest between conservationists, indigenous people, and developers such as governments or transnational companies (for more details, see Galvin and Haller 2008).

7.2.3.10 Transdisciplinarity

Transdisciplinarity is used to describe the co-production of knowledge between academic and non-academic communities. Transdisciplinarity is an adequate way for academic and non-academic communities to jointly engage in collective action aimed at applying the principles of sustainability to concrete socio-environmental situations. Indeed, transdisciplinary co-production of knowledge is increasingly considered a prerequisite for research that is directed towards finding more sustainable development pathways (Pohl et al. 2010). When it comes to addressing and solving complex problems with a large number of stakeholders – such as, for example, land degradation or natural resource management – combining diverse types of knowledge and research is a fruitful approach (Russell et al. 2008). Transdisciplinary research allows tackling questions about the causes of current problems, the further development of these problems, relevant values and norms, and desired transformation processes. The purpose of transdisciplinarity is to combine diverse perceptions of an existing problem in order to gain a better understanding of it. According to Hirsch Hadorn et al. (2006, p. 127), transdisciplinary research should explore empirical, pragmatic, normative, and purposive aspects of the problem in question. As this is done on the basis of dialogue and cooperation between scientific and non-scientific ontological communities, transdisciplinarity is not only an adequate approach to including the cultural dimension of problem definition and problem framing; it also provides an excellent opportunity for establishing an intercultural dialogue about the ontological foundations underlying scientific and other forms of knowledge.

7.3 Sustainability, Culture, and Forestry

7.3.1 The Cultural Dimension of (Rural) Development

In 1987, the Brundtland report pointed out that global economic development could only be sustainable if societies and their economies managed to grow without exhausting natural resources. This was the first time that the ecological, economic, and social dimensions of sustainable development were addressed together in the political arena. The report provided a key statement on sustainable development, defining it as "development that meets the needs of the present without compromising the ability of future generations to meet their own needs" (World Commission on Environment and Development 1987).

Although UN member states recognised sustainable development as a guiding principle in 1992, two decades later progress towards achieving it continues to be fairly slow. We argue that the lack of progress is related – among many other things, indeed – to a lack of inquiry into the cultural foundations (ontology) underlying the principles of sustainable development. This omission has made it difficult to explore synergies with other cultural systems that could help to make the general principles of sustainability work within the various existing cultural systems.

A look at African worldviews can help to illustrate potential synergies with the worldview underlying to sustainable development. In many African communities, attitudes towards nature and natural resources as well as natural resource management are linked to people's religions, spirituality, and worldviews. Forests and certain kinds of trees, for example, were preserved in the name of traditional African religion (Omari 1990). Values and norms regarding the use of natural resources are passed on from generation to generation through songs, myths, and stories, as well as feasts and ceremonies. In such cases, land is traditionally considered a common property that belongs to all beings, both living and dead. This type of environmental ethic rests on the principle of communalism (Peterson 2004, p. 169). Accordingly, all owners of land must be aware of its being a limited public good, regardless of the form of property rights attached to it (private, public, collective, etc.). To hold the rights of ownership for particular resources means to assume great responsibility at both the individual and the community levels, as well as towards the communities of beings that lived in the past or coexist with nature in the present (ancestors, spirits).

A similar situation was observed when discussing the (cultural) meaning of sustainability with indigenous people in Bolivia. To them, a major motive for engaging in debates about sustainable development is the insight that these debates are linked to indigenous people's current struggle for recognition as full, but culturally different, citizens of the state to which they belong. They are striving for recognition that is not merely legal but includes their right to self-governance in their historically existing territories based on their own forms of organisation, political representation, body of knowledge, and understandings of "development". The latter is often labelled as "vivir bien", or "living well" (Delgado et al. 2011). The overlaps between such forms of "ethnodevelopment" and the basic principles of sustainability are a key factor that can make indigenous or native people important allies who reinvent, revitalise, promote, and practise an emancipatory understanding of sustainable development (Chartock 2011). A specific novel feature of such ethnically based types of sustainable development is that they not only strive to satisfy the needs of present generations without compromising future generations' ability to meet their needs: indigenous people in South America and Africa go one step further to include past generations, as well. They understand sustainable development as a mode of co-evolution of humans and nature that fosters the relationship between present generations, past generations (ancestors and related living entities of nature), and future generations (Rist 2010).

Many more examples suggest that indigenous forms of governing natural resources overlap more or less strongly with the principles of sustainable development (Ostrom and Nagendra 2006). However, people practising these forms of governance are not necessarily aware of this. Far more often, they consider their ways of governing natural resources as an expression and a source of cultural resistance to the homogenising tendencies of current economic globalisation – a process that many non-Western cultural groups perceive as an expression of Western cultural hegemony. This focus on resistance among indigenous communities, along with the global community's lack of attention to indigenous forms of sustainable development, impedes a proper conceptualisation of sustainable development and an exploration of its potentials and limitations from a cross-cultural perspective.

From a culturally inclusive perspective, the aim of sustainable development has to be defined in a culturally neutral way. Accordingly, we propose to say that the aim of sustainable development is *to improve any culturally defined inherent qualities of life* – rather than just its economically measurable aspects. Obviously, inherent qualities of life differ between diverse cultures, depending on their respective systems of values, norms, beliefs, and traditions. Although economically measurable aspects of life, expressed as "standard of living", are widely assumed to represent an universal main objective of "development", some cultures may give other qualities of life a higher priority. In this sense, considering the cultural dimension of sustainable development does not mean simply adding a fourth dimension to the "magic triangle" of sustainability; instead, it means focusing on how different cultures define, combine and weight the three classical dimensions of sustainable development (that is, its economic, social, and ecological dimensions). The specific way in which a given culture combines and weights these dimensions of sustainability reflects this culture's set of values, institutions, traditions, and ontological assumptions concerning "reality" and the relations between the social, biological-material, and spiritual spheres of life (see Fig. 7.2). When comparing different situations in which more or less sustainable development is taking place, we must therefore examine to what degree these situations express different "cultures of sustainability".

Defining a group's specific understanding of sustainable development as the culture-specific order in which that group prioritises the social, economic, and ecological aims or implications of a given form of "development" raises the question of whether this order is based on an underlying (culture-specific) normative hierarchy that give coherence to these dimensions. Drawing on the framework of Bargatzky (1986), Darr (2011) convincingly showed that this question is worth answering in the affirmative. She demonstrated that cultural systems contain a clear hierarchy ranging from sacred assumptions about the world to abstract principles to concrete norms, regulations and practices (see Fig. 7.3). Abstract assumptions and concrete rules are linked through ethical and moral principles that are guiding behavior of actors in everyday life.

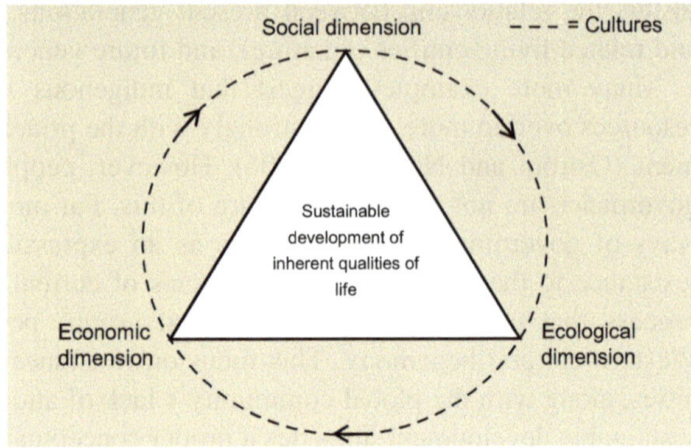

Fig. 7.2 Culture shapes the relationships between three dimensions of sustainable development (Source: Own elaboration)

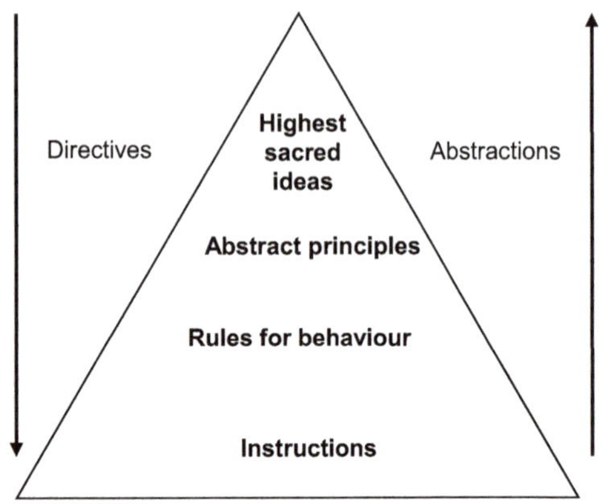

Fig. 7.3 Control hierarchy in orderly adaptive structures (Source: Bargatzky (1986, p. 190), modified according to Irrgang (2005))

The core components of cultural knowledge at the top of the pyramid – which are generally of a sacred nature – give meaning and legitimacy to the normative domains of the cultural system, such as, for example, rules and regulations defining access to and ownership of forests. These rules, in turn, are directly related to socio-productive and technological practices and determine, for example, the extent of slash-and-burn activities or the balance between timber and non-timber uses of forests.

A culturally sensitive understanding of sustainable development – in this case, of forests – has the advantage that it allows shedding light on the processes of cultural construction of multiple functions that forest and trees fulfil in relation to different social groups. Forests or trees do not normally fulfil a single specific function at one certain point in time and space; they serve multiple purposes such as, for example, timber and non-timber production, soil fertility restoration, water cycle maintenance, and provision of habitats in support of genetic, species, and ecosystem diversity. In addition, they fulfil certain functions in human communication with living beings belonging to past generations (ancestors) or to animated nature (spiritual beings associated with minerals, plants, animals, and other natural phenomena). A cultural perspective on forestry that makes it possible to reconstruct the multidimensionality of forests and trees as perceived by different social groups is thus an important link between policies and politics of sustainable development and local forest users' practices. Indeed, research demonstrates that multifunctionality of forests is a promising conceptual and practical alternative to approaches focusing on conservation and sustainable use of forest resources (Pinedo-Vasquez and Sears 2011).

7.3.2 The Link Between Culture and Forest Governance

The significance of "cultures" in global forest management becomes evident when we consider the important role of community forests – especially in developing countries, where their proportions have increased considerably over the last decade or so (Sunderlin et al. 2008). Community forests are managed locally by mostly non-Western peasant or indigenous communities and nations. They provide a livelihood for more than 350 million rural people and, in addition, produce local and global public goods, for example related to biodiversity conservation, water storage, soil conservation, and carbon sequestration (Eliasch 2008; Barbien et al. 2009).

Today it is widely recognised that community-based forestry systems are not only local expressions of a high diversity of indigenous cultures, but that they also embrace a high potential for sustainability if the following basic conditions are fulfilled: tenure security, clear rules, shared norms, appropriate leadership, multi-level decision-making, and adapted forms of sanctions (Ostrom and Nagendra 2006).

Due to the importance of global forests in biodiversity conservation, about 12 % of them have been placed under different forms of protection (FAO 2010). Regardless of whether this is sufficient, it is important to highlight that excluding such a significant area from commercial forestry or agro-industrial use also represents a cultural valuation of forests: protecting forests at such scales means placing the intrinsic values of forests higher than the monetary assets available in the short term. This means that nature protection or conservation, like many indigenous forms of natural resource governance, must also be understood as a cultural expression that has strong arguments against the dominant trend of giving priority to capital-intensive, mainly capitalist forms of using forest land and resources.

The cultural dimension gains even greater importance in achieving more sustainable forest governance and management in light of the growing emphasis placed on the intrinsic links between biological and cultural diversity. The emerging field of research, development, and policy referred to as the "biocultural approach" aims to simultaneously understand and act upon the links between the world's linguistic, cultural, and biological diversity as different manifestations of the general diversity of life (Maffi 2005). The emphasis on these links confirms that in the long run, biological diversity can only be conserved if cultural diversity is maintained and enhanced.

Another important argument for integrating "culture" into the search for more sustainable governance and management of forests is that both community-based institutional arrangements and biocultural diversity are being translated into concrete management practices that are widely based on traditional ecological knowledge (in the comprehensive sense of the term, including beliefs and practices along with knowledge). In order to better understand traditional ecological knowledge and put it to use in enhancing indigenous forestry development we propose considering the links between (1) worldviews, (2) social institutions, (3) land and resource management systems, and (4) endogenous knowledge on the environment and ecosystems (ethnoecological knowledge) (Berkes 2006; Rist and Dahdouh-Guebas 2006; Pretty et al. 2009).

7.4 Making Approaches to Rural Development Culturally Sensitive: Case Studies from Kenya and Bolivia

7.4.1 Introduction

The following section presents two case studies from Kenya and Bolivia that show what it means to achieve greater mutual understanding and, on this basis, to develop solutions for more sustainable and culturally sensitive forest governance. The two cases also illustrate how disciplinary and sectoral boundaries become blurred when the shared understanding of forestry and agroforestry systems is expanded to include the dimension of cultures. The dissolution of such boundaries is a precondition for moving from disciplinary towards inter- and transdisciplinary knowledge production. This implies to engage in a processes where knowledge is produced jointly by scientists and non-scientific actors such as indigenous or peasant communities, extensionists, non-governmental organisations, businesspeople, representatives of the public administration, and policymakers (Pohl et al. 2010).

The case study from Kenya allows to see how identity, traditions, gender relations, spirituality, and institutional, political, economic, and ecological factors are interacting and how this relates to local forms of multidimensionality of forests.

The Bolivian case describes how two culturally different indigenous groups have developed very different institutions (understood as norms, rules, and regulations) for managing forest and associated natural resources even though they live in the same ecosystem in the lowlands of Bolivia; thereby, it demonstrates that cultural backgrounds, traditions, and history have a fundamental influence on how forests are managed. This poses a particular challenge to social forestry policies. "One-fits-all" solutions are bound to fail, as they are too abstract and thus ignore local actors' agency and the resulting high and dynamic biocultural diversity.

7.4.2 The Cultural Forests of the Tiriki in Kenya: An Example of Joint Forest Ownership by Past, Present, and Future Generations

How are identity, traditions, gender relations, spirituality, and institutional, political, economic, and ecological factors linking in a concrete of forest management in Africa? This is shown on the basis of the Tiriki ethnic group in Kenya. They are one of sixteen ethnic subgroups of the Luhya, who belong to the Bantu language family which represents Kenya's second largest ethnic group. The Luhya live mainly in Western Kenya between Lake Victoria and Mount Elgon.

The Tiriki speak a dialect distinctly different from the other Luhya dialects. They adhere either to traditional African religion or to Christianity and live mainly as farmers in the south-eastern part of the Kakamega forest, which is administered by the Kenyan Forest Service. The whole forest shows clear signs of overuse and degradation (Bleher et al. 2006). Yet, more than 40 smaller, species-rich and biologically less disturbed "cultural forests" of the Tiriki (Onyango and Nyanja 2004) exist in the same densely populated area. These "cultural forests" vary in size between 0.5 and 8 ha and generally consist of indigenous tree species. Numerous trees are said to be more than 100 years old.

The Tiriki are well known locally for their initiation ceremonies in these "cultural forests". Every 5 years, they celebrate a traditional passage ceremony for the boys that marks their transition from childhood to adulthood. It is celebrated in these forests and includes circumcision as well as other rituals that strengthen the Tiriki's community spirit. Indeed, a man is only accepted in society if he has undergone the ritual of circumcision and initiation inside the forest.

Since the circumcision ceremonies require the availability of water, cultural forests are normally situated in valleys. The forests need to be densely stocked, so as to provide a good shelter and hiding place for the boys during the festivities, which nowadays last 4–5 weeks. Although women are not allowed to enter the forests, they also participate in the traditions: circumcision ceremonies are accompanied by cheerful dances and beer drinking outside the forest. During these activities the forest is not entered.

Tiriki culture strictly prohibits use of the cultural forests for common purposes such as fuelwood or fodder collection. In every village, at least one of the village elders is formally responsible for overseeing the cultural forest. In addition, the entire village community feels responsible for reporting misuse of these sacred forests by villagers or strangers.

The importance of these cultural forests to the Tiriki is expressed in the following statements by Tiriki interviewees (Darr 2011):

> "Those forests or shrines are considered holy places for all those circumcision things and so they are totally protected by the culture because from our forefathers [ancestors] it was said if you cut a tree from my shrine you will die! You will die. So that fear has grown from age to age. [...] And people are happy to protect the shrines."

→ *The cultural forests are protected by traditional taboos. The fear of doing something wrong is rooted in traditional beliefs about the ancestral spirits.*

"The sacred forests give us happiness [. . .] I am happy when my boys come safe from the circumcision."

→ *Speaking about her feelings towards these forests, this mother reveals her close connection to this natural environment.*

"The place [sacred forest] is only and specifically for circumcision – all trees found in that place they are considered special. It is not a planted forest, it is naturally grown."

→ *This statement emphasises the uniqueness that is attributed to the cultural forests. The combination of vivid rituals and a traditionally largely natural environment is highly valued.*

The Tiriki perceive their "sacred forests" as an important part of their rich cultural heritage. Until today, they value the cultural forests not only for contributing to their identity and the maintenance of their traditions, but also for the indigenous tree species they contain. The species combination in the cultural forests ensures that they serve the local people and their cultural uses. Various specific tree species are valued as homes of the ancestral spirits, as well as for their multiple traditional uses, including aesthetic aspects and biodiversity conservation mainly of medicinal plants, and for their characteristic sizes and forms that give the whole forest a closed canopy (Darr 2011). Endogenous knowledge serves the safeguarding of the "cultural forests". It represents the infinite bond and long-lasting relationship that unites past, current, and coming generations. The vital rituals celebrated in the cultural forests contribute to the continuance of this perception. Territoriality also plays an essential role in the process of forest protection and management: only people with a close relation to the local sites know about the history and meanings of important people and places.

In the case of the Tiriki and their cultural forests, it is obvious that greater importance is attached to the combination of social and ecological aspects (see Fig. 7.2) than to any economic use. The fact that this is a long-standing tradition lends the rules and governance of these forests stability.

A look at the Tiriki's value structure with regard to the cultural forests (Fig. 7.4, middle column) shows that highest sacred ideas and abstract principles are rooted in the Tiriki's understanding of their Tiriki identity and in traditional African religion. The cultural forests are perceived as places for vivid spiritual life. These values are the main motivation for the Tiriki to uphold their strong traditional taboos; they also help the community to develop and maintain their typical quality of life. Adherence to the protective rules is motivated by strongly rooted and inherited fears. This is one example of the effectiveness of traditional norms and social pressures in ensuring conservation. Additionally, the small size of the sacred forests and their location in central places helps the communities to easily monitor them via social control in their daily life.

This study also explored how the Tiriki use and perceive the government-administered Kakamega forest. Findings showed that the abstract values they attach to the Kakamega forest differ from the ones ascribed to the sacred forests: for example, they value the Kakamega forest as a space of reciprocity between humans and ancestors and as a resource (Fig. 7.4, right column). Overall, the Tiriki's

Orderly adaptive structure (Bargatzky, 1986)	Tiriki sacred forests	Tiriki Kakamega forest
Highest sacred ideas	Cultural identity Traditional African religion	Reciprocity of humans and ancestors
Abstract principles	Forest as a place for vivid spiritual life	Forest as resource (climate, health, income)
Rules for behaviour	Strong traditional taboos	State and traditional rules
Instructions (examples)	• Do not collect fire wood (except for ceremonies within the forest) • Women, children, and strangers must never enter the sacred forest • Tiriki men may use the forest every five years for circumcision and initiation	• Do not plant exotic tree species along rivers • Collect only dry firewood, and only after payment of fees • All humans are allowed to enter the forest
Directives ↓ / Abstraction ↑		

Fig. 7.4 Value hierarchy of the Tiriki (Source: Own elaboration)

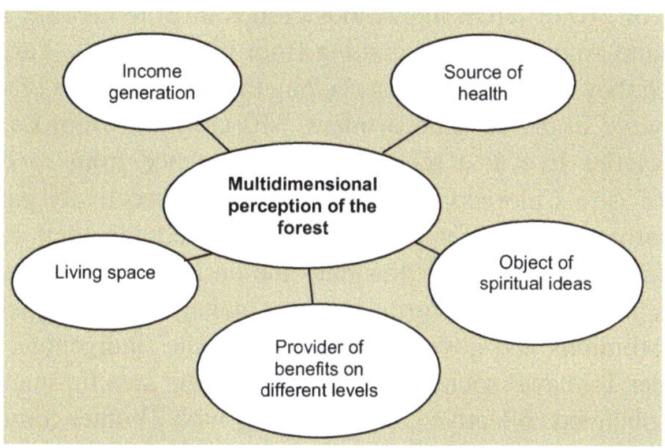

Fig. 7.5 The Tiriki's multidimensional perception of forest (Source: Own elaboration)

perception of forest in general –including both the sacred forests and other forests – is multidimensional: they view the forest not exclusively as an object of spiritual ideas but – depending on the type of forest – attribute other functions to it as well. For example, they also perceive the forest as a living space for medicinal and other plants, a source fuelwood, or a place to relax and enjoy the cool and fresh air (Fig. 7.5). These non-spiritual functions of the forest are valued more from the perspective of everyday use.

Behaviour towards Kakamega forest is governed mainly by official rules. The Tiriki's respect for the forest service as the institution defining the rules of use is not nearly as high as for traditional African religion as the source of rules governing their behaviour towards the sacred forests. Accordingly, it seems worthwhile to know the values and abstract ideas guiding forest users' behaviour, as a basis for understanding the driving forces of development and for fostering the more sustainable ones among them.

The example shows that culturally sensitive sustainable forest management cannot be made just by relying on the scientific view of forests. Scientific contributions might be meaningful, but only to the degree that local people can integrate them into their own multidimensional view of the forests. For that purpose intra and interontological dialog and cooperation between the different ontological communities implied can lay the ground for defining culturally sensitive and inclusive forms of sustainable forest governance.

7.4.3 *How Two Different Indigenous Cultures in the Bolivian Lowlands Influence Forest Governance and Conservation*

In this section we present a comparative study that shows how important elements in the cultures of two different indigenous groups in Bolivia – the Tsimane' from the lowlands, and Andean settlers coming from the highlands – are affecting the forests in which they live. These forests belong to the buffer zone of the Pilón Lajas Biosphere Reserve in the Beni department, 370 km north from La Paz. The two indigenous societies live at a relatively short distance from each other in the communities of Alto Colorado and Villa Imperial, respectively partly inside the reserve and partly in its buffer zone (Fig. 7.6). Studying their rather different social–ecological systems clearly illustrates the variability of cultural aspects and how they are related to different forms of forest management and their implications in terms of institutions and governance. The Tsimane' indigenous community is part of the wider Tsimane' society that has lived in the area for many generations. The Tsimane' obtained collective land titles along with 21 other communities in the early 1990s in the context of a national joint indigenous and conservationist movement (Bottazzi 2008, 2009). The indigenous Andean settlers coming from the highlands were granted individual land titles in the context of the government's colonisation programme launched in the 1970s.

7.4.3.1 The Indigenous Tsimane'

Tsimane' Amerindians settled in the area during the pre-Hispanic period. Today, they make up a total population of about 15,000 people who share a distinct language system and distinct cultural patterns. Alto Colorado is the oldest Tsimane' community and was established in the Pilón Lajas Biosphere Reserve 50 years ago. It has a total population of about 150 individuals. Their livelihoods are based on hunting-gathering as well as on small-scale shifting cultivation. Generalised collective ownership of land allows for great flexibility in terms of agrarian practices.

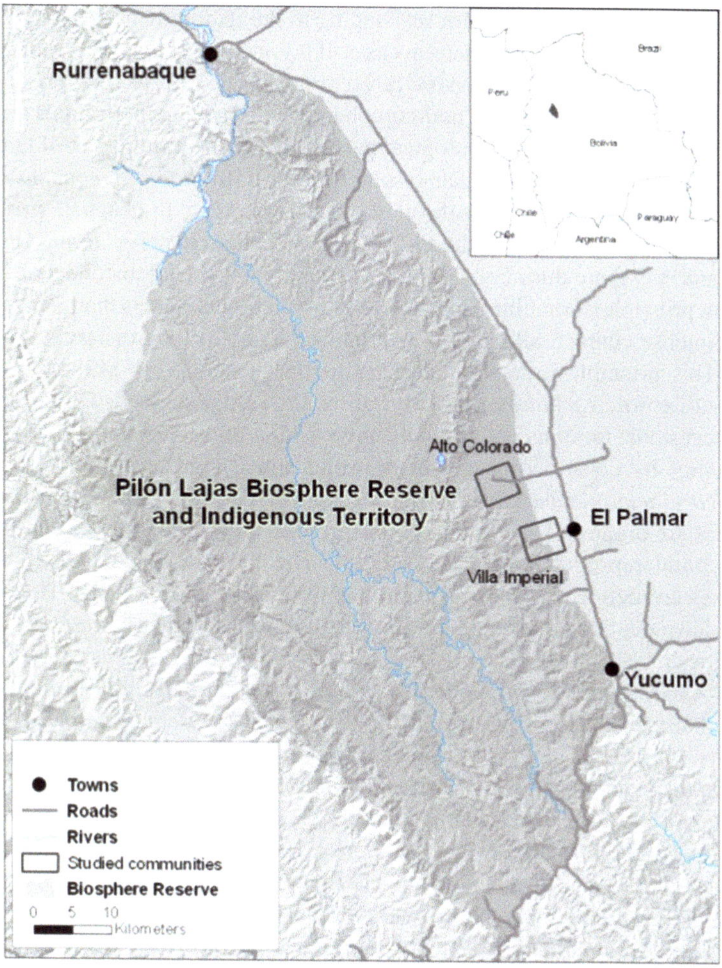

Fig. 7.6 Localisation of Alto Colorado and Villa Imperial in the Pilón Lajas Biosphere Reserve (Source: Own elaboration)

Land is usually cultivated individually by each household; use of forest products, however, is collective. A small number of cattle are kept in the community, although the Tsimane' generally dislike the activity of cattle-breeding: for them, hunting is the most significant source of protein (Chicchón 1995). Hunted game is divided between family members following kinship rules. Tsimane' have also developed a highly specific knowledge about plant diversity in terms of the plants' economic, practical, and cultural values (Reyes-García et al. 2006, 2007).

Forest extraction is organised by multiple rules based on Tsimane' cosmology, according to which non-human entities such as animals and spirits control the use of hunting resources (Riester 1993; Daillant 1998, 2003). The spiritual chief, called *cocojsi'*, is in charge of establishing links between the spiritual and the material dimensions. He is responsible for responding to the requirements of spirits such as the *jëjëbe*, who can punish humans in cases of non-compliance with certain spiritual principles. Excessive extraction of forest products or the destruction of certain sacred wood species as well as non-compliance with fundamental rituals are highly prohibited according to the ontological perspective of the Tsimane'. All families in Alto Colorado extract non-timber resources, especially *jatata* (*Geonoma diversa*) for roof construction or for sale at the local markets. In August, part of the community move deeper into the Pilón Lajas Biosphere Reserve, to the Quiquibey River, and stay there during several weeks, mainly to fish and to collect turtle eggs. A strong principle of mobility, referred to as *sóbaqui,* guarantees that families are in highly intense contact, although they might be living at long distances from each other. This principle is motivated by extractive benefits, but also by the social benefits of enforcing family ties and sharing knowledge or seeds (Ellis 1996). On some occasions, families practise collective fishing in the Colorado River near the community by constructing a dam of wood and using a natural poison called *barbasco* (*Jacquinia barbasco*) that can stun or paralyse fish. These extractive activities are organised according to kinship rules and respond to several logics of sociocultural reproduction (Castillo 1988). The localisation and distribution of resources are defined by the spiritual chief for each family or clan. Hunting and fishing have declined over the past 20 years due to the damage caused by logging companies.

The Alto Colorado community is part of a larger set of 21 communities whose representatives elect a regional authority called *Consejo Regional Tsimane' Moseten'* (CRTM). The CRTM is in charge of representing Tsimane' communities living in the shared territory, in regional political arenas and processes, but has no control over resource extraction (Bottazzi and Dao 2013). At the village level, the Tsimane's main formally elected authority is called the *corregidor*; this is a function that was introduced by Jesuit and Franciscan missionaries and is practised in extremely variable ways from one community to another, depending on the context, history, and individuals involved. Amazonian societies are known for their horizontal power distribution and subdivision into family clans (Clastres 1974).

Over the last decades, indigenous Tsimane' societies have undergone considerable acculturation, which has manifested itself, among other things, in the loss of ontological principles that used to guide their forest management. This development is mainly due to increasing contacts with external actors related to the logging sector, participation in product and labour markets, and the increasing proximity to growing urban centres. Significant financial gains from logging led to a concentration of power in the hands of several small groups of individuals. At the same time, the Tsimane' were hardly able to adjust their own principles to the institutional rationale of the national forest regime, which focuses on the needs and interests of large-scale logging companies (Bottazzi 2009). The practically unaffordably high transaction costs for obtaining a legal forest management plan forced them to completely bypass the formal rules. As a result, they now sell cheap logs extracted far from their community, thereby breaching many of their own principles of indigenous sustainable forest management. In several cases, Tsimane' have become genuine *entrepreneurs*, even though their territories are subject to a collective property regime. They are selling access rights to resources and gradually investing their income from illegal forestry into large-scale commercial agriculture, sometimes even employing Andean migrants as wage labourers. This development paves the way for forest degradation and might result in rapidly increasing deforestation within their territory (WCS 2005). Even though some families oppose the cutting of their rainforest due to their symbolic and religious beliefs (Riester 1993), 64 % of them are strongly involved in owing to market pressure and the societal influence of external agents.

7.4.3.2 The Indigenous Andean Settlers

The "colonisation zone" located on the eastern part of the Pilón Lajas Biosphere Reserve is a large project started in 1974 by military dictator Hugo Bánzer. The newly established National Institute of Colonisation distributed land in the area to migrants over several years. Villa Imperial is one of the resulting so-called *colonias*, a settlement unit composed of 40 parcels of 25 ha each, on a total surface of 1,500 ha. The first migrants of Villa Imperial arrived in 1983, mostly from the Andean city of Potosí, which they had been forced to leave due to the collapse of the mining industries. The "colony" of Villa Imperial is organised in a "syndicate" composed of the parcel owners (mainly men). The syndicate functions on the basis of annually renewed community posts (*cargos*) in a way that practically all owners constantly take part in the community's governance, which is coordinated by the syndicate's executive board. A larger parcel situated at the centre of the colony is managed by the executive board, its main purpose being to generate the resources required for the functioning of the community organisation. Family plots are worked individually and managed depending on the families' productive interests; at the same time, all families are required to participate in collective work and to pay monthly fees to the executive board. The syndicate has the right to seize the parcel of an uncooperative owner and sell it to another community member. All resources found on the individual parcel belong to its owner and are not subjected to any use constraings imposed by the community. The main conditions on which families are guaranteed the right of ownership are their participation in community work, attendance of monthly assemblies, and payment of the self-determined fees required for maintaining the community organisation. The syndicate is part of a broader organisation, the *Central*, which, in turn, forms part of a *Federación* at the regional and national levels.

Compared to the Tsimane's agricultural plots, the Andean settlers' cultivation areas are much larger. They are mainly used for intensive cultivation of rice, maize, and cassava. As settlers do not know most of the highly diverse plants and other properties of their forests, they rely mainly on agriculture for their commercial and subsistence needs. Since the beginning of the 1990s, parcel owners have been receiving credits from small financial agencies.

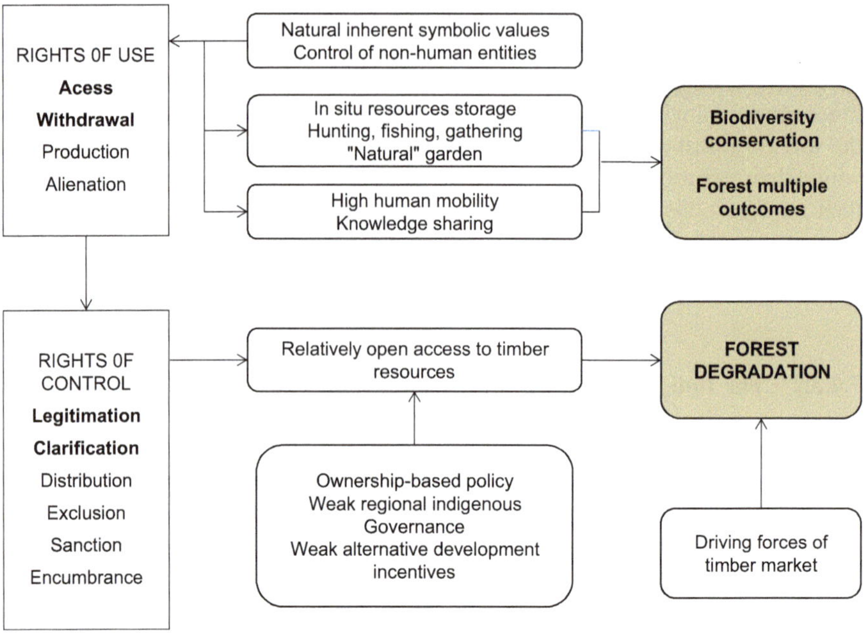

Fig. 7.7 The cultural and institutional drivers of change in Tsimane' forest use (Source: Own elaboration)

These loans are invested in off-farm activities, such as commerce and providing transportation services. Benefits from agricultural and non-agricultural exploitation are mainly invested in cattle as a form of financial savings (Fig. 7.7). Most peasants keep one to five animals due to the limited size of their family parcels of 25 ha, which means that breeding is not profitable in commercial terms. At this small scale, cattle-breeding makes sense exclusively as a saving strategy. Given that this strategy is widespread among community members, the number of cattle has become a sign of prestige. Consequently, peasants tend to find ways for increasing their number of cattle. As the syndicate forbids its members to own more than one parcel, it is impossible for them to keep large numbers of livestock on their own land. This has led to the development of an internal leasing market. Owners of a larger number of cattle lease parcels from other peasants according to a rotational scheme. The related deforestation happens along steadily expanding networks of highways and local roads and has created a unique fishbone pattern of deforestation in the landscape which is characteristic of Amazonian migrant settlements (Walker 2003) (Fig. 7.8).

The Bolivian case study shows how, in a similar forest ecosystem, contrasting cultural patterns structure institutional as well as productive settings, and how these settings, in turn, shape forest land cover in different ways. The Tsimane's animist perception of the nature–society relationship is part of a wider reciprocity that includes interaction between humans and non-human beings and forms the basis for

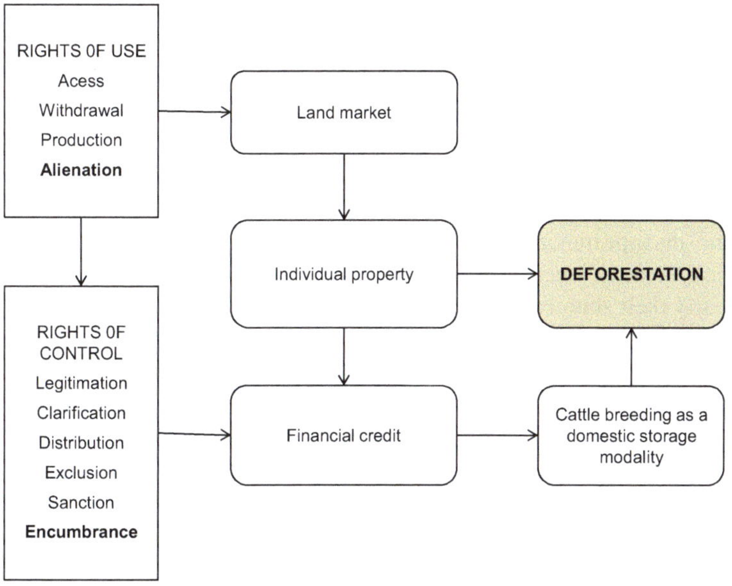

Fig. 7.8 The cultural and institutional drivers of change in Andean settlers' forest use (Source: Own elaboration)

the symbolic representation of norms of resource regulation. Such cultural patterns, combined with social institutions such as collective access to forests, are not prepared to adapt to accelerated societal changes triggered by timber market influences or by the increasing loss of hunting-gathering resources. The clash of such indigenous forms of resource appropriation with the capitalist rationale of resource management results in forest degradation due to the lack of organisational capacities among indigenous communities, which manifests itself in rather uncontrolled logging practices. However, growing interaction and comparison with Andean settlers has made the Tsimane' aware that have to find ways of strengthening the coherence of their social organisation as a means to regulate the use of their forest resources in more coherent ways. They have learnt from Andean settlers that improving forest governance is not an isolated organisational aspect but should be part of a wider effort to reconnect worldviews, indigenous identities, and socio-political organisation. A possible way of advancing such a project is to facilitate communication of the principles discussed at the level of their regional political organisation – the CRTM – and the principles guiding forest management at the level of communities and families. These local visions of life and agency are constantly challenging present forms of external domination, fuelling a process of reflection that reveals potential ways of resisting the dominant externally driven forms of resource appropriation.

The Andean settlers form a complete contrast to this development. They have brought their habitual principles of intensive land use on family plots from their communities of origin in the Andes into contact with a new vernacular valuation of

Mother Earth's generosity (called *Pachamama*). The concept of *pacha*, meaning totality, is divided into three levels: *urco* (highlands), *taypi* (centre or middle), and *uma* (lowlands). This concept of totality (*pacha*) underlying the concept of *Pachamama* is expressed in the Andean cosmology characterized in Fig. 7.1. From the point of view of agricultural production, the *pacha* concept is central in the highlands because it includes a concept of complementarity between multiple ecological zones (Murra 1972). Migration to the lowlands has not resulted in a complete disappearance of this concept but rather in its adaptation to a new ecological context, where it helps producers to reduce the risk of harvest loss and to diversify their sources of revenue. This is combined with an ethics of hard work and self-sacrificing efforts that will be compensated by the human community and the related beings that are all part of Mother Earth. The fact that Andean migrants' invest their labour in land that is a quasi-individual property regime is a necessary consequence of the long-standing integration of highlanders into capitalist labour and product markets. However, in the case of these Andean migrants, private property does not mean a complete individualisation of production but, instead, is used as an additional means of control by local political bodies. The capitalist trends in external market influences are adopted, but governed by local entities so as to protect social cohesion and maintain indigenous ontological values. These "hybrid" institutions stimulate constant collective reflection, leading Andean migrant communities to continuously reinterpret the relations between their own cultural roots and the external context in which their culture evolves. Within this dialectic, deforestation can be viewed as a temporary expression of the interfaces created between the two ontological poles determining that in the type of their interrelation define the trajectory of the socio–ecological coexistence of humans and nature.

7.5 Lessons Learned and Outlook

The case studies described above showed that working at the interface of culture and development in search of more sustainable management of forests – for example in Kenya and Bolivia, but certainly elsewhere, too – requires, first of all, a clarification of the role of culture in the context of sustainable development. The findings demonstrated that it is useful in this context to understand culture as the expression of specific relations between the three classical dimensions of sustainable development: economy, society, and ecology. However, this research has also demonstrated the use of understanding that these culturally defined relationships involve a hierarchy between basic ontological assumptions underlying the different cultural systems and the related institutions, norms, and regulations that shape the scope of human behaviour in the context of managing forest resources. The model of control hierarchy in orderly adaptive structures developed by Bargatzky (1986) was helpful in visualising this fundamental aspect. Together, these insights not only facilitated a sound conceptualisation of the role of culture in the context of a broader understanding of sustainable development, but also set the stage for a visualisation of the multiple aspects involved in the social construction of the perceived multifunctionality of trees and forests belonging to native, indigenous, and other local communities.

The case study from Bolivia revealed possible ways in which the interdisciplinary insights resulting from such processes can be fed back into action and policymaking with a view to creating synergies between different forms of knowledge. Both case studies support the conclusion that sustainable development in non-Western cultures can be complemented with exogenous or scientific knowledge. Endogenous development pathways offer the possibility of reconstructing traditions, taking into account local communities' internal cultural settings as well as the broader external political, economic, and cultural context.

We see the following basic ways in which cultural foundations can contribute to endogenous-driven forms of sustainable development:

- *First*, it is important to create spaces for cultural invigoration and joint construction of governance regimes – processes which, instead of disregarding the cultural dimension, use it as a source of normative guidance in devising new types of development projects that go beyond expressing narrow sectoral or unreflected scientific views of "development". If the starting point is an outside actor, such as, for example, a university, a non-governmental organisation, or a public authority, we recommend achieving this by rearticulating research and rural extension based on transdisciplinarity. This leads to a dialogue between endogenous and scientific knowledge and to co-production of knowledge for more sustainable development. Local cultural characteristics need to be integrated on an equal footing with external cultural assumptions when planning and land use has to obtain long-lasting support from the local people. The coexistence of forests and people is rooted in many different cultural perceptions of the natural environment.
- *Second*, this requires establishing or strengthening a society–nature relationship that links past, present, and future generations based on the creation of adequate overall conditions that are propitious for the enhancement and invigoration of indigenous identities, traditions, institutions, and the related environmental knowledge and management practices that express territorialised principles of the endogenous world. Traditions such as those of the Tiriki are nothing fixed – they evolve and change depending on how the given society understands them in the light of contemporary developments. Traditions need to be practised so that they can develop, because only constant development allows them to persist. This form of endogenous knowledge deserves respectful acceptance and has to be considered as equal to other forms of knowledge. From a holistic view it is part of life and not just a "backward attitude".
- *Third*, it is important to recognise that indigenous, peasant, and other native communities, as well as the principles of sustainability as it is commonly understood, tend to give priority to common-pool resource management; this requires supporting efforts to foster the recognition of existing or the

establishment of new institutions that allow endogenous communities to increase their agency as territorial actors. This seems essential, as many traditional societies view natural resources as a heritage handed down from past to current generations. It is nothing to be owned individually, and it has to be maintained and handed down to future generations. The long held values that guide modest and careful treatment and use of forests are rooted in the cultural memory of societies. This can be stronger than short-term political decisions.

- *Fourth*, considering the cultural dimension of forest governance in a context of sustainable development provides a basis for making explicit the different types of multidimensionality that result from culture-specific ways of perceiving and valuating forest resources. This provides a particularly important entry point to sustainable development. In more advanced interpretations of sustainable forest development, multifunctionality is viewed as one of the most promising approaches to forest governance which is said to have a high potential for optimally integrating tangible economic and intangible sociocultural and political functions of trees and forests.

References

Adorno T (1977) Über Tradition. In: Adorno TW, Tiedemann R (eds) Gesammelte Schriften, 1st edn. Suhrkamp, Frankfurt am Main

Agrawal A (2001) Common property institutions and sustainable governance of resources. World Dev 29:1649–1672

Andrade I, Shweder R, Le Vine R (1993) Cultural theory – essays on mind, self and emotion. Cambridge University Press, Cambridge

Antweiler C, Mersmann C (1996) Local knowledge and cultural skills as resources for sustainable forest development. Bundesministerium für wirtschaftliche Zusammenarbeit und Entwicklung (BMZ) and Deutsche Gesellschaft für Technische Zusammenarbeit (GTZ), Eschborn

Arora V (2006) The forest of symbols embodied in the Tholung sacred landscape of North Sikkim, India. Conserv Soc 4(1):55–83

Barbien E, Baumgärtner S, Chopra K, Costello C, Duraiappah A, Hassan R, Kinzing AP, Lehman M, Pascal U, Polasky S, Perrings C (2009) The valuation of ecosystem services. In: Naeem S, Bunker DE, Hector A, Loreau M, Perrings C (eds) Biodiversity, ecosystem functioning, & human wellbeing. An ecological and economic perspective. Oxford University Press, Oxford, pp 248–262

Bargatzky T (1986) Einführung in die Kulturökologie. Umwelt, Kultur und Gesellschaft. Reimer, Berlin

Barnard A, Spencer J (eds) (2003) Encyclopedia of social and cultural anthropology, Reprint. Routledge, London

Berkes F (1999) Sacred ecology: traditional ecological knowledge and resource management. Taylor & Francis, Philadelphia

Berkes F (2006) From community-based resource management to complex systems. Ecol Soc 11 (1):45

Berkes F (2007) Community-based conservation in a globalized world. Proc Natl Acad Sci 104:15188–15193

Bleher B, Uster D, Bergsdorf T (2006) Assessment of threat status and management effectiveness in Kakamega Forest, Kenya. Biodivers Conserv 15(4):1159–1177

Boillat S (2007) Traditional ecological knowledge, land use and ecosystem diversity in the Tunari National Park (Bolivia). An ethnoecological approach for dialogue between traditional and scientific ecological knowledge. Ph.D. thesis of the Faculty of Sciences. Inauguraldissertation (Ph.D.). Supervision by Urs Stephan Rist, Wiesmann, Hans Hurni, Centre for Development and Environment (CDE), Institute of Geography of the University of Bern, Bern, p 487

Bottazzi P (2008) Linking “"Socio” and “"Bio” diversity: the stakes of indigenous and non-indigenous co-management in the Bolivian Lowlands. In: Galvin M, Haller T (eds) People, protected areas and global change: participatory conservation in Latin America, Africa, Asia and Europe. NCCR Perspective/University of Bern/Geographica Bernensia, Bern, pp 81–109

Bottazzi P (2009) Indigenous governance, protected areas and decentralised forestry: a comparative analysis of two Tsimane' territories in the Bolivian Lowlands. In: Rist S, Geiser U (eds) Decentralisation meets local complexity: local struggles, state decentralisation and access to natural resources in South Asia and Latin America. NCCR Perspective/University of Bern/ Geographica Bernensia, Bern, pp 155–189

Bottazzi P, Dao H (2013) On the road through the Bolivian Amazon: a multi-level land governance analysis of deforestation. Land Use Policy 30:137–146

Brechbühl U, Krieger D, Lesch W, Rey L, Thomas C (1995) Ökologie und Kulturwandel: Wort, Bild, Wert und Glaube als Vermittler zwischen Individuen und Gesellschaft. In: Fuhrer U (ed) Ökologisches Handeln als sozialer Prozess. Ecological action as a social process. Birkhäuser, Berlin

Caro V, Ewert A (1995) The influence of acculturation on environmental concerns: an exploratory study. J Environ Educ 26:13–21

Castillo FG (1988) Chimanes, Cambas Y Collas' Relaciones Interetnicas En las Tierras Bajas Tropicales Del Beni (Chimanes, Cambas and Collas Interetnical Relations in the Beni Tropical Lowlands). Escuela profesional Don Bosco, La Paz

Chartock S (2011) How movement strength matters: social movement strength and the implementation of ethnodevelopment policy in Ecuador and Peru. Stud Comp Int Dev 46:298–320

Chicchón A (1995) Faunal resource use by the Chimane of Eastern Bolivia: policy notes on a biosphere reserve. In: Sponsel LE (ed) Indigenous peoples and the future of Amazonia: an ecological anthropology of an endangered world. The University of Arizona Press, Tucson/ London, pp 225–243

Clastres P (1974) La société contre l'Etat. Recherches d'anthropologie politique. Minuit, Paris

Cowen M, Shenton RW (1996) Doctrines of development. Routledge, London/New York

Daillant I (1998) Ils sont comme nous, mais ... Relations de parenté et de genre entre Chimane et gens de dedans (They look like us, but ... kinship and gender between Chimane and inside people). Anthropol Soc 22:75–97

Daillant I (2003) Sens dessus dessous. Organisation sociale et spatiale des Chimane d'Amazonie bolivienne (Meaning above and below. Social and spatial organisation among Chimane of Bolivian Amazon). Société d'ethnologie, Nanterre

Darr B (2011) Die kulturelle Bedeutung des Waldes – eine Untersuchung anhand ausgewählter Kulturelemente von drei Ethnien in Westkenia. Ph.D. Technische Universität Dresden

Dax T (2001) Endogenous development in Austria's mountain regions: from a source of irritation to a mainstream movement. Mt Res Dev 21(3):231–235

Delgado F, Rist S, Escóbar C (2011) Desarrollo endógeno sustentable: camino para re-actualizar el “Vivir Bien” en el contexto de la revolución democrática y cultural de Bolivia. In: Farah I, Vasapollo L (eds) Vivir bien: ¿Paradigma no capitalista? Postgrado en Ciencias del Desarrollo de la Universidad Mayor de San Andrés (cides-umsa) de Bolivia y el Departamento de Economía de la Universidad de Roma “La Sapienza” de Italia., La Paz, pp 401–422

Descola P (2005) Beyond nature and culture. University of Chicago Press, Chicago. Par-delà nature et culture (English trans: Descola P). Gallimard, Paris

Devisch R, Crossman P (2002) Endogenous knowledge in anthropological perspective. In: Odora Hoppers C (ed) Indigenous knowledge and the integration of knowledge systems: towards a philosophy of articulation. New Africa Books, Claremont, South Africa, pp 96–125

Dittmann K (no year) Über den Begriff der Tradition. Vorüberlegungen zu den Schwierigkeiten mit einem Begriff. Available online http://www.holmespeare.de/tradition/etexte/ubertrad.html. Accessed on 9 Jan 2009

Eliasch J (2008) The Eliasch review: climate change – financing global forest. UK Government, London

Ellis R (1996) A taste for movement: an exploration of the social ethics of the Tsimanes of Lowland Bolivia. Ph.D. dissertation. St. Andrews University, St. Andrews

Ember CR, Ember M (2002) Cultural Anthropology, Tenth Edition. Upper Saddle River, NJ: Pearson Prentice Hall

Escobar A (1995) Encountering development: the making and unmaking of the third world. Princeton University Press, Princeton

FAO (Food and Agriculture Organization) (2010) Global forest resources assessment. 2010 main report. Forestry Department, FAO, Rome

Frow J, Morris M (1993) Australian cultural studies a reader. University of Illinois Press, Urbana

Galvin M, Haller T (eds) (2008) People, protected areas and global change: participatory conservation in Latin America, Africa, Asia and Europe. Geographica Bernensia, Berne

Göhler D, Cashore B, Blom B (2013) Forest governance and sustainable rural development. In: Pretzsch J, Uibrig H, Darr D, Auch E (eds) Forests and rural development. Springer-Verlag, Berlin, pp.

Habermas J (1970) Towards a theory of communicative competence. Inquiry 13:360–375

Haller T, Galvin M, Meroka P, Alca J, Alvarez A (2008) Who gains from community conservation? Intended and unintended costs and benefits of participative approaches in Peru and Tanzania. J Environ Dev 17:118–144

Hamilton P (2003) The enlightenment and the birth of social science. In: Hall S (ed) Formations of modernity. Polity Press, Cambridge, pp 17–58

Harmon D (2004) Intangible values of protected areas: what are they? Why do they matter. George Wright Forum 21(2):9–22

Harris M (1989) Kulturanthropologie – Ein Lehrbuch. Campus Verlag, Frankfurt am Main

Haverkort B, Rist S (2007) Endogenous development and bio-cultural diversity. The interplay of worldview, globalization and locality. COMPAS-CDE, Leusden, p 445

Haviland WA (1993) Cultural anthropology. Wadsworth/Thompson Learning, Orlando, USA

Hirsch Hadorn G, Bradley D, Pohl C, Rist S, Wiesmann U (2006) Implications of transdisciplinarity for sustainability research. Ecol Econ 60:119–128

Hobsbawm EJ, Ranger TO (1983) The invention of tradition. Cambridge University Press, Cambridge

Hoffmann A (2009) Auxilium-online. Lateinwörterbuch-Deutsch-Latein. CommTec-Softwareentwicklung. Available online http://www.auxilium-online.net/wb/formenanalyse.php. Accessed on 12 Oct 2009

Ingold T (2000) The perception of the environment. Essays on livelihood, dwelling and skill. Routledge/Taylor& Francis, London/New York

Irrgang B (2005) Work Ethics in Judaism, Christianity, Islam and African Religion. Paper presented at the DAAD workshop 28 May to 06 June, Tharandt

Kingsbury N (2001) Impacts of land use and cultural change in a fragile environment: indigenous acculturation and deforestation in Kavanayen, Gran Sabana, Venezuela. Interciencia 26:327–336

Lele S, Norgaard RB (2005) Practicing interdisciplinarity. Bio Sci 55:967–975

Löwy M (2005) What is ecosocialism? Capital Nat Soc 16:15–24

Maffi L (2005) Linguistic, cultural and biologic diversity. Annu Rev Anthropol 34:599–617

Magdoff F, Bellamy Foster J (2010) What every environmentalist needs to know about capitalism. Mon Rev 61:1–21

Martínez-Alier J (2012) Environmental justice and economic degrowth: an alliance between two movements. Capital Nat Soc 23:51–73

Mathez-Stiefel S, Boillat S, Rist S (2007) Promoting the diversity of worldviews: an ontological approach to bio-cultural diversity. In: Haverkort B, Rist S (eds) Endogenous development and bio-cultural diversity. The interplay of worldview, globalization and locality. COMPAS-CDE, Leusden, pp 67–81

McLaren P, Houston D (2004) Revolutionary ecologies: ecosocialism and critical pedagogy. Educ Stud 36:27–45

Mead GH, Morris CW (2000) Mind, self, and society. From the standpoint of a social behaviorist. University of Chicago Press, Chicago

Meadows DH (1972) The limits to growth; a report for the Club of Rome's project on the predicament of mankind. Universe Books, New York

Meadows DH (2005) Limits to growth the 30-year update, 2005th edn. Earthscan, London

Muhlinghaus S, Walty S (2001) Endogenous development in Swiss mountain communities: local initiatives in Urnasch and Schamserberg. Mt Res Dev 21(3):236–242

Murra J (1972) El "control vertical" de un maximo de pisos ecologicos en la economía de las sociedades andinas. Essai publié dans le tome II de la Visita de la provincia de Leon Huanuco (1562). Universidad Hermilio Valdizan, Huanuco

Omari CK (1990) Traditional African land ethics. In: Engel JR, Engel JG (eds) Ethics of environment and development. Global challenge, international response. The University of Arizona Press, Tucson

Onyango JC, Nyanja R (2004) Conservation of biodiversity in the East African tropical forest. Lyonia J Ecol Appl 7(2):151–157

Opotow S, Brook A (2003) Identity and exclusion in rangeland conflict. In: Clayton S, Opotow S (eds) Identity and the natural environment. The psychological significance of nature. MIT Press, Cambridge, pp 249–272

Ostrom E, Nagendra H (2006) Inaugural article: insights on linking forests, trees, and people from the air, on the ground, and in the laboratory. Proc Natl Acad Sci 103:19224–19231

Peterson RB (2004) Central African voices on the human – environment relationship. In: Gottlieb R (ed) This sacred earth. Religion, nature, environment, 2nd edn. Routledge, New York/ London

Pinedo-Vasquez MA, Sears RP (2011) "Várzea" forests: multifunctionality as a resource for conservation and sustainable use of biodiversity. In: Pinedo-Vasquez M, Ruffino M, Padoch LC, Brondízio ES (eds) The Amazon Várzea, Springer, Netherlands, pp 187–206

Pohl C, Rist S, Zimmermann A, Fry P, Gurung GS, Schneider F, Ifejika Speranza C, Kiteme B, Boillat S, Serrano E, Hirsch Hadorn G, Wiesmann U (2010) Researchers' roles in knowledge co-production: experience from sustainability research in Kenya, Switzerland, Bolivia and Nepal. Sci Public Policy 37:267–281

Pretty J, Adams B, Berkes F, Ferreira de Athayde S, Dudley N, Hunn E, Maffi L, Milton K, Rapport D, Robbins P, Sterling E, Stolton S, Tsing A, Vintinner E (2009) The intersections of biological diversity and cultural diversity: towards integration. Conserv Soc 7(2):100–112

Ray C (1999a) Endogenous development in the era of reflexive modernity. J Rural Stud 15:257–267

Ray C (1999b) Towards a meta-framework of endogenous development: repertoires, paths, democracy and rights. Sociol Rural 39:521–537

Reyes-García V, Huanca T, Vadez V, Leonard WR, Wilkie D (2006) Cultural, practical, and economic value of wild plants: a quantitative study in the Bolivian Amazon. Econ Bot 60:62–74

Reyes-García V, Vadez V, Tanner S, Huanca T, Leonard WR, McDade T (2007) Ethnobotanical skills and clearance of tropical rain forest for agriculture: a case study in the Lowlands of Bolivia. Ambio 36:406–408

Riester J (1993) Universo mítico de los Chimane. Hisbol, La Paz

Rist G (2001) Le développement: histoire d'une croyance occidentale. Presses de Sciences Polities. Paris, France
Rist S (2010) Que es la materia y el espíritu y como se relacionan? Desarrollo endógeno en Bolivia. In: Viceministerio de Medio Ambiente Biodiversidad y Cambio Climático y de Gestión de Desarrollo Forestal (ed) Construcción de la Sustentabilidad desde la visión de los pueblos indígenas de Latinoamérica. Ministerio de Medio Ambiente y Agua de Bolivia, La Paz, pp 179–187
Rist S, Dahdouh-Guebas F (2006) Ethnosciences – a step towards the integration of scientific and non-scientific forms of knowledge in the management of natural resources for the future. Environ Dev Sustain 8(4):467–493
Rist S, San Martín J, Tapia N (1999) Andean Cosmovision and self-sustained development. In: Haverkort B, Hiemstra W (eds) Food for thought – ancient visions and new experiments of rural people – COMPAS. Zed Books, London, pp 177–190
Rist S, Chiddambaranathan M, Escobar C, Wiesmann U, Zimmermann A (2007) Moving from sustainable management to sustainable governance of natural resources: the role of social learning processes in rural India, Bolivia and Mali. J Rural Stud 23:23–37
Rist S, Boillat S, Gerritsen P, Schneider F, Mathez-Stiefel S, Tapia N (2011) Endogenous knowledge: implications for sustainable development. In: Wiesmann U (ed) with an international group of coeditors Research for Sustainable Development: Foundations, Experiences, and Perspectives. Perspectives of the Swiss National Centre of Competence in Research (NCCR) North–South. University of Bern/Geographica Bernensia, Berne, pp 119–146
Russell AW, Wicksona F, Carew AL (2008) Transdisciplinarity: context, contradictions and capacity. Futures 40:460–472
Sack RD (1987) Human territoriality and space. Graduate School of Geography/Clark University, Worcester
Schech S, Haggis J (2000) Culture and development a critical introduction. Blackwell, Oxford
Schreurs MA (2012) Rio +20: assessing progress to date and future challenges. J Environ Dev 21 (2):19–23
Seeland K (1997) Nature is culture. Indigenous knowledge and sociocultural aspects of trees and forests in non-European cultures. Intermediate Technology Publications, London
Sneddon C, Howarth R, Norgaard RB (2006) Sustainable development in a post-Brundtland world. Ecol Econ 57:253–268
Sunderlin WD, Hatcher DJ, Liddle M (2008) From exclusion to ownership? Challenges and opportunities in advancing forest tenure reform. Rights and Resources Initiative, Washington, DC
Taringa N (2006) How environmental is African traditional religion? Exchange 35(2):191–214
UNCED (United Nations Conference on Environment and Development) (1992) Report of the united nations conference on environment and development. Rio de Janeiro, United Nations Department of Economic and Social Affairs (DESA), New York, 3–14 Jun 1992
UNESCO (2005) Convention on the protection and promotion of the diversity of cultural expressions. UNESCO, Paris
van Ploeg JD, van Dijk J (1995) Beyond modernization: the impact of endogenous rural development. Van Gorcum, Assen, Netherlands, 297 pp
Walker R (2003) Mapping process to pattern in the landscape change of the Amazonian Frontier. Ann Assoc Am Geogr 93:376–398
WCED (World Commission on Environment and Development) (1987) Our common future. Reprint edn. Oxford University Press, Oxford/New York
WCS (2005) Actualisacion del Plan de Manejo PL 2005–2009 (Actualized management plan PL 2005–2009). Wildlife Conservation Society, La Paz
Werth L, Mayer J (2008) Sozialpsychologie. Spektrum, Akademischer Verlag, Deutschland

Chapter 8
Extension Services for Rural Development

Dietrich Darr, Volker Hoffmann, and Simone Helmle

Abstract At the beginning of the twenty-first century, rural populations are faced with a number of opportunities that generations of farmers have only dreamed of. Rural extension services play a pivotal role in providing smallholders and other client groups with the information, knowledge, and qualifications required to exploit these emerging opportunities. Starting with a review of the evolution of concepts and paradigms in extension science and practice over the past decades, this chapter elaborates on the major cornerstones of successful rural extension work; viz. extension contents that help farmers understand the functioning of markets and improve market information and transparency; initiating and institutionalizing communication and horizontal exchange among farmers through group-oriented extension approaches; answering the manifold of information requirements and consulting needs of the diverse client groups; and designing organizational arrangements for extension that are equally effective and cost-effective. The chapter concludes by providing an outlook on the opportunities and limitations that extension services face in relation to rural development.

Keywords Advisory services • Agricultural innovation • Farmer groups • Market-oriented extension

D. Darr (✉)
Faculty of Life Sciences, Hochschule Rhein-Waal, Marie-Curie-Str. 1, 47533 Kleve, Germany
e-mail: dietrich.darr@hochschule-rhein-waal.de

V. Hoffmann
Institute of Social Sciences in Agriculture, Universität Hohenheim, Goethestr. 16, 72654 Neckartenzlingen, Germany
e-mail: vhoff@t-online.de

S. Helmle
Department of Rural Sociology and Extension, Justus-Liebig Universität Gießen, Senckenbergstr. 3, 35390 Gießen, Germany
e-mail: simone.helmle@agrar.uni-giessen.de

J. Pretzsch et al. (eds.), *Forests and Rural Development*, Tropical Forestry 9,
DOI 10.1007/978-3-642-41404-6_8, © Springer-Verlag Berlin Heidelberg 2014

8.1 Introduction

At the beginning of the twenty-first century, rural populations are faced with a number of opportunities that generations of farmers have only dreamed of: In most parts of the tropics and subtropics, smallholders are realising increasing exposure to market opportunities born from the increasing demand for high-value agricultural and forestry products in the growing urban centres. Specialized production technologies and advanced technical equipment have been developed that match the requirements of very specific ecological, economic and social conditions and that make farming and forest operations more safe, environmentally sound and profitable activities. Globalizing information flows, growing levels of international cooperation and exchange, and the increasing mobility of populations have contributed to strengthened civil societies that encourage political participation of previously marginalized population groups and foster entrepreneurial freedom. Yet, most of these opportunities come at the price of increased ambiguity and complexity. To name but few, smallholders that attempt to exploit the growing market opportunities are confronted with an unprecedented volatility of agricultural and forestry commodity prices, strict production standards, and tightened quality requirements prescribed by global wholesalers and processing companies; climate change and the more frequent incidence of extreme weather events lead to an increased risk of crop failures and require farmers to adapt their farming practices to these conditions; resource scarcity and degradation increasingly impact the potential for intensifying or expanding primary production; and the coalescing world results in changing lifestyles and preferences that tend to accelerate out-migration from rural areas and deeply affect the way of rural life. Navigating these challenges puts high demands on smallholders in terms of the contextual knowledge, specialized skills, and methodological competencies they require.

Rural extension services can play a pivotal role in providing smallholders and other client groups with the information, knowledge, and qualifications required to exploit the emerging opportunities while simultaneously balancing risks. Instructing farmers on more sustainable production technologies – or helping farmers to generate and disseminate these technologies; educating smallholders on the functioning of markets and strategies to mitigate the risks of increased market production; supporting women farmers to organize in marketing cooperatives or other groups that link them with input providers and wholesalers are some of the eminent contributions rural extension services can make. But the role of extension by far transcends the immediate effect of such services: In a broader sense, extension can help foster entrepreneurship thereby initiating and guiding the transformation of largely subsistence-based communities to more diversified rural economies (Kennedy and Thomas 1996; Saether 2010). It can enhance the capacity of farmers for critical inquiry and for taking action in the pursuit of a more democratic and liveable world (Kroma 2006); and is thus an important accelerator of rural and community development (Hoffmann et al. 2009b).

So far, most of the work in extension science and practice has focussed on the improvement of agricultural practices and the promotion of cropping and livestock technologies. Extension advice specifically tailored to forest management or farm forestry development has received comparatively little interest. Kress (2001) explicates a number of distinctive features that differentiate forestry extension from other extension work, among them:

- The long time horizon that the extension advice needs to take into account, as the lifetime and growing cycle of forests usually extends over generations;
- Legal restrictions imposed on the utilization and management of forest resources by local communities;
- The predominantly collective ownership of forest resources that requires extension services to develop and employ methods suitable to address groups of farmers and entire communities rather than individuals;
- The central role of women in utilizing and maintaining forest resources in rural areas;
- The difficulties in motivating resource-poor farmers to participate in long-term forest management activities;
- The low economic value of most forest products and marginal profitability of forest management, explaining why forest management activities are of only secondary importance for most smallholders;
- The necessity for extension agencies to also consider the non-productive forest functions;
- The large number of societal demands and influences on the forest resources;
- The requirement for forest extension agents to undergo trainings that enable them to appropriately address the multi-disciplinarity of topics.

Despite these very specific frame conditions of forestry extension, the strict separation of forestry from agricultural extension work is not very helpful. Many pertinent questions in the extension sphere are more general in character and thus refer to agricultural and forestry extension alike. Such questions comprise, inter alia, the quest for extension approaches and systems that are equally effective and cost-effective, culturally adapted, and exploit the benefits of modern communication technologies. More fundamental questions concerning strategic objectives refer to operating models and delivery modes, breadth of contents and client groups, and the sustainability of extension interventions per se.

Moreover, there are a number of arguments in support of integrated extension services that take a more holistic and cross-sectoral perspective: the composite nature of most smallholder farm systems that integrate various undertakings and activities; the complex relationships, interactions and repercussions among the single elements of the ecosystem; the inter-sectoral nature of the rural development challenge; and the highly diverse range of service needs of the various client groups and strata of the farming population, for example. The points raised by Kress, then, no longer represent features unique to forestry extension but constitute the basic conditions that are pertinent for rural extension services in general. Yet, much of the traditional work on agricultural extension fails to appropriately address these aspects.

This chapter aims to trace the evolution of concepts and paradigms in extension science and practice over the past decades to provide the foundation of the contemporary understanding of predominating extension contents, approaches and organizational arrangements appropriate to address the challenges of rural development. The chapter concludes with an outlook on the opportunities and limitations that extension services face in relation to rural development.

8.2 Some Definitions

Except in the sections that describe the more conventional understanding of these concepts, the following comprehension shall be used throughout this book:

Extension denotes the process whereby the extension agent seeks to enhance the client's capability to solve his acute problems. This largely involves providing content expertise and process facilitation during the cycle of joint problem definition, identification of alternative options, selecting, and implementing the most appropriate solution. This also involves providing information and facilitating linkage with markets and stakeholders of the value chains in order to enhance rural entrepreneurship. This requires a trusted and reciprocal relationship between the advisor and the client, with the advisor being solely committed to the welfare of his client. The client preserves the freedom to make decisions and to assume personal responsibility for any actions, because he or she alone must ultimately bear the consequences of these actions (adapted from Hoffmann et al. 2009b).

Innovations are products, equipments, services, practices, methods, ideas, and socio-organizational or political arrangements that often emerge in response to a distinct problem and that usually imply new forms of co-ordination among stakeholders (Leeuwis and van Ban 2004). Innovations are dynamic and undergo a constant process of change and gradual improvement by various stakeholders (cf. Kroma 2006; Reij and Waters-Bayer 2001). Innovations are not necessarily entirely new, but novel in a particular place and context (OECD 2005).

Social networks denote a set of social relationships of different strength that link a set of actors, and effectively convey role model effects, persuasive forces, opportunities for confirmation, and other forms of social influence required to promote the diffusion of innovations. Communication networks are the patterns of contact that are created by the flow of messages among communicators through time and space (Monge and Contractor 2003).

Farmer organizations denote formal organizations representing small-scale farmers who produce a range of commodities (ISNAR 1994). They mainly comprise national or regional level federations, associations and producer organizations that aim to handle the supply of farm inputs and/or marketing of particular commodities.

Farmer groups represent one distinct category of farmer organizations at the local and community level (Heemskerk and Wennink 2006; ISNAR 1994). Following the definition of groups, farmer groups are social aggregates of a minimum of 5–7 (Ellis and Fisher 1994) individuals who possess a common identity, with the same goals and objectives, share a common fate, exhibit structured patterns of interaction and modes of communication, hold shared perceptions about group structure, are personally and instrumentally interdependent, reciprocate interpersonal attraction, and consider themselves to be a group (Carron and Hausenblas 1998, cited in Carron and Brawley 2000).

8.3 Evolution of Concepts and Lines of Thought

8.3.1 From Linear to Complex: Concepts of Innovation Generation and Diffusion

The conception of a linear research-development chain prevailed during most of the previous century. This view postulates a clear division of tasks between the various stakeholders engaged in technology generation and application: Researchers generate technologies through fundamental and applied research, extension agencies persuade farmers to adopt the technology packages by creating awareness, interest and know-how, and smallholders and commercial enterprises ultimately apply the technologies generated (Leeuwis and van Ban 2004; Röling 1995). Inherent in this conception was the thinking that innovations are largely technical improvements of current practices, spreading among target audiences largely mechanistically as described in the diffusion of innovation theory (Rogers 1962). This theory postulates that the innovation diffusion process, represented by the cumulative number of adopters over time, typically takes a sigmoid growth curve. Based on the assumption that innovativeness, as other human traits, is normally distributed among humans, populations are classified into distinct adopter categories[1] that typical characteristics are ascribed to. In addition to these characteristics, the properties of the innovations, type of the innovation-decision, nature of the social system, and type of extension interventions have been postulated to influence the diffusion process.

Yet, accumulating evidence suggested that the predictions made by the linear model were largely not supported by empirical observations. For instance, there exist many examples where farmers change, adjust and improve technologies and practices without the involvement of scientists thereby independently generating locally adapted innovations (see, for example, Reij and Waters-Bayer 2001; Hoffmann et al. 2007). It is now largely undisputed among innovation scholars that rather than in a linear fashion, innovations emerge through an iterative process of technical and institutional change that necessarily involves a mix of stakeholders,

[1] I.e., innovators, early adopters, early majority, late majority, and laggards (Rogers 2003).

requiring new forms of coordination (e.g. Leeuwis and van Ban 2004; Rametsteiner and Weiss 2004; Kroma 2006; McIntyre 2009; Röling 2009). Innovations are no longer perceived as being merely technical in nature but as containing novel social, organizational, and political arrangements as well. Mumford (2002), for example, describes ten social innovations that emerged between 1726 and 1757 in Philadelphia. More recent examples of innovative socio-political and organizational practices and arrangements from rural development comprise, inter alia, the enactment of participatory planning procedures, the establishment of micro-finance institutions and grain banks, or the inception of village resource user groups. Finally, the diffusion of innovation theory proved to be an over-simplistic model of reality (Geroski 2000; Frenzel Baudisch and Grupp 2006). Vanclay and Lawrence (1994) pointed out the fact that the diffusion of innovation is not a suitable framework for environmental innovations due to the external effects usually associated with this type of innovation. Hoffmann (2007) criticises the largely empiricist nature of the framework that lacks a sound theoretical foundation. Table 8.1 provides an overview on the evolution of models of innovation during the past 60 years.

In contemporary rural extension science, system models of innovation, of which Röling and Engel's agricultural knowledge information system (AKIS) is probably the currently most pre-eminent example, have superseded linear models of innovation (Röling and Engel 1991). The concept denotes the set of stakeholders involved in the generation, transformation, dissemination, and utilization of relevant knowledge and information, and their links and interactions (Röling 1990, cited in Röling and Engel 1991). Innovation systems grant conditions conducive for innovation generation and dissemination by fostering cooperation and managing conflict among stakeholders; providing a platform for learning and experimentation; reducing uncertainty; developing an infrastructure for strategy and vision development; and creating incentives (Kubeczko and Rametsteiner 2002; Rametsteiner 2010).

In line with the advancement of innovation theories, the diffusion of innovation theory has also been replaced by more sophisticated conceptions of innovation spread. Building on accumulating evidence that human behaviour, attitudes, values, and norms are socially constructed, modified and transferred through interpersonal relationships (e.g., Erickson 1991; Friedkin 2001; Stahr 2001), social network theory and related concepts have been increasingly used to investigate and explain the diffusion of innovations. Communication network analysis (e.g., Monge and Contractor 2003; Valente 2006) represents the sub-section of social network analysis commonly applied to the diffusion of innovations. Recent work in the field of agricultural and agroforestry innovations comprise, inter alia, Conley and Udry (2001, 2005), Darr (2008), Raini et al. (2005), and Spielman et al. (2010). The diffusion of forestry-related innovations has, to our knowledge, not been studied using the social network framework.

Table 8.1 Evolution of models of innovation

Innovation model	Impetus	Agenda	Theoretical foci and predominant concepts
1950s Black box models	Explain innovation as an outcome of resource use – but without explaining the process of innovation itself	Input and output of R&D sector	'Big science': scientific autonomy and independence through public funding
1960–1970s Linear models	Explain innovation as a step-by-step process that culminates in adoption by final users	Success factors and failure causes for innovations; diffusion of innovations; market failure as justification for public investments in R&D	Technology-push and need-pull models; barriers to innovation
1980s Interactive models	Replace sequential conception of innovation process by a more complex understanding stressing interacting stages and feedback loops	Factors and players involved in innovation; communication and collaboration during the innovation process	Innovation cycles, technological gaps
Early 1990s System models	Explain the success of small firms in innovation as a result of their interactions and cooperation with other stakeholders	Innovation networks; interactions, interconnectedness and synergies in innovation networks	National systems of innovation (e.g., Freeman 1995); Agricultural knowledge information system (Röling and Engel 1991)
Late 1990s Evolutionary models	Explain success and failure of technologies as a consequence of selection processes and the influence of the surrounding environment	Variety of innovations; fitness, adaptation, and selection of innovations; competition and collaboration; influence of the external environment	Innovation avenues, technological trajectories
1990–2000s Innovative milieu models	Explain success of highly innovative small and medium enterprises that do not possess resources to maintain aggressive R&D strategies	Influence of territorial and local factors on innovation process, e.g. support networks, working environment, and social, cultural and natural characteristics (e.g. climate, quality of life, etc.)	Innovation clusters/geographic proximity (Porter 1990); collective learning

Source: Adapted from Marinova and Phillimore (2003)

8.3.2 From Production Focussed to Facilitating Linkage: Extension Paradigms, Strategies, and Systems

Driven by the linear conception of the innovation process, the transfer of technology[2] paradigm has been the dominant view in extension science and practice for decades. The concept is rooted in (1) a realist and positivist stance in science, i.e. the belief in scientific progress; (2) the premise of a continuing productivity increase, which leads to scale enlargement, intensification, and specialization; and (3) diffusion of innovations as a key mechanism for scaling up (Röling 2002). Hence in this thinking, the primary purpose of rural extension services has been to aid general economic development and industrialization policies, such as the increase of primary sector production through the promotion of modern farm and forest management technologies, or the increase of governmental revenues from export commodities such as coffee, rubber, timber, or palm oil (Rivera 1991). Heavy investments in agricultural and forestry research and technology development; the creation of large governmental extension services; and the promotion of large-scale and inflexible extension systems such as 'T&V'[3] by international donors were direct manifestations of this paradigm.

Growing concerns with the cost of financing public extension and its apparent failure to meet broader environmental and development goals sparked increasing criticisms of the paradigm, starting from the 1980s (McIntyre 2009; Rivera 1991). The marginalization and ignorance of farmers' local indigenous knowledge (Vanclay and Lawrence 1994) were increasingly regarded as the cause of the limited effectiveness of public extension efforts. The structural adjustment programs implemented by World Bank and IMF additionally increased the pressure to abandon ineffective public agricultural and forestry extension services and the paradigm they had been based on in a number of developing countries (e.g., Farrington 1994). Table 8.2 provides an overview on the evolution of the extension paradigms and predominant concepts.

Based on the conviction that farmers are active problem-solvers who continuously change, improve, and adjust technologies to local conditions; that successful technological innovations usually need to be accompanied by appropriate social and institutional arrangements; and that innovations emerge not from research alone but through the interaction of a multitude of stakeholders, extension paradigms have increasingly acknowledged the contribution of farmers and other

[2] Also called 'linear model' (e.g. Kline and Rosenberg 1986), 'pipeline model' (e.g. Biggs 2007), 'persuasive transfer' (Leeuwis and van Ban 2004), or 'technology supply push pathway' (Röling 2009).

[3] 'Training and Visit', hierarchically organized extension system mainly promoted by the World Bank and adopted by many governmental agricultural extension services throughout the developing world. The system was based on a rigid schedule of extension staff trainings and fortnightly farmer visits, mainly of innovative contact farmers, strong linkages with research stations, and supported by agricultural input and credit supply (Benor et al. 1984).

Table 8.2 Evolution of extension paradigms

Extension paradigm	Impetus	Agenda	Theoretical foci and predominant concepts
Pre-1960s Transfer of technology	Extending research findings from laboratories to farmers' fields	Increasing agricultural and forestry production through introducing modern technologies	Message transmission – transmission model (Shannon and Weaver 1949)
1960s Improving technology transfer	Increasing farmers' awareness in order to improve poor uptake of technology	Educating farmers to establish market-orientated farm enterprises	Adoption behaviour – diffusion of innovations theory (Rogers 1962), adult learning
1970s Farming systems research	Improving fit of technologies to farmers' heterogeneous environments	Creating innovations oriented to farming systems	Holistic thinking, systems thinking
1980s Participatory technology development	Tackling inappropriate technologies and inequity among target group by involving farmers in the innovation process	Exploring and building on indigenous knowledge, ensuring participation of powerless population strata (e.g., women, poor)	Power, community development, gender, organizational learning, group work, team building
1990s Facilitating learning	Improving understanding of ecological and social processes and general principles	Fostering ecological sustainability and social justice by changing institutions and relationships	Communicative rationality, systems thinking, cognitive processes
2000s Facilitating linkage	Facilitating emergence of locally adapted (social and technological) innovations by managing interactions at interpersonal and institutional level	Improving linkage, coordination, and (horizontal) knowledge exchange among extension stakeholders	Agricultural knowledge information systems (Röling and Engel 1991), social networks, farmer groups

Source: Adapted after King (2000), cited in Holding Anyonge (2002)

stakeholders in the innovation process since the 1980s (Leeuwis and van Ban 2004; Röling 1995). Building on the systems model of innovation, the focus of the current paradigm is on facilitating the linkage and knowledge exchange at the interpersonal level in farmer groups and organizations, as well as at the level of stakeholders and institutions of the entire innovation system.

After disbanding the hegemony of the traditional public extension services, an increasingly complex and pluralistic extension landscape has emerged. This consists of a multitude of private sector and not-for-profit extension service providers and in some countries also includes revitalized public extension agencies (Rivera and Alex 2004). This development followed a diversification of extension objectives, exceeding the narrow focus on production and productivity increase promoted under the traditional paradigm. Examples include promoting learning and knowledge exchange among farmers, enhancing the general problem-solving ability of smallholders, fostering general human resource development, and building institutions and institutional capabilities (Rivera 1991; Swanson 2008). Leeuwis and van Ban (2004) distinguish six major goals that extension interventions can pursue, and identify appropriate strategies that suit the particular purpose (see Table 8.3). Numerous extension systems have been developed and implemented in the field. With their emphasis on facilitating farmer problem solving, improving the link between stakeholders, or enhancing farmer empowerment, these systems fundamentally differ from the extension work under the traditional paradigm of persuasive technology transfer.

Providing cutting-edge knowledge and expertise to farmers who actively seek specialist advice for a specific problem; guiding their problem-solving process; and permanently enhancing their problem-solving ability by imparting suitable methodologies and frameworks are the major goals of advisory extension interventions. Typical examples comprise, inter alia, recommending to a woman farmer a number of tree species that she could grow on her farm-yard; instructing a forest user group how to apply basic forest inventory methods to determine the allowable cut and monitor the amount of timber extracted by the group members; or advising the manager of a forest holding on how to increase revenues from timber sale by optimizing grading of timber. The strategy is thus equally appropriate in situations where clients are individual farmers who are highly specialized and technologically advanced (Röling 1995), or where groups of farmers are concerned with the management of communal resources.

The core objective of the facilitation strategy is to encourage farmers' experimentation and to support them to exchange their knowledge, experiences, and observations with their colleagues thereby promoting the spread of locally adapted innovations (Gerster-Bentaya and Hoffmann 20011). This typically refers to group extension settings. The Farmers' Forest Management Schools established in Nepal as a response to the farmers' interest in learning more about community forest management practices in groups is one example of this strategy (Singh 2003). Through facilitating participatory and discovery learning approaches, group dynamic exercises, and regular interactions, the extension worker aims to build the farmers' capacities and enhance their self-confidence. This ultimately makes them less dependent on outside expert knowledge and external inputs when managing their resources. Thus, the strategy is particularly appropriate in circumstances where the solution is highly site-specific, requiring profound local knowledge that external experts usually do not possess (Röling 1995; Leeuwis and van Ban 2004). An example of this would be the conversion from conventional logging to community forest management or even watershed management.

In situations where appropriate solutions to the problem at hand do not readily exist, building coherent innovations becomes the main purpose of the extension intervention. For example, the short supply of high quality seedlings of native high value tree species at affordable prices is a common problem for many forestry projects. Developing simple propagation methods that can easily be applied in village nurseries, where methods of modern germplasm research are usually

Table 8.3 Different intervention goals, appropriate strategies, and typical extension systems

Intervention goal	Extension strategy	Role of extension worker	Role of client(s)	Key processes involved	Typical extension systems
Aid policy objectives mainly related to productivity increase	Persuasive transfer of innovations	Strategically manipulate the clients' behaviour	'Unexpecting' receiver (initially)	Adoption, acceptance	T&V
Assist client to solve problem/ enhance problem solving ability; focus on whole farm	Advisory communication	Consultant/counsellor (expertise and/or process guidance)	Active problem owner	Problem solving, counselling	Socio-economic advisory work in German smallholder forestry
Enhance knowledge exchange; enhance diffusion of innovations	Facilitation/support horizontal knowledge exchange	Expert, facilitator	Active learners/ source of experience	Learning, networking, problem solving	Farmer field/forest management schools, Farmer-to-farmer extension
Create coherent innovations	Facilitate process of innovation generation	Facilitator, resource person, supporting vertical knowledge flow	Active participants	Problem solving, social learning, network building, negotiation	Participatory technology development, Farmer research groups, Farmer-led experimentation, Farmer innovation approach
Form organizations that take over initiative; strengthen position of a group	Support organization development and capacity building	Organizer, trainer, facilitator	Active participants	Social learning, negotiation	Participatory action research
Manage pre-existing conflict	Conflict management	Mediator, facilitator	Stakeholder participant	Negotiation, social learning	

Source: Adapted from Leeuwis and van Ban (2004); Hoffmann et al. (2009a)

unavailable, represent a significant challenge. Extension workers could organize the process of generating these methods. In addition to solving the technical aspects of how best to produce the seedlings, the innovation also requires new organizational, socio-cultural and/or political arrangements, such as funding of research work by donors, village-level collective action to establish and operate the nursery, or cooperation with private-sector stakeholders for input supply and marketing. Generating this innovation would therefore require the interplay of multiple actors – such as, farmers, researchers, development practitioners, operators of private nurseries and local authorities. Ensuring the farmers' participation in experimentation and testing the germination practices; facilitating the interaction of the stakeholders, e.g. by forging links between the community and private-sector tree nurseries for knowledge exchange; agreeing implementation plans and time-lines; overseeing the adequate provision of resources; and creating platforms that foster stakeholder coordination and mutual understanding are some of the critical tasks to perform by the extension worker in such context. Group as well as individual extension approaches can apply in the scope of this strategy.

Based on the conviction that most innovations require adequately functioning farmer and community organizations, such as forest user groups, farmer innovation circles, marketing cooperatives, etc., establishing the organizations that are required to make the intervention successful or sustainable, or strengthening the position of a particular group towards other stakeholders, e.g. traders or exporters, can also be the objective of an extension intervention. Organization development and group capacity building is the appropriate extension strategy in such contexts. By working with farmer groups, this strategy aims at strengthening a particular group's capacity to innovate, to help themselves, and/or to make claims towards local authorities or more powerful stakeholder groups. Extension workers in such situations typically initiate the establishment of groups and organizations, contribute to group activities and processes, provide training in organizational skills, and facilitate processes of organizational change (Leeuwis and van Ban 2004).

Managing pre-existing conflict is the appropriate intervention strategy in circumstances where unresolved disputes, e.g. over access to natural resources, have potentially negative effects on the well-being of the community and/or the state of the natural resources. This contemporary extension strategy has the least in common with the traditional understanding of extension services. Typical conflicts arise between sedentary farmers and pastoralists over the use of communal grasslands or water resources; over the extraction of firewood and timber from the village forest by members of a neighbouring community or forest user group; or between the forest authority and forest dwellers around the legitimacy of their traditional forest use rights. In such situations, the neutral extension worker facilitates in creating a negotiation platform, institution building and organizational development, and mediation that aims to support the stakeholders in finding innovative and more productive solutions to the conflict. Usually this involves working with groups of stakeholders rather than with individual farmers. However, in order to be successful mediators, extension workers require specific skills and insights that are not yet widely available (Leeuwis and van Ban 2004).

Regardless which of these strategies is identified as most appropriate in a particular context, there is widely shared consensus among extension researchers and practitioners that extension interventions need to address the farmers' demands – or even stimulate the demand for the innovation[4]; respond to changing knowledge needs; and assume a systemic perspective that considers the economic, social and environmental consequences of the innovation along the entire value chain.

8.4 Cornerstones of Successful Rural Extension Work

8.4.1 Extension Contents: The Need for Market-Oriented Advice

Farmers and tree growers usually face a number of challenges that concern technical, environmental, managerial, and other aspects of their management activities. While they successfully overcome many of these problems themselves, some might be rather complex, requiring specific expertise to solve. Determining growth rates and the maximum allowable cut of natural forests and timber plantations; preparing detailed management plans that meet the legal requirements; selecting the tree species most appropriate to the prevailing environmental conditions and management objectives; identifying the necessary management practices and most economic working methods for thinning, pruning, timber harvest, and other operations; or optimizing the grading of timber and other forest products after harvest are some examples. While such needs create a strong demand on the side of farmers for clearly defined advisory services and extension contents, past rural extension work has not always been successful in responding to these specific needs.

Extension advice traditionally focused on introducing more productive and sustainable farm and forest management practices, where technological aspects of these innovations have been the focus of extension work for decades. Relatively little attention was paid to the economic context the farmers were operating in, neither with regard to the micro-economics of the farm households nor regarding the market demand and quality requirements for farm products. Yet, economics have a significant impact on farmers' management objectives and choice of practices. This is evident in the influence on farmers' livelihood strategies, through

[4] Pointing at the trade-off between demand-led and persuasive extension, Garforth (2004) holds the view that extension advice needs to be demand-driven, as long as farmers pay the full price for the services; and that if governments or other public stakeholders provide extension services as quasi public goods, it is legitimate for the extension service to support specific policy objectives. Yet, the challenge remains that farmers are unlikely to adopt innovations on a sustainable basis that do not match their needs. Stimulating the demand for the innovation, e.g. by creating successful demonstration examples, helping farmers to access markets, or by creating legal frame conditions that enable – or if necessary also stipulate – innovation adoption, is likely to be the more successful approach.

determining the optimal combination of assets, resources, and off-farm activities in a certain farming system; the optimal balance of risks and security measures; and the appropriate level of intensification and market integration. Most field advisors are not familiar with, or trained to help farmers in, solving these kinds of problems (Hoffmann et al. 2009b).

Aspects such as the size of the market for the introduced products, the availability of storage and processing facilities, or negotiation and marketing skills required by farmers to profitably sell their produce have rarely been addressed by rural advisors. Extension services that operated under the transfer of technology paradigm, and likewise progressive intervention projects with a strong local or sector focus, were particularly prone to insufficient market orientation. Case in point is where farmers could not sell their produce due to lacking markets, products spoil because of a lack of transport infrastructure, or markets collapse after too many farmers had jumped on the bandwagon (cf. Swanson 2008). Non-adoption has often been the rational response by smallholders to the introduction of such innovations that largely disregarded the economic realities.

Decades of trade liberalization have significantly increased the exposure of most smallholders to market forces. The globalizing agribusiness and forestry industries and trade in these sectors' commodities have created demand for farm and forestry products in regions largely cut off from the global economy until recently. Lifting farmers out of subsistence economies through encouraging market production has become a common rural development strategy (World Bank 2007). Smallholders are thus increasingly exposed to market risks, face volatile input and commodity prices, and are confronted with the bargaining power of global market players. In order to enable farmers to take proper advantage of the arising opportunities, the need for extension services emerges that help them to better understand the functioning of markets, assess the economic risks and opportunities of certain products, link farmers to insurance and other risk-reducing mechanisms, and improve market information and transparency.

Market-oriented advisory services (Neuchatel Group 2008; Swanson 2008) aim at unlocking rural entrepreneurship and creating income opportunities for smallholders through market development and improving their market access. Given the poorly developed value chains for most agriculture and forestry products in the tropics, these services primarily focus on facilitating and brokering between different value chain actors, establishing transparency, and enhancing the reliability and efficiency of the interactions. This requires extension agencies to expand their traditional scope from working primarily with farmers to addressing the stakeholders along the entire value chain that can contribute to overcoming deficiencies and bottlenecks. Such actors include producers, input and commodity traders, wholesalers, processors, export companies, providers of infrastructure, or financial institutions. Market-oriented advisory services also require the provision of more integrated extension advice that encompasses enhancing farmers' technical, economic and marketing know-how including, inter alia, post-harvest handling, contract negotiation, and brand development; imparting managerial skills and initiating the development of self-help organizations such as producer groups or marketing cooperatives; providing market information; and facilitating linkage and negotiation among the stakeholders of the entire value chain. This is not to say that a single unified extension service will have to cover all the client groups and topics mentioned. What is necessary is an increasing awareness of the need for market-oriented advice among all stakeholders of the extension sector.

With the objective of establishing functioning markets and improving the efficiency of agricultural and forestry value chains, thereby fostering rural development, market-oriented extension services can be regarded as a core domain of governmental action. A variety of organizational models, ranging from public, public-private, entirely private sector, or community extension services can contribute to this aim. In their most comprehensive form, these services can trigger the transformation of the entire regional or even national agricultural or forestry sector.

8.4.2 Extension Approach: Group Versus Individual Extension

Traditional and modern methods of mass communication such as television and radio programs, print products, mass events like trade fairs and field days, as well as internet and mobile phone services continue to be useful in extension work as far as simple messages are communicated. Yet, in a globalizing and ever growing complex world, extension messages become increasingly differentiated and complicated. Extension agents can best support farmers to successfully cope with uncertainty and volatility, to interpret abstract and partly contradictory information, and to derive concrete actions that apply to their own situation through personal face-to-face interaction with farmers in individual or group extension settings.

8.4.2.1 Why Group Extension Is the Approach of Choice

As a consequence of the commodity orientation of agricultural extension services during the colonial era, and of the productivity growth and technology focus after independence, vast agricultural and forestry estates as well as progressive and large-scale farmers have been the primary targets and recipients of rural extension services for decades. Extension systems such as T&V have been built on the contact farmer concept, according to which, extension agents primarily work with few individual farmers who were expected to share their experiences and knowledge with other farmers (Benor et al. 1984). However, the secondary transfer of information from the contact farmers to the communities has been much less successful than predicted, and adoption rates have commonly been very low among non-contact farmers (Röling and Pretty 1997). Addressing groups of peasants rather than individual farmers, therefore, has been one modification to the T&V, which aimed at increasing the number of farmers receiving face-to-face service from extension, thus alleviating the limited technology diffusion from contact farmers (e.g. Wuyts-Fivawo 1996; Anderson et al. 2006). Yet, the unified and rather simple messages were inappropriate to the highly diverse world of rainfed farming and did not fit the specific demands of the small-scale farmers (Hoffmann et al. 2009a).

With the onset of system models of innovation and the shift to facilitative extension systems, rural extension services increasingly focused on small-scale farmers and their traditional institutions. Initiating, intensifying and institutionalizing communication and horizontal knowledge exchange among farmers have become core tasks of extension work (e.g. Scarborough et al. 1997; Selener et al. 1997). The following advantages are commonly attributed to group-oriented extension approaches (e.g., Belay and Abebaw 2004; Feder et al. 2010; Hagmann et al. 1999; Hoffmann et al. 2009b):

- Reduced transaction cost of providing extension by exploiting the scale effects and farmer linkage mechanisms that these groups offer;
- Increased extension effectiveness based on a higher level of farmer participation and the reinforcing effects of group learning and group action;
- Improved accountability of extension providers by making them directly accountable to the members of the group;
- More sustainable extension interventions as a consequence of organizational development and capacity building that help institutionalize communication and collaboration patterns among farmers;
- Opportunity to address the free rider problem of public-good extension by enabling some cost recovery at local level.

Historical cases of technology diffusion, as well as contemporary empirical evidence, suggest that many rural innovations disseminate through the farmers' ordinary social relationships and networks. These relationships and networks usually are not a target of extension interventions. Through their interactions, farmers are exposed to information, attitudes, and the behaviour of their peers, thus increasing the likelihood that they will develop beliefs, assumptions, and attitudes similar to those of their fellow farmers (Monge and Contractor 2003).[5] Darr (2008) for example, found that relatively simple agroforestry innovations, such as live fencing, spread without the intervention of extension agents. This is merely a product of day-to-day farmer interactions. The comparative advantage of group-oriented extension particularly lies in activities requiring collective action, such as natural resource management and pest management (Feder et al. 2010). Many accounts of group-oriented rural extension interventions have been successful (e.g. Heemskerk and Wennink 2004; Noordin et al. 2001; Simpson and Owens 2002; Wambugu et al. 2001). Yet, these also face a number of challenges, which include weak dissemination of technologies within and beyond the targeted population (e.g., Davis 2006; Tripp et al. 2005) or only a marginal increase of agricultural

[5] This mechanism of innovation diffusion is commonly referred to as 'contagion by cohesion' (e.g., Valente 1999) or 'relational proximity' (Rice 1993).

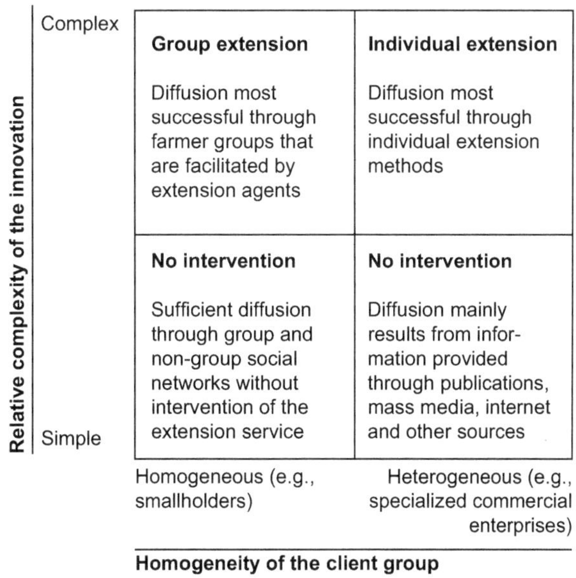

Fig. 8.1 Most appropriate extension approach as determined by the client group and innovation complexity (Source: Authors' elaboration)

productivity (e.g., Feder et al. 2004); the failure to benefit low-resource and powerless farmers (Carney 1996); and high demands in terms of the interpersonal and social skills and competencies required by the extension agents, in order to manage group processes and mediate conflicts, for example (see Box 8.1).

Group-oriented rural extension is deemed less suitable in situations where there is a complex technological problem to solve, or a technologically progressive and highly specialized client group require advanced and specific expert advice (Feder et al. 2010); where competition among the enterprises limits the scope for cooperation in groups[6]; or where a dispersed location or other socio-cultural and economic limitations constrain collective interactions. Thus, with increasing farm development and specialization of smallholders in the tropics their need for situation-specific and individually-tailored extension advice will certainly increase. Individual extension may be the more appropriate choice under such circumstances (see Fig. 8.1).

Thus, group extension is clearly not a panacea and 'one size fits all' solution. Identifying the most appropriate intervention, which is in accordance with the prevailing socio-cultural, environmental, and economic conditions, remains a difficult task in rural extension work. Rural extension work in most parts of the developing world is still mainly about promoting the adoption of comparatively basic technologies by culturally and socio-economically homogenous subsistence

[6] Yet, Rosenfeld (2001) reports that, promoting inter-firm networks through working with groups of enterprises has been a successful approach to foster rural development in a number of European countries and the US. Among the most important benefits of cooperation for members were reduced costs, access to information and ideas, and solutions to common problems.

farmers, rather than about giving specialist advice to highly advanced farming entrepreneurs. Given the obvious advantages and empirically proven success of the group approach under these circumstances, extension services should aim to work with existing farmer groups, actively promote group establishment in settings where they are lacking, and facilitate farmer-to-farmer knowledge exchange across social and spatial divides. Ensuring that the majority of rural households have access to basic extension services through their farmer groups, additional specialist advice could be provided, on request, to individual households belonging to minor or marginalized subsections of the population, such as particularly vulnerable households or highly advanced farming entrepreneurs.

Box 8.1 Failing to Reap the Benefits from Group-Oriented Extension: The Case of the 'Integrated Natural Resource Management in Ukambani' Project in Kenya (2005)

The Forest Department of Kenya and the Belgian Technical Cooperation jointly implemented the 'Integrated Natural Resource Management in Ukambani' Project (INRMU) in four districts of the Eastern Province between 1997 and 2005. Through sustainable management and use of natural resources, the project aimed to increase food security and income of the farmers living in the project area (INRMU 2005; van den Abeele and Macharia 2001). To this end, INRMU addressed the development and application of advanced technologies for intensive dry-land farming through efficient water use and soil conservation techniques. Extension packages consisted of technical advice, provided through regular farm visits by the extension officers, a number of highly advanced technical inputs such as soil conditioners and water cisterns made of corrugated iron sheets, and an attractive cost-sharing arrangement. Through these packages the project aimed to encourage farmers to adopt the project innovations and to commercialize their activities (INRMU 2005). Farmer-owned businesses, such as a sawmill, a honey processing centre and a horticulture marketing cooperative, had been set up in order to allow farmers to sustain their commercial activities after project conclusion.

36 extension officers of the Forest Department and a number of untrained frontline extensionists constituted the field-level personnel of the project. The area of responsibility of each extension officer extended over approximately three administrative divisions. Each extension officer had been equipped with a project motorcycle.

Due to the long tradition and cultural importance of farmer groups in the area (Tiffen et al. 1994), self-help groups, mainly of women, originally had been identified as the most suitable vehicle for extension (INRMU 2005). 46 groups had been provided with training, material and credit for project implementation during the pilot phase (Ngatiah 2000). However, the project

(continued)

Box 8.1 (continued)

has failed to fully reap the benefits of group extension. Extension agents of the project have not been able to appropriately manage the groups and mediate the conflicts that frequently arose over group members' individual land dedicated to group purposes. In the eyes of the project staff, the groups failed to effectively coordinate joint work, to responsibly administer the communal funds, and to properly maintain the group-owned infrastructure (van den Abeele, personal communication). Driven by concerns of project success, the project managers ultimately decided to dismiss the group extension approach and rather focus their efforts on a small number of individual farmers who were more easily to work with. These farmers eventually implemented the project activities on their individual land holdings.

The apparent failure of this case is the inadequacy of the project to equip its extension officers with the methodological competencies and interpersonal skills required to successfully manage farmer groups and facilitate intragroup processes. For group extension, a new type of sensitive extension agent with good communication skills and social empathy is required, which was – and, unfortunately, partly is – not readily available (Leeuwis and van Ban 2004). At a more abstract level, the case also illustrates the incongruity of group-oriented extension with interventions pursuing a traditional transfer of technology. This is inherently the case in the technocratic design of INRMU's extension packages.

8.4.2.2 How Innovations Disseminate Through Farmer Groups

Hoffmann et al. (2009b) recommend starting the promotion of local innovation by forming a group of farmers who will study it, and decide who of them will put it to test. Theoretically this has a number of advantages and reads as a plausible suggestion. Yet, studies that investigate the performance of farmer groups, and success in terms of diffusion of rural innovations in particular, are rare (exceptions include Alvarado 1980; Davis et al. 2004; Place et al. 2002; Were et al. 2006). Hence, existing theoretical conceptions of innovation dissemination in farmer groups are generally not well elaborated and are partly inconsistent.

Based on a profound review of four theories that explain the diffusion of innovations from the perspective of different scientific backgrounds and research traditions, Darr (2008) derived the multiple-pathway model of innovation diffusion (Fig. 8.2). The model is founded on the premise that the spread of innovations is influenced by direct contact between individuals.[7] In line with the propositions

[7] Apart from pure information on the innovation, other aspects such as trust, advice, social influence, and role model effects simultaneously transfer through social relationships. In addition to creating knowledge and awareness of an innovation among potential adopters, social exchange

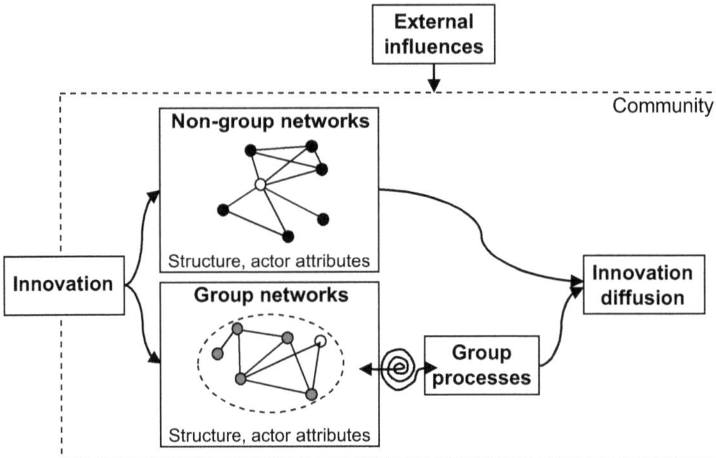

Fig. 8.2 The multiple-pathway model of innovation diffusion (Source: Darr (2008), modified)

made by social network theory (e.g., Coleman et al. 1957; Kilduff and Tsai 2005; Monge and Contractor 2003; Scott 1998; Valente 1999; Wasserman and Faust 1998), the information flow and, hence, diffusion of innovations within the community is presumed to take place simultaneously through multiple pathways, i.e. via different types of relationships in various network types such as formal farmer groups, informal communication networks, family bonds, or relationships of mutual assistance and exchange of tools, farm inputs and other supplies among households. Network structure as well as individual attributes of the network actors are conceptualized as key variables in determining innovation diffusion at the network level. In accordance with functional group theory (e.g., Ellis and Fisher 1994; Hollingshead et al. 2005; Wittenbaum et al. 2004), the diffusion of innovations among the members of a farmer group is thought to be influenced by group processes and variables such as homogeneity, activity, and group climate in a non-linear manner. Innovation-related and context-specific variables are additional factors influencing the diffusion of innovations according to the model.

From an empirical point of view, the proposition of innovation diffusion by relational proximity is in line with the social reality of most rural regions of the tropics and subtropics. Given the remote location, poor road access and limited availability of modern communication media (see Chap. 3), face-to-face conversations represent the primary form of information exchange amongst farmers. The innovation-related information flows are two-step, moving from modern media to advisors and other disseminators, who then introduce the contents into farmers'

plays an important part in persuasion and social imitation that induce individual decision-making, implementation, and confirmation of the adoption decision.

communication networks. Rural life in socio-economically homogeneous communities provides plenty of opportunities for daily interaction and the formation of relationships between neighbours, kin, members of age cohorts etc. An important aspect of these relationships, support networks, and indigenous organizations is that they allow farmers to better cope with the hardship of rural life.

8.4.2.3 Why Groups Are More Effective

Innovations that require collective actions, such as the sustainable management of community forests or the implementation of soil protection structures on communal slopes, can most effectively be promoted by extension agents working with groups of farmers and using group methods (Feder et al. 2010). But also for innovations that require farmers to take independent adoption decisions, farmer groups are important and effective vehicles to promote the diffusion of these innovations.

The strength of interpersonal social relationships is likely the attribute most critical for the diffusion of innovations in social networks. Infrequent interactions with socially and spatially distant others are typically referred to as 'weak' ties, while frequent and regular interactions with people who are similar and socially close, i.e. share the same attitudes and behaviours, are usually labelled as homophilous, cohesive, or 'strong' ties. Both weak and strong ties have shown to be necessary for successful innovation diffusion. Weak ties are stronger at carrying information about new ideas (Granovetter 1973; Rogers 2003). The argument here is that social circles of weakly tied actors tend to overlap less (Strang and Soule 1998), providing non-redundant information at a comparatively low cost (Hansen 1999). On the contrary, strongly related partners share many ties to third parties and so have little news to report to each other. By creating more and shorter links between distant individuals in a network, weak ties accelerate the rate of diffusion (Valente 1999). A lack of weak ties can therefore act as a diffusion barrier (Rogers 2003). The positive effect of strong ties on innovation diffusion, on the other hand, has been attributed to more rewarding interpersonal communication (Rogers and Kincaid 1981), higher levels of trust and reciprocity (cf. Monge and Contractor 2003), compounded by social pressures and persuasive forces exerted by cohesive networks (Strang and Soule 1998).

More than most other forms of social interaction in rural places, farmer groups combine the features and benefits of both strong and weak ties. This fact forms the first argument in explaining the effectiveness of the group approach. Farmer groups, particularly if facilitated by extension agents, usually foster the formation of weak ties and the exchange of information across social and spatial community strata, i.e. among farmers who would not be connected to each other without the groups. Simultaneously, farmer groups provide the opportunity, if not necessity for group members to regularly interact and cooperate in order to perform the group activities and to attain a common group objective. These interactions create relationships of mutual dependency and trust that effectively convey role model effects, persuasive forces, opportunities for confirmation, and other forms of social influence that facilitate the promotion and diffusion of innovations.

Further illuminating the effects of strong vs. weak ties on innovation diffusion, knowledge sharing has been conceptualized as a dual process of knowledge 'search' and knowledge 'transfer' (Hansen 1999), or of 'innovation' and 'imitation' (Liu et al. 2005): On the one hand, actors look for and identify relevant knowledge and information through their social relationships in order to innovate or adopt innovations earlier than others. Due to their capacity to link distant actors and to provide better information access, weak ties are more useful during this process. On the other hand, mimicking the adoption behaviour of others and incorporating new knowledge into the activities of the recipient typically requires extra effort. Due to their higher interaction intensity, the strength of behavioural pressures, the actor's emotional commitment to the relationship, and the restricted access to potentially contradicting information, strong ties are more useful here, particularly when complex knowledge is concerned.[8]

Darr and Pretzsch (2008) show that the relative contribution of the 'search' and 'innovation' versus 'transfer' and 'imitation' mechanisms to the diffusion process varies with the prevailing extension approach. In situations of group extension, where a relatively large number of farmers have direct access to information, innovations disseminate more successfully through strong ties because they convey behavioural pressures and role model effects that shape individual innovation adoption decisions more effectively. Due to the abundant information available in this context, weak ties cannot capitalize on their search advantage. In contrast, in the context of individual extension and information scarcity when looking for the knowledge sources is relatively more important, actors who possess extended weak ties are in a better position to access limited and dispersed information; and actors who additionally possess strong links to these sources are in the best position to ultimately adopt the innovation.

The second argument explaining the effectiveness of the group approach in rural extension refers to the coincidence of information abundance and strong social relationships. Successful group extension interventions, through directly addressing a large number of farmers, significantly increase the abundance of information in the community. Simultaneously, through supporting group formation and facilitating group processes, group extension enhances the formation of dense and cohesive relationships in farmer groups (e.g., Darr 2008). Thus, successful group extension interventions promote the emergence of – or even overtly create – those social relationships that most effectively disseminate the innovations given the level of information available.

[8] In a similar vein, Levin and Cross (2004) introduce the notion of 'trusted weak ties', which combine the advantages of both strong and weak ties; Swan et al. (2003) show the relevance of local vs. global networking during the various stages of the innovation process; and Uzzi (1996) concluded that optimal networks consist of both types of ties.

The third argument refers to the mediating influence of external facilitation and increased group activity on the relationship of group homogeneity and innovation diffusion. Groups composed of members who are dissimilar in terms of major socio-economic or cultural characteristics and therefore linked by sparse and weak ties, usually face barriers to innovation diffusion that are caused, for example, by less satisfactory modes of communication among members, less effective interaction patterns and lower degrees of mutual trust (cf. Rogers and Kincaid 1981; Monge and Contractor 2003). Darr (2008) shows that external facilitation and increased group activity are key mechanisms to overcome diffusion barriers inherent to unfavourable group compositions. Innovations, in fact, disseminate more effectively in heterogeneous farmer groups if these groups are characterized by high levels of group activity. Directly and indirectly enhancing group activity and facilitating the emergence of such cohesive relationships among the members through, for example, intensifying member interaction, aiding in conflict resolution, and fostering a climate of trust and mutual obligation, constitute the central contribution of group-oriented extension services to enhance the diffusion of innovations among farmers. Likewise, external facilitation and increased group activity can also improve the performance of overly homogeneous groups. Rural extension, thus, has the potential to simultaneously facilitate and balance the two opposing mechanisms of structural cohesiveness and relational weakness of group networks by encouraging member interaction and group activity. This fosters a climate of trust, mutual dependency and cohesive relationships. Furthermore, extension services facilitate and encourage linkages with other groups and external sources of information, exposing the members to new ideas, as well as increasing group diversity through purposively bringing farmers together, from across many social and spatial divides.

Darr (2008) makes a fourth argument for the group extension approach: There is evidence that group-oriented extension, in particular if it is strong and persuasive, can at least in parts compensate for any obstruction of the innovation diffusion process that may result from insufficient group processes, such as poor communication among group members or low levels of member commitment. In addition, the involvement of extension agents in group decision-making that, for example, helps to expedite lengthy discussions and tough processes of consensus building, that presses for necessary decisions, and that responsibly steers the development of the group, seems to be one important success factor of the group extension approach. Figure 8.3 summarizes the arguments made.

The superiority of group-oriented rural extension interventions in popularizing the adoption of comparatively basic technologies among culturally and socio-economically largely homogenous populations of subsistence farmers is thus grounded on three facts: group-oriented extension services (a) facilitate the emergence of cohesive relationships among the group members, which in turn foster effective innovation spread; they (b) have the potential to stimulate the activity of the farmer groups, which can compensate for less effective diffusion under unfavourable group composition; and they (c) partially compensate for diffusion disadvantages that result from an unfavourable group climate. This proposition further augments the case made by Liu et al. (2005) and Newman and Dale (2007),

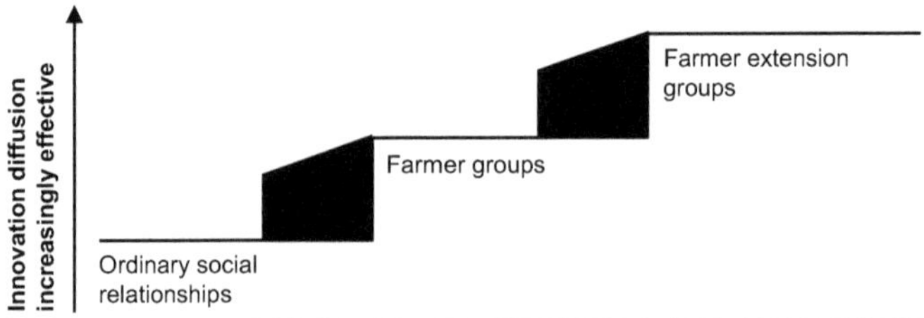

	Ordinary social relationships	Farmer groups	Farmer extension groups
Examples	Communication networks, exchange relationships, age cohort and family bonds	Formal and non-formal farmer organizations not adequately addressed by extension agents - e.g., self-help groups	Farmer groups actively supported and facilitated by extension agents in the context of group-oriented extension
Predominant mechanisms of diffusion	**'Contagion by cohesion'**: adoption initiated by information, knowledge, persuasive forces, and role model effects conveyed in social relationships between farmers	As before plus **Combined strong and weak tie effect**: groups provide opportunity to create relationships of trust and mutual dependency as well as linkage across usually isolated community strata	As before plus **Coincidence** of group cohesiveness and information abundance; **Offset diffusion barriers** caused by unfavourable group composition, lack of group activity and insufficient group processes

Fig. 8.3 Effectiveness and predominant mechanisms of diffusion of rural innovations in various types of farmer networks (Source: Authors' elaboration)

who suggest that diffusion processes and group performance can be promoted by actively manipulating network structure and group composition.

8.4.3 Diversity of Client Groups: Serving Distinct Needs

Individuals vary with regard to their capabilities, resources, and preferences; and they make decisions based on their personal experiences and evaluation of the future, the expectations of their social environment, and other subjectively perceived opportunities and constraints. Farmers who face comparable natural and socio-economic conditions nevertheless are dissimilar by way of their livelihood strategies, the farming practices they employ, the risks they take, or with regards to the aims and levels of the timber production they pursue, for example.

Small-scale and mostly subsistence-oriented farmers constitute, by far, the largest client group for providers of rural extension services in the tropics and subtropics. Although the degree of their integration in local markets may vary, these farmers typically grow staple crops, rear small domestic animals, or collect forest products mostly for their own sustenance. Consequently, the farmers usually lack the cash and capital resources required to adopt more productive or more profitable farm activities such as farm forestry and high-value agricultural production. Adherence to traditional farming practices maintains this feedback mechanism by limiting their ability to generate marketable surpluses and thus cash income. Extension services can help these farmers to break the vicious circle by increasing their ability to participate in local markets, invest in their farm businesses, and take the risks of increased market involvement. Providing advice on agroforestry practices, establishing micro-credit schemes, and linking the farmers to input and output markets are some critical contributions extension services can make.

Women-headed households are usually more resource constrained. They often face significant challenges concerning farm labour that result from the lack of the male workforce once the husband has moved or passed away. Commonly, these women are not able to summon the physical labour needed to cultivate their entire farm. This usually results in lower yields, marketable surplus production and hence cash income. Consequently, these households often need to lease out parts of their farmland in order to purchase their needs. Given their lack of land and financial resources these women farmers often prefer to grow a small number of trees, primarily for firewood. Extension advice would therefore need to focus on intercropping and other agroforestry practices. Affluent households, in contrast, may be able to acquire the land, labour, and capital resources needed to establish and operate a decent farm forestry undertaking. Producing firewood and timber for household consumption and sale may be their typical objectives; and the extension services would focus on the establishment and management of woodlots and small-scale plantations.

Forest dwellers and farmers living in the vicinity of forests typically derive a significant part of their livelihood from harvesting timber and collecting non-timber forest products such as resins, gums, vines, nuts, or mushrooms. During hunger periods forests are an important safety net that provide alternative sources of supply for foodstuff and farm income. Yet, the unregulated over-exploitation of some of these products, the adverse effects of forest degradation and deforestation, and natural fluctuations of availability and supply of the NTFPs threaten the livelihood of these communities. Extension interventions could mitigate these impacts through a focus on improving local value-added processing and marketing of these products, facilitating the establishment of local institutions that enforce sustainable management practices, or developing technologies that allow for domestication of these products on farmsteads.

Absentee landlords may primarily be interested in land use options that require a minimum of supervision and labour. The cultivation of perennial tree crops under a contract farming arrangement (see Sect. 8.4.4) is therefore a very attractive alternative for this client group. With the objective of optimal returns on their land, owners of large estates may predominantly be interested in establishing a commercial plantation for high quality export timber production, or softwood production for pulp and paper industries. In this instance, extension services would focus on mechanized plantation management or the mediation of conflict between the plantation and adjacent communities, inter alia.

Finally, commercial farmers and operators of large-scale plantations of trees or perennial agricultural crops such as coffee, tea, or oil palm typically require general business consultancy and customized technical advice on specific aspects of their production process. The first may include, for example, strategy and marketing topics, assistance in accounting, financial assessment of investment options, or provision of legal advice. Customized technical advice usually concerns aspects such as the training of operating staff on new harvesting equipment, soil sample analysis to determine the appropriate rate of fertilizer application, and recommendations on pest management among others. Thus, as diverse the realities and forest production objectives of these archetypes of farmers, so too are their information requirements, demand for technical services and business advice, assistance in linking to markets and value chain stakeholders, and thus the consulting and extension needs of these client groups.[9]

Besides the various types of small-scale and commercial tree growers, there exist a number of other potential client groups for rural extension services – especially where the role of extension is understood to transcend traditional advisory communications. Noteworthy of these include forest contractors, farm input and commodity traders, operators of warehouses and storage yards, timber transport companies or processing plants and other value chain stakeholders that may be significantly involved in, and benefit from, the provision of advisory services aimed at increasing market and value chain efficiency. Unauthorized users of contested natural resources, such as pastoralists who graze their herds on a sedentary farmer's land; pharmaceutical companies and biotech start-ups that encroach on natural forests for bioprospecting; or migrant settlers who do not obey local customary rules are potential client groups if extension agents engage in the mediation and resolution of pre-existing conflicts. Public authorities, NGOs, research institutions, donor agencies, churches, and other stakeholders of civil society may also be important client groups if the purpose of the intervention is to facilitate linkage and exchange, initiate joint action, and institutionalize forms of mutual cooperation that lead to the generation of innovative and locally adapted socio-technical arrangements or other forms of societal progress.

Despite the variety of client groups and the diversity of their specific extension needs, addressing clients in groups rather than individually is the preferred approach for rural extension in most tropical and subtropical countries (see Fig. 8.1). Relative homogeneity of the groups with regard to the major determinants of the farmers' extension needs can be achieved by directing the extension services towards groups that aggregate individuals in pursuit of a common interest, plagued with identical environmental or socio-economic constraints, or willing to exploit the same opportunities despite all the remaining individual, socio-cultural, or economic differences between the farmers.

8.4.4 Extension Organization: No Single Best Solution

The plurality of extension stakeholders and the diversity of extension systems, strategies, and objectives they pursue have led to the emergence of new organizational arrangements of advisory service provision during the past years.

[9] The diversity of farmers, in particular with regard to their economic performance and the required extension services, has occasionally been viewed as a major justification of, as well as discriminator for, charging selected farmers for the extension services they obtain.

Governments through their line agencies, input suppliers and traders, private sector processing or export enterprises, churches, community based organizations, local NGOs, and donor-funded projects constitute just a few of the diverse set of extension providers that coexist or work in collaboration.

Providing extension services that address environmental issues, other public interests, or target poor or non-commercial farmers, are traditionally regarded as governmental responsibility, and that these should be provided free of charge. Conversely, providing commodity-oriented advice, or targeting commercial farmers, is perceived to be a task of the private sector, with partial or full cost recovery as the standard model. Yet, this distinction may no longer hold, as private sector companies are realizing the economic potential and business opportunities that lie in providing poor and non-commercial farmers with the products and services they need (Prahalad 2010).

Traditionally, agricultural and forestry extension has been provided through public extension service organisations, financed by and organized under the respective Ministries of agriculture, forestry, fisheries, natural resources, or rural development. This continues to be the case in most Asian and African as well as in a number of Latin American and European countries (Neuchatel Group 2008). Usually organized hierarchically from national, down to communal levels, these services, often through a number of specialized advisors, provide advice in multiple sectors such as agriculture and horticulture, livestock, natural forest management, farm forestry, primary processing, or health. Common problems of public extension services often result from the sheer size of these organizations, which frequently leads to bureaucratic and inefficient processes; role conflict between serving farmers' or the government's interest; poorly motivated and equipped frontline extension staff; and a lack of accountability of extension agents. The paradigm shift from traditional towards more contemporary extension strategies and methods requires a shift in the mindsets of long-serving extension staff, which is sometimes challenging. As one example, Swanson cites the continued tendency of public extension services to work with progressive high-resource farmers. Organizing smallholders into farmer or producer groups can help to alleviate this problem (Swanson 2008).

Recognizing that other actors can impart certain types of extension services more efficiently and more effectively, public sector agencies are increasingly promoting some form of partnership with these new providers (Collion 2004). Nevertheless, public extension services are often the only advisory services available to farmers, particularly in marginal areas where service provision is not deemed profitable from the viewpoint of the private sector.

Organizational arrangements particularly important for market-oriented extension services are contract farming and outgrower schemes.[10] Under such arrangements, trading or processing companies provide technical assistance, credit, or farm

[10] Outgrower schemes connote the involvement of a public enterprise or parastatal, while contract farming refers to private sector arrangements (Baumann 2000).

inputs to smallholders in return for a contractually fixed quantity and quality of an agricultural or forest product delivered at a predefined date and at a price fixed in advance of the growing season (Binswanger et al. 1995, cited in Setboonsarng 2008; see Box 8.2). While such arrangements are usually only of marginal importance if markets and value chains function well, they can be essential during the building and growth of a processing industry and considerably facilitate the transformation from subsistence to commercial economies (Setboonsarng 2008). For the purchaser, contract farming and outgrower schemes help to increase the reliability and quality of the raw material supply and reduce the cost for land leases and labour. Smallholders usually benefit from guaranteed market access, reduced risk of price fluctuations, and the availability of appropriate technical advice and inputs. Empirical evidence in relation to the benefits and economic viability of contract farming is mixed (Baumann 2000; for some examples see Arnold 1998; Gibbon et al. 2009; Kudadjie-Freeman et al. 2008; WRI 2005). Disadvantages of the arrangement stem from the power imbalance between the smallholders and the commercial entities (Neuchatel Group 2008). Through strength in numbers, farmer groups can therefore play an important role to improve the bargaining power of farmers and contribute to the success of contract farming arrangements (Setboonsarng 2008).

The restructuring of public extension organizations and the partial shift of advisory services from the public to the private domain has led to the increasing involvement of farmer groups and organizations in rural service provision (Heemskerk and Wennink 2004). At the community level, extension service providers address farmer interest or producer groups for participatory research and technology development, to facilitate experience sharing and enhance innovation diffusion. The establishment of marketing and export cooperatives serve as vehicles for improving the farmers' market access by aggregating geographically dispersed supply and demand to commercially viable levels. At the regional or national level, farmer unions, federations, or industry associations are involved in defining priorities and planning of research and extension interventions. Yet, the role of farmer organizations goes much beyond that of simply participating in, and contributing to, research, planning, and innovation diffusion. Farmer groups and organizations also act as independent extension facilitators that, either through employed advisors or through linking with external advisors, actively provide farm input, credit, market information, transport, training, or technical advice to their members or other farmers (Heemskerk and Wennink 2006; Neuchatel Group 2008). These groups further serve to increase political influence and negotiating power of smallholders. Shortage of funding and lack of management skills probably represent the greatest challenges for farmer organizations to successfully and sustainably provide these services to smallholders.

For an overview on other typical organizational arrangements in extension see, for example, Moris (1991) and Hoffmann et al. (2009b).

In view of the manifold of local contexts and challenges of rural extension work, and given the success of diverse organizational extension arrangements that suit particular circumstances, there is clearly not a single most appropriate organizational form of advisory services. Characteristic of any successful arrangement, however, is that the organization of extension services fits with the requirements of the client groups and their economic and environmental realities; that the organization is flexible and allows for cooperation and partnerships with multiple stakeholders whenever appropriate; and that it is robust and ensures continued and sustainable service provision even if external environments are changing and unstable. A countrywide monopolistic governmental extension service, even when providing services free of charge, is no longer appropriate to serve the different functions and meet the diverse needs of the various client groups in today's world. Largely concentrating on topics of public interest and doing so in collaboration with partners that have comparative advantages should be the primary strategy of governmental extension agencies. Services that primarily target private interests should be provided through a pluralistic extension and support system, where various providers compete for the farmers' preference through an array of possible solutions. The role of governments in such situations should be limited to providing the regulatory conditions for the extension system to function well, rather than acting as a service provider itself.

Box 8.2 Smallholder Contract Tree Farming: Experiences from Thailand

Of all Southeast Asian countries, Thailand likely has the most extensive experiences in contract farming in both the agriculture and forestry sectors. Contract farming of trees, mainly of eucalyptus for the production of woodchips, has been practiced for more than 30 years. A minimum of 30,000 smallholders are estimated to manage approximately two thirds of the country's 460,000 ha of eucalyptus plantations under some form of contract farming. Purchases of logs from contracted and non-contracted farmers represent a primary supply strategy for every Thai particleboard and pulp and paper producer.

The early contract farming schemes have frequently failed as a consequence of misconception by both contractual parties on the responsibilities, benefits and limitations of the system; the competitive markets that either tempted growers or contractors to breach the contract and pursue the business with other potential partners; and an oversupply of woodchips and stagnant farm gate prices for logs in the course of the economic crash of 1997. Restrictions on timber extractions from forest concessions, civil resistance to large-scale commercial tree plantation projects, and the construction of additional processing facilities have intensified the competition for wood fibre, improved the position and bargaining power of smallholders, and thus led to a revival of contract tree farming.

Contract tree farming projects vary widely in terms of their contractual arrangements, their ability to equally benefit farmers and the processing companies, and the sustainability of the forest management operations.

(continued)

Box 8.2 (continued)

Nevertheless, contract tree growing has frequently proved to be a more profitable option to smallholders than competing cash crops. Typically, contracts are concluded between a pulp and paper, plywood or particleboard producer and individual farmers within a radius of 100–150 km from the factory gate. Depending on the plant capacity, the area harvested annually may well amount to 10,000 ha. The size of the individual smallholder plantation ranges from less than 1 ha to 50 ha, with an average size of 5–8 ha. Consequently, companies must deal with several thousand contract farmers, partially organized in farmer groups, in order to reduce the transaction costs.

Eligibility for the contract farming scheme usually requires farmers to possess a minimum holding size and land area available for planting, as well as proof of land rights. Previous experience in tree cultivation is occasionally also required. Contract periods typically range between 10 and 20 years with repeated cuttings after 3–5 years. After being accepted, farmers normally receive tree seedlings free of charge or at reduced cost, technical advice on tree planting and maintenance, and a minimum guaranteed timber price at harvest. In some cases, farmers also receive fertilizers and pesticides at cost price, labour for tree harvesting, and trucks for transportation. In return, the smallholders must strictly follow the instructions of extension officers of the company, and they are usually obliged to sell their timber exclusively to the company or its middlemen.

Overall, contract tree farming arrangements have proved to be beneficial to smallholders in terms of greater and more profitable market involvement, especially if these arrangements involve mixed cropping systems that provide farmers with a diversified income source in the early years, before trees are harvested. Furthermore, participation in a contract farming scheme can be the best way for smallholders to access financial inputs, quality tree seedlings and technical expertise required to use or rehabilitate degraded lands, such as those covered by spear grass *(Imperata cylindrica)*, that they otherwise would not have the capital to clear, plant and manage. Although most arrangements are still too inequitable and unstable to be considered partnerships in any normative sense, they can provide communities with the building blocks necessary for economic and political empowerment in the longer term.

Source: Compiled from Barney (2004); Baumann (2000); FAO and CIFOR (2002); Makarabhirom and Mochida (1999); Sriboonchitta and Wiboonpongse (2005); Vermeulen et al. (2008)

8.5 Outlook

Extension and advisory services never before in history have been as necessary as today. No single farming family and no forestry or agricultural enterprise will be able to fully exploit the upcoming opportunities and successfully navigate the risks and challenges that lie ahead without relying on professional advisory support in one way or another. Intensifying labour division, specialisation and global dependencies are improving the standard of living for a majority of people worldwide. However, embedded in this are decision making, collaboration, and management considerations that are becoming ever more complex.

In light of these developments, extension is not at all an outdated concept. Providing farmers and tree growers with specific advice related to their land management practices has been necessary and relatively successful for centuries. Yet, extension services have to adapt to the changing situation and needs of their client groups. Given the specific requirements in industrialized and developing countries, in countries in transition, and in the upcoming giant economies, simply transferring blue-print approaches, methods, and organisational arrangements in extension is futile.

Based on a joint understanding of current problems and their root causes, actors in the rural development arena have to analyze the situation on a case-by-case basis and identify potential solutions. Based on an understanding of the respective visions and individual contributions of all stakeholders, suitable coalitions and cooperation partners need to be selected, concurrently, with appropriately designed approaches, methods, organizational structures, and working programs. Particularly in the field of rural extension, the learning organisation is the model for the future, as joint learning of clients and advisory service providers paves the way for success. The challenge of rural development comprises a plethora of important tasks, and rural extension services need to develop the differentiated offers that can assist problem solving in these areas.

Rural advisors, once they no longer are part of governmental structures, should organize themselves in professional unions that drive the development of binding values and quality standards, codes of conduct, training programs, and quality management systems. Explicating and adhering to these values can provide a transparent baseline for clients operating in an otherwise increasingly pluralistic and chaotic market of extension service offers (Hoffmann et al. 2009b).

References

Alvarado ML (1980) Genossenschaften als Träger der Diffusion von Innovationen Die Bedeutung der Agrargenossenschaften für die Diffusion von Innovationen in Entwicklungsländern dargestellt am Beispiel Boliviens, vol 10, Schriften zur Kooperationsforschung: C, Berichte. Mohr, Tübingen

Anderson JR, Feder G, Ganguly S (2006) The rise and fall of training and visit extension: an Asian mini-drama with an African epilogue, Policy research working paper. World Bank, Washington, DC

Arnold M (1998) Trees as out-grower crops for forest industries. Experience from the Philippines and South Africa. Rural Development Forestry, London
Barney K (2004) Re-encountering resistance. Plantation activism and smallholder production in Thailand and Sarawak, Malaysia. Asia Pac Viewpoint 45(3):325–339
Baumann P (2000) Equity and efficiency in contract farming schemes. The experience of agricultural tree crops. Working paper. Overseas Development Institute (ODI), London
Belay K, Abebaw D (2004) Challenges facing agricultural extension agents: a case study from south-western Ethiopia. Afr Dev Rev 16(1):139–168
Benor D, Harrison JQ, Baxter MWP (eds) (1984) Agricultural extension. The training and visit system. International Bank for Reconstruction and Development, A World Bank Publication. The World Bank, Washington, DC
Biggs S (2007) Building on the positive: an actor innovation systems approach to finding and promoting pro poor natural resources institutional and technical innovations. Int J Agric Res Gov Ecol 6(2):144–164
Binswanger HP, Deininger KW, Feder G (1995) Power, distortions, revolt and reform in agricultural land relations. In: Behrman JR, Srinivasan TN (eds) Handbook of development economics vol. 3, Part 2. Elsevier, Amsterdam, pp 2659–2772
Carney D (1996) Formal farmers' organisations in the agricultural technology system: current roles and future challenges. Natural resource perspectives. Overseas Development Institute (ODI), London
Carron AV, Brawley LR (2000) Cohesion: conceptual and measurement issues. Small Group Res 31(1):89–106
Coleman JS, Katz E, Menzel H (1957) The diffusion of an innovation among physicians. Sociometry 20(4):253–270
Collion M (2004) Introduction to revitalization within public sector services. In: Rivera WM, Alex G (eds) Revitalization within public sector services, case studies of international initiatives, vol 4, Extension reform for rural development. World Bank, Washington, DC, pp 1–5
Conley T, Udry C (2001) Social learning through networks: the adoption of new agricultural technologies in Ghana. Am J Agr Econ 83(3):668–673
Conley T, Udry C (2005) Learning about a new technology: pineapple in Ghana, Center discussion paper. Economic Growth Center, Yale University, New Haven
Darr D (2008) Effective even when neglected: farmer groups and the diffusion of agroforestry innovations in rural communities of Eastern Africa, vol 89, Kommunikation und Beratung. Margraf, Weikersheim
Darr D, Pretzsch J (2008) Mechanisms of innovation diffusion under information abundance and information scarcity – on the contribution of social networks in group vs. individual extension approaches in semi-arid Kenya. J Agric Educ Ext 14(3):231–248
Davis KE (2006) Farmer field schools: a boon or bust for extension in Africa? J Int Agric Ext Educ 13(1):91–97
Davis KE, Franzel S, Hildebrand P, Irani T, Place NT (2004) Extending technologies among small-scale farmers in Meru, Kenya: ingredients for success in farmer groups. J Agric Educ Ext 10(1):53–61
Ellis DG, Fisher BA (1994) Small group decision making. Communication and the group process. McGraw-Hill, Boston
Erickson BH (1991) The relational basis of attitudes. In: Wellman B, Berkowitz SD (eds) Social structures. A network approach. Cambridge University Press, Cambridge, pp 99–121
FAO, CIFOR (2002) Towards equitable partnerships between corporate and smallholder partners. Relating partnerships to social, economic and environmental indicators. Bogor, Indonesia, 21–23 May 2002. Food and Agriculture Organization of the United Nations (FAO), Rome
Farrington J (1994) Public sector agricultural extension: is there life after structural adjustment? Natural resource perspectives. Overseas Development Institute (ODI), London
Feder G, Murgai R, Quizon JB (2004) Sending farmers back to school: the impact of farmer field schools in Indonesia. Rev Agric Econ 26(1):45–62

Feder G, Anderson JR, Birner R, Deininger K (2010) Promises and realities of community-based agricultural extension, IFPRI discussion paper. International Food Policy Research Institute (IFPRI), Washington, DC

Freeman C (1995) The 'National System of Innovation' in historical perspective. Camb J Econ 19 (1):5–24

Frenzel Baudisch A, Grupp H (2006) Evaluating the market potential of innovations: a structured survey of diffusion models, Jenaer Schriften zur Wirtschaftswissenschaft. Universität Jena, Wirtschaftswissenschaftliche Fakultät, Jena

Friedkin NE (2001) Norm formation in social influence networks. Soc Netw 23(3):167–189

Garforth CJ (2004) Introduction. In: Rivera WM, Alex G (eds) Demand-driven approaches to agriculture extension, case studies of international initiatives, vol 3, Extension reform for rural development. World Bank, Washington, DC, pp 3–6

Geroski PA (2000) Models of technology diffusion. Res Policy 29:603–625

Gerster-Bentaya M, Hoffmann V (2011) Rural extension, volume 3: Training concepts and tools. Margraf, Weikersheim

Gibbon P, Lin Y, Jones S (2009) Revenue effects of participation in smallholder organic cocoa production in tropical Africa. A case study, vol 6, IIS/GI. Kongevej working paper. CDR, Copenhagen

Granovetter MS (1973) The strength of weak ties. Am J Soc 78(6):1360–1380

Hagmann J, Chuma E, Murwira K, Connolly M (1999) Putting process into practice: operationalizing participatory extension, Agricultural research and extension network paper. Overseas Development Institute (ODI), London

Hansen MT (1999) The search-transfer problem: the role of weak ties in sharing knowledge across organization subunits. Adm Sci Q 44(1):82–112

Heemskerk W, Wennink B (2004) Building social capital for agricultural innovation. Experiences with farmer groups in Sub-Saharan Africa, vol 368, KIT development policy and practice. KIT Publishers, Amsterdam

Heemskerk W, Wennink B (eds) (2006) Farmers' organizations and agricultural innovation. Case studies from Benin, Rwanda and Tanzania, vol 374, KIT development policy and practice. KIT Publishers, Amsterdam

Hoffmann V (2007) Book review. Five editions (1962–2003) of Everett Rogers's diffusion of innovations. J Agric Educ Ext 13(2):147–158

Hoffmann V, Probst K, Christinck A (2007) Farmers and researchers. How can collaborative advantages be created in participatory research and technology development? Agric Hum Values 24(3):355–368

Hoffmann V, Christinck A, Lemma M (2009a) Handbook: rural extension, vol 2, Examples and background material. Margraf, Weikersheim

Hoffmann V, Gerster-Bentaya M, Christinck A, Lemma M (2009b) Handbook: rural extension, vol 1, Basic issues and concepts. Margraf, Weikersheim

Holding Anyonge C (2002) Forest extension: equitable partnerships for sustainable multifunctional forestry. Paper presented to the IUFRO division 6 meeting, IUFRO, Valdivia, 11–17 Nov 2002

Hollingshead AB, Wittenbaum GM, Paulus PB, Hirokawa RY, Ancona DG, Peterson RS, Jehn KA, Yoon K (2005) A look at groups from the functional perspective. In: Poole MS, Hollingshead AB (eds) Theories of small groups. Interdisciplinary perspectives. Sage, Thousand Oaks, pp 21–62

INRMU (2005) General information and major activities of INRMU. Unpublished project document, 'Integrated Natural Resource Management in Ukambani' Project, Machakos

ISNAR (1994) Report of a workshop: strengthening the role of farmers' organizations in technology development and transfer, International Service for National Agricultural Research. Briefing paper no. 15. International Service for Agricultural Research (ISNAR), The Hague

Kennedy JJ, Thomas JW (1996) Evolving forestry and rural development belief at midpoint and close of the 20th century. In: Glück P, Weiss G, European Forest Institute (eds) Forestry in the context of rural development. Future research needs. European Forest Institute, Joensuu, pp 27–44

Kilduff M, Tsai W (2005) Social networks and organizations. Sage, London
King CA (2000) Systematic processes for facilitating social learning: challenging the legacy. Agraria 233, Swedish University of Agricultural Sciences, Uppsala
Kline S, Rosenberg N (1986) An overview of innovation. In: Landau R, Rosenberg N (eds) The positive sum strategy. Harnessing technology for economic growth. National Academy Press, Washington, DC, pp 275–306
Kress A (2001) Rahmenbedingungen für partizipative Beratungsprozesse. Eine Analyse des Einflusses externer Faktoren auf die nachhaltige Wirkung forstlicher Programme in Bolivien, vol 40, Kommunikation und Beratung. Margraf, Weikersheim
Kroma MA (2006) Organic farmer networks: facilitating learning and innovation for sustainable agriculture. J Sustain Agric 28(4):5–28
Kubeczko K, Rametsteiner E (2002) Innovation and entrepreneurship – a new topic for forest related research. EFI-INNOFORCE discussion paper I and IFSPE discussion paper P/2002-1, University of Agricultural Sciences, Vienna
Kudadjie-Freeman C, Richards P, Struik PC (2008) Unlocking the potential of contract farming. Lessons from Ghana, Gatekeeper series. International Institute for Environment and Development (IIED), London
Leeuwis C, van den Ban AW (2004) Communication for rural innovation. Rethinking agricultural extension, 3rd edn. Blackwell Science, Oxford
Levin DZ, Cross R (2004) The strength of weak ties you can trust: the mediating role of trust in effective knowledge transfer. Manage Sci 50(11):1477–1490
Liu BS, Madhavan R, Sudharshan D (2005) DiffuNET: the impact of network structure on diffusion of innovation. Eur J Innov Manag 8(2):240–262
Makarabhirom P, Mochida H (1999) A study on contract tree farming in Thailand. Bull Tsukuba Univ For 15(5):1–157
Marinova D, Phillimore J (2003) Models of innovation. In: Shavinina LV (ed) The international handbook on innovation. Pergamon, Oxford, pp 44–53
McIntyre BD (2009) Global report, vol 6, Agriculture at a crossroads. Island Press, Washington, DC
Monge PR, Contractor NS (2003) Theories of communication networks. Oxford University Press, Oxford
Moris J (1991) Extension alternatives in tropical Africa, Occasional paper. Westview Press, Boulder
Mumford MD (2002) Social innovation: ten cases from Benjamin Franklin. Creativity Res J 14 (2):253–266
Neuchatel Group (2008) Common framework on market-oriented agricultural advisory services. Swiss Centre for Agricultural Extension and Rural Development, Lindau
Newman L, Dale A (2007) Homophily and agency: creating effective sustainable development networks. Environ Dev Sustain 9(1):79–90
Ngatiah JM (2000) Self-help groups supported by the [Support to the District Forestry Development Programme] project: internal evaluation of results so far. Unpublished project report, Nairobi
Noordin Q, Niang A, Jama B, Nyasimi M (2001) Scaling up adoption and impact of agroforestry technologies: experiences from western Kenya. Dev Pract 11(4):509–523
OECD (2005) Oslo manual. Guidelines for collecting and interpreting technological innovation data. Organisation for Economic Co-operation and Development (OECD), Statistical Office of the European Communities (Eurostat), Paris
Place F, Kariuki G, Wangila J, Kristjanson P, Makauki A, Ndubi J (2002) Assessing the factors underlying differences in group performance: methodological issues and empirical findings from the highlands of central Kenya, Collective action and property rights working paper series. CGIAR Systemwide Program on Collective Action and Property Rights, International Food Policy Research Institute (IFPRI), Washington, DC
Porter M (1990) The competitive advantage of nations. Free Press, New York
Prahalad CK (2010) The fortune at the bottom of the pyramid. Eradicating poverty through profits, 5th edn. Wharton School Publishing, Upper Saddle River, 2. print

Raini R, Hoffmann V, Zebitz CPW (2005) Integrated pest management (IPM) and information flow case study. Tomato stakeholders' practices in Kenya. Paper presented to the conference Deutscher Tropentag 2005, held in Stuttgart-Hohenheim, 11–13 Oct 2005
Rametsteiner E (2010) Integrating innovation and development policies for the forest sector: an introduction. In: Rametsteiner E, Weiss G, Ollonqvist P, Slee B (eds) Policy integration and coordination. The case of innovation and the forest sector in Europe. COST Action E51. Publications Office of the European Union, Luxembourg, pp 1–10
Rametsteiner E, Weiss G (2004) Innovation and entrepreneurship in forestry in central Europe. Draft paper presented at the conference sustain life – secure survival II, Prague, 22–25 Sept 2004
Reij C, Waters-Bayer A (eds) (2001) Farmer innovation in Africa. A source of inspiration for agricultural development. Earthscan, London
Rice RE (1993) Using network concepts to clarify sources and mechanisms of social influence. In: Richards W Jr, Barnett G (eds) Advances in communication network analysis, Progress in communication sciences. Ablex, Norwood, pp 43–62
Rivera WM (ed) (1991) Agricultural extension. Worldwide institutional evolution and forces for change. Elsevier, Amsterdam
Rivera WM, Alex G (2004) Revitalization within public sector services, case studies of international initiatives, vol 4, Extension reform for rural development, Agriculture and rural development discussion paper. World Bank, Washington, DC
Rogers EM (1962) Diffusion of innovations. Free Press, New York
Rogers EM (2003) Diffusion of innovations. Free Press, New York
Rogers EM, Kincaid DL (1981) Communication networks. Toward a new paradigm for research. Free Press, New York
Röling NG (1995) What to think of extension? A comparison of three models of extension practice. AERDD bulletin. University of Reading, Reading
Röling NG (2002) Issues and challenges for FFS: an introductory overview. Paper presented to the international workshop on Farmer Field Schools (FFS): emerging issues and challenges, held in Yogyakarta, 21–25 Oct 2002
Röling N (2009) Pathways for impact: scientists' different perspectives on agricultural innovation. Int J Agric Sustain 7(2):83–94
Röling NG, Engel PGH (1991) The development of the concept of agricultural knowledge information systems (AKIS): implications for extension. In: Rivera WM (ed) Agricultural extension. Worldwide institutional evolution and forces for change. Elsevier, Amsterdam, pp 125–137
Röling NG, Pretty J (1997) Extension's role in sustainable agricultural development. In: Swanson BE (ed) Improving agricultural extension. A reference manual. Food and Agriculture Organization of the United Nations (FAO), Rome, pp 181–191
Rosenfeld SA (2001) Networks and clusters: the Yin and Yang of rural development. In: Centre for the Study of Rural America (ed) Exploring policy options for a new rural America. Federal Reserve Bank of Kansas City, Kansas City, pp 103–120
Saether B (2010) Agricultural extension services and rural innovation in inner Scandinavia. Nor J Geogr 64(1):1–8
Scarborough V, Killough S, Johnson DA, Farrington J (eds) (1997) Farmer-led extension. Concepts and practices. Intermediate Technology Publications, London
Scott J (1998) Social network analysis. A handbook. Sage, London
Selener D, Chenier J, Zelaya R, Endara N, Fadherbe P, Jacques A (eds) (1997) Farmer-to-farmer extension. Lessons from the field. IIRR, New York
Setboonsarng S (2008) Global partnership in poverty reduction. Contract farming and regional cooperation, ADBI discussion paper. Asian Development Bank Institute (ADBI), Tokyo
Shannon CE, Weaver W (1949) A mathematical model of communication. University of Illinois Press, Urbana
Simpson BM, Owens M (2002) Farmer field schools and the future of agricultural extension in Africa. J Int Agric Ext Educ 9(2):29–36

Singh HB (2003) Farmer field school in community forest management in Nepal. An approach to group learning and experimentation by forest user groups. In: Farmer field schools: from IPM to platforms for learning and empowerment. Users' Perspectives with Agricultural Research and Development, Los Baños, pp 60–61
Spielman DJ, Davis K, Negash M, Ayele G (2010) Rural innovation systems and networks: findings from a study of Ethiopian smallholders. Agric Hum Values 27:1–18
Sriboonchitta S, Wiboonpongse A (2005) Analysis of contract farming in Thailand. Chiang Mai Univ J 4(3):361–385
Stahr K (2001) Dörfliche Kommunikationsnetzwerke, vol 42, Kommunikation und Beratung. Margraf, Weikersheim
Strang D, Soule SA (1998) Diffusion in organizations and social movements: from hybrid corn to poison pills. Annu Rev Soc 24:265–290
Swan J, Scarbrough H, Robertson M (2003) Linking knowledge, networking and innovation processes. A conceptual model. In: Shavinina LV (ed) The international handbook on innovation. Pergamon, Oxford, pp 680–694
Swanson BE (2008) Global review of good agricultural extension and advisory service practices. Food and Agriculture Organization of the United Nations (FAO), Rome
Tiffen M, Mortimore M, Gichuki F (1994) More people, less erosion. Environmental recovery in Kenya. Wiley, Chichester
Tripp R, Wijeratne M, Piyadasa VH (2005) What should we expect from farmer field schools? A Sri Lanka case study. World Dev 33(10):1705–1720
Uzzi B (1996) The sources and consequences of embeddedness for the economic performance of organizations: the network effect. Am Sociol Rev 61(4):674–698
Valente TW (1999) Network models of the diffusion of innovations, 2nd edn, Quantitative methods in communication. Hampton Press, Cresskill
Valente TW (2006) Communication network analysis and the diffusion of innovations. In: Singhal A, Dearing JW (eds) Communication of innovations. A journey with Ev Rogers. Sage, New Delhi/Thousand Oaks, pp 61–82
van den Abeele J, Macharia JK (2001) Identification report for intervention in natural resources project in Kitui, Machakos, Makueni and Mwingi (Kenya). Unpublished project report, Nairobi
Vanclay F, Lawrence G (1994) Farmer rationality and the adoption of environmentally sound practices: a critique of the assumptions of traditional agricultural extension. Eur J Agric Educ Ext 1(1):59–88
Vermeulen S, Nawir AA, Mayers J (2008) Rural poverty reduction through business partnerships? Examples of experience from the forestry sector. Environ Dev Sustain 10:1–18
Wambugu C, Franzel S, Tuwei P, Karanja G (2001) Scaling up the use of fodder shrubs in central Kenya. Dev Pract 11(4):487–494
Wasserman S, Faust K (1998) Social network analysis. Methods and applications, vol 8, Structural analysis in the social sciences. Cambridge University Press, Cambridge
Were E, Swallow B, Roy J (2006) Water, women, and local social organizations in the western Kenya highlands, Collective action and property rights working paper series. International Food Policy Research Institute (IFPRI), Washington, DC
Wittenbaum GM, Hollingshead AB, Paulus PB, Hirokawa RY, Ancona DG, Peterson RS, Jehn KA, Yoon K (2004) The functional perspective as a lens for understanding groups. Small Group Res 35(1):17–43
World Bank (2007) Agriculture for development, World development report, 30/2008. World Bank, Washington, DC
WRI (2005) The wealth of the poor. Managing ecosystems to fight poverty. World Resources Institute (WRI) in collaboration with UNDP, UNEP, and World Bank. WRI, Washington, DC
Wuyts-Fivawo A (1996) Linkages between research, farmers, and farmers' organizations in Kenya: a summary of findings, Briefing papers. International Service for Agricultural Research (ISNAR), The Hague

Chapter 9
Payment Schemes for Environmental Services: Challenges and Pitfalls with Respect to Effectiveness, Efficiency and Equity

Francois Jost and Ingo Gentes

Abstract Payments for environmental services (PES) is an expanding mechanism of environmental conservation that is becoming increasingly relevant, especially when the projected REDD + activities are taken into account. Therefore, particular attention is being focussed on the design and implementation of current and future programmes so that they are carried out in the most effective and equitable manner, and in a way that is as efficient as possible. However, market mechanisms alone cannot provide sufficient guarantees for environmental sustainability and local development subject to equity conditions, as there are many gaps in these mechanisms as yet and they need time to mature. Investing in physical, human and social capitals can secure land tenure, reduce vulnerability, strengthen local institutions and capacity building and, in many cases, be more equitable than is the case with money incentives. As objectives pertaining to efficiency, equity and effectiveness cannot all be fully achieved in the same measure, a trade-off has to be considered. Consequently, the role of decision makers, and the influence stakeholders exert over them, is a deciding factor. A good starting point is to set a minimum acceptable equity level that is to be achieved (enough to be of advantage for the participants) and to define from there the most efficient way to achieve this level, so that ultimately it is effective enough to ensure the benefits of the programme for this and the coming generation. PES is an instrument of significant potential and should be seen as a mechanism helping in the attainment of part of the objectives of ecosystem conservation and rural development, but not as a final solution in itself.

F. Jost (✉)
Institute of International Forestry and Forest Products, Technische Universität Dresden, Tharandt, Germany
e-mail: francoisjost@gmail.com

I. Gentes
European Union, Program for Public Administration and Regional Integration (PAAPIR), Honduras
e-mail: ingo.gentes@gmail.com

J. Pretzsch et al. (eds.), *Forests and Rural Development*, Tropical Forestry 9,
DOI 10.1007/978-3-642-41404-6_9, © Springer-Verlag Berlin Heidelberg 2014

Keywords Effectiveness • Efficiency • Equity • Payments for environmental services • PES • Environment • Ecosystem services • Trade-off • Livelihood assets • Market mechanisms • Biodiversity • Reducing emissions from deforestation and degradation • REDD • Valuation methods • Decision-making • Additionality • Targeting • Deforestation • Transaction costs • Opportunity costs • Compliance • Monitoring • Sustainability • Baseline • Fairness • Tenure rights • Developing countries

9.1 Introduction

Payments for environmental services (PES) is a relatively new economic instrument for environmental conservation and rural development that is becoming more and more relevant, especially in the Global South. During the last two decades PES agreements have mainly been focussed on hydrological services and forests. The leading argument for their promotion is that classical state regulation measures together with public programmes and development aid policies have been failing with regard to water protection and forest conservation initiatives, especially in rapidly developing countries. PES schemes are seen as an opportunity to bridge the shortcomings of the protection efforts through the implementation of market mechanisms.

The emphasis on water and forest protection in PES programmes has been driven mainly by an increasing scarcity of adequate water supply in many areas, as well as by the fact that rights to forests can be allocated to – and so benefitted from by – groups operating at a distance. Development agencies have promoted such agreements as a successful strategy for the conservation of forest and water resources (see Cordero et al. 2008). While water quality and quantity is essential for human and economic development, the constant growth of the human population and its economic activity has dramatically increased the water demand, resulting in growing pressure on the ecosystems providing these hydrological services. Maintaining a sufficient supply of clean water for human, agricultural and other uses requires healthy ecosystems (Stanton et al. 2010). Through payments and other rewards, PES agreements help to establish a link between the consumers and producers of the hydrological services, thereby incentivising forest users and owners to effectively conserve their resource to ensure the ongoing provision of water.

Generally, two types of PES schemes can be distinguished. On the one hand, PES agreements are used as an economic tool to enhance regulating environmental services, such as water quality and availability, carbon sequestration and disaster prevention. A typical characteristic of this type of arrangement is the consumptive or indirect use values of these services, where their values are available on the market. Alternatively, there are PES programmes with a focus on biodiversity conservation, tourism development and the promotion of aesthetics or cultural values. However, the number of such projects is relatively small, as these values are being treated as externalities and, consequently, the benefits provided are not being transmitted through prices. The main reasons behind this are (1) the lack of appropriate institutional structures to ensure the local delivery of these services and (2) that most of these services remain public goods without a well-defined private demand.

Although generally conceived as a mechanism to improve the efficiency of natural resource management, and not primarily as a tool for poverty reduction (Pagiola et al. 2005; Pagiola 2008), PES schemes are also increasingly regarded as a means to create new opportunities for local income and employment and, thus, to contribute to reducing poverty in the area of implementation (FAO 2004). However, it has been claimed that this 'triple-bundling' of goals – environmental conservation and protection, the creation of local development opportunities and poverty alleviation – cannot be left to the markets themselves, as market development is driven by certain groups, and the corresponding improvement of human welfare is not certain (Landell-Mills 2002). While many development agencies and think tanks around the globe are deeply convinced that PES can create enough financial flows from the users to the providers of environmental services that they constitute an incentive to protect and to maintain those natural resources generating these flows, this deflects PES from their main purpose and casts them as an instrument expected to solve or ease certain social and/or political problems, which usually have other, more fundamental, backgrounds. While the overall consensus is that compensation schemes for natural resource services are at least not detrimental for the participants, most PES agreements require special efforts to assure that poor people have access to the opportunities created by the programme (Pagiola et al. 2005).

Despite the fact that some environmental services in the forest and hydrology sectors have been quantified, a broader assessment of the structural impact of PES projects on forest management, poverty alleviation and social equity and their overall benefit-sharing for local actors is still required (Pattanayak et al. 2010; Porras et al. 2011). It has yet to be carefully assessed over longer periods of time, whether PES projects can actually help reduce overall poverty, including an equitable distribution of the benefits generated within a selected group, community or country as a whole.

Although it is much too early for a conclusive assessment of PES agreements, the overall purpose of this chapter is to examine whether PES programmes are effective enough to achieve equity and efficiency for the users and providers of environmental services. The chapter begins with some definitions of essential terms and an analysis of the evolution of the PES concept over time. An overview of the current PES projects and the different approaches to determining the value of their services is provided and the effectiveness, efficiency and equity concerns surrounding the promotion and implementation of 'fixed and individualised' environmental property rights to natural resources in PES regimes are discussed. The chapter concludes with an outlook on the PES schemes and stresses the limitations of the contribution of PES to worldwide environmental sustainability and rural development objectives.

9.2 Some Definitions

Ecosystem services denote the benefits people obtain from ecosystems (MEA 2005). These benefits are divided into four categories:

- Provisioning services, e.g., food, fuel-wood and medicinal resources
- Regulating services, e.g., climate regulation, air purification, water regulation and pollination
- Supporting services, e.g., conservation of biodiversity and the gene pool
- Cultural and amenity services, e.g., recreation, tourism, cultural and spiritual uses

Environmental services refer to the often intangible benefits that people obtain from the environment. While there is a significant overlap with the concept of ecosystem services, environmental services typically do not include provisioning services, i.e., the tangible products and services that often have a clear market value (Swallow et al. 2009).

The term **payments for environmental services (PES)** refers to the full set of environmental services and includes payments for the regulating, supporting and cultural services provided by ecosystems. PES are defined as a transfer of monetary or non-monetary resources between social actors, which aims to create incentives to align individual and/or collective land use decisions with the social interest in the management of natural resources, and which is being articulated through a market and/or other mechanisms such as incentives or public subsidies (Muradian et al. 2010). In a more restrictive definition, PES is described as a "voluntary transaction where a well-defined environmental service is being bought by a buyer from a service provider if—and only if—the provider secures the service provision" (Wunder 2005). However, in practice PES programmes rarely live up to this restrictive definition and can best be described as 'PES-like' programmes. Typical problems are often a not entirely voluntary nature of the programme, difficulties establishing a clear causal link between a certain land use practice and a quantity of environmental service provided, and problems ensuring a long-term monitoring of the resource and the level of service provision (e.g., Corbera 2011; Muradian et al. 2008, 2010; Rapidel et al. 2011; Rico et al. 2011; Swallow et al. 2009).

Effectiveness is described here as the ability of the PES scheme to achieve the environmental objective. In a watershed-based PES project, for example, this would refer to the extent to which the project contributes to the improvement of the water quality or water flow regulation. The main characteristics implied in terms of PES effectiveness are (1) the optimisation of the payments; (2) the high transaction costs; and (3) the targeting of land that contributes the highest value to the desired services.

Efficiency refers to the level of achievement of the PES goals at a minimum cost. For example, a PES project that sequesters carbon at a lower cost is more efficient than a project where the cost per tonne of carbon sequestered is higher.

Low transaction costs, a focus on large tracts of land and untargeted payments are the main characteristics considered with respect to the efficiency of a PES scheme.

Equity is determined by the level of representation and involvement of different stakeholders in the design and implementation of the PES programme, for example, providers, intermediaries and/or users of the environmental services. Equity also refers to the fair distribution of the costs and benefits created by the programme. Therefore, the terms *fairness* and *justice* are often used as a complement or substitute for equity. The main characteristics with respect to the equity of a PES scheme are high transaction costs, untargeted payments and a focus on small land users.

9.3 Evolution of the PES Concept and Its Strengths and Weaknesses

Throughout the world, environmental conservation regimes for a long time mainly used rigid mechanisms and a *command-and-control* logic to achieve their objectives, such as legislation and regulations, the demarcation of protected areas, restrictions on property rights, and subsidies, taxes and other traditional financial instruments. It is only relatively recently that, along with a more general paradigm shift towards neo-classical economic approaches, many more flexible and market-based policy instruments have been developed and implemented in the context of environmental conservation programmes. Among these are, for example, cap-and-trade systems for carbon emissions, environmental offset schemes such as the voluntary carbon markets, eco-labelling such as Fairtrade and FSC, reverse auctions and agreements between beneficiaries and providers of environmental services (for deeper analyses see Mayrand and Paquin 2004; Broughton and Pirard 2011; Swallow et al. 2009).

Ecosystem services as a concept were introduced in the 1980s by ecologists as a 'communication tool' with the aim of making the societal value of ecosystem functions more explicit and publicly known (Ehrlich and Ehrlich 1981). Since the late 1990s the concept has slowly entered the policy and market arena, and research and development projects have increasingly focused on determining the monetary value of ecosystem services and designing market-based instruments to create economic incentives for conservation. Moreover, the Millennium Ecosystem Assessment project (MEA 2005) added significantly to fostering policy support for the conservation of ecosystem services, and to creating awareness for the link between the provision of these services and human well-being. The process of ecosystem function commodification went through three stages (cf. Gomez-Baggethun et al. 2010):

- The framing of ecological functions as services in utilitarian terms during the 1970s and 1980s. At this stage, the ecosystem service concept was mainly used

as a communication instrument without yet associating it to the economic processes of valuation, appropriation and exchange.
- The monetisation of environmental services mainly in the 1990s, when numerous studies revealed the economic importance of environmental services, for example, through monetary valuation of the benefits and costs associated with natural ecosystems and their loss or conversion.
- The increasing appropriation and exchange of environmental services, where providers and users of these services were linked in a market place. Building on the process of appropriation of intangible environmental services, institutional structures and markets for these services were created during the 2000s, which allowed for transactions and market exchanges addressing environmental services.

PES schemes operate on the premise of the Coase theorem, which postulates that economic efficiency can be achieved by the allocation of property rights through bargaining in the absence of transaction costs. By identifying externalities in the environment, and by internalising these effects so that prices transmit the true cost and benefits of any product or service, economic resources can be allocated most efficiently. Yet, rather than the *polluter pays* principle, PES is based on the friendlier 'beneficiary pays' principle (cf. Engel et al. 2009). When forest users obtain payments to compensate for the external benefits their land use decisions generate, the opportunity cost of forest conversion reflects the full environmental costs, making this a less attractive land use option.

PES has been evolving logically to incorporate a broader promotion of markets for property rights in the context of the *tragedy of the commons*. The objective here being the avoidance of the underutilisation of a resource in common property regimes.

In comparison to government regulations, PES schemes are said to have the potential to be more efficient, effective and flexible (Mayrand and Paquin 2004). While PES agreements tend to concentrate on areas that generate higher benefits from environmental services, traditional conservation mechanisms often have a more limited impact on the provision of environmental services. In addition, command-and-control mechanisms are often poorly enforced, mainly due to a lack of institutional and financial resources in public administrations. There exist severe limitations with regard to the establishment of protected areas in or close to densely inhabited spaces, increasing the costs of implementation and creating social and political distress. Furthermore, there are many ethical and practical implications when limiting access to natural resources for people whose livelihoods directly depend on these (Mayrand and Paquin 2004). PES schemes, in contrast, directly address environmental conservation by generating financial resources for users who foster the sustainable provision of environmental services. These revenues also have the potential to contribute to rural development and the alleviation of poverty.

However, experiences so far have shown that there are a number of critical issues that require special attention when implementing PES programmes. Properly identifying and quantifying the environmental services provided, and monitoring the effective and fair distribution of revenues among the service providers are core prerequisites for the delivery of the agreed services, which typically need to be put in place by independent intermediary agents. As a consequence of the lack of effective monitoring institutions, many PES programmes do not ensure the delivery of the desired services due to the absence of additionality, or the inability to clearly demonstrate the actual service provision (Bond et al. 2009). Further criticism stems from the following: (1) the actual complexity of the system is often oversimplified, by breaking down the intricate relationships between nature and humans, thus neglecting the attributes of the environment and its role; (2) an inability to adequately reflect all facets of the actual value, worth and meaning of environmental services, as trading uses a single exchange value; and (3) the development of market-based mechanisms for environmental conservation may foment underlying power asymmetries related to access to these resources and services rather than reduce them (Kosoy and Corbera 2010).

9.4 Local and Global Challenges for PES

9.4.1 Types and Current Scope of PES Projects

PES schemes are highly heterogeneous, as they differ with regard to the environmental service(s) being provided, the geographical scale, the governance model and the source(s) of financing. It is worth briefly distinguishing the different existing categories of PES programme before drawing inferences on the mechanisms. A more detailed description of examples and categories of PES scheme are provided in Bond et al. (2009), Landell-Mills (2002), Mayrand and Paquin (2004) and Porras et al. (2011).

PES schemes differ essentially with regard to the *type of service* they provide. Of the broad spectrum of environmental services, only a small subset is currently of relevance for PES schemes. Watershed protection, carbon sequestration, biodiversity conservation and preservation of landscape beauty (mainly used for ecotourism purposes), either addressed individually or as a bundle, are the services that presently dominate these mechanisms. Most other environmental services are currently not amenable to PES due to a lack of a demonstrable and defined private demand, as well as the lack of appropriate institutions promoting and ensuring the provision of these services. With a further advancing international environmental regulatory framework these services may also be internalised in PES schemes focused in bundled services. If well intended, PES schemes could become an instrument embracing the values of these supporting and cultural services along with other services, making them economically attractive and bringing them to the market in a landscape approach (see Rosa et al. 2004).

The *geographic scale* is another defining variable of PES schemes. These schemes operate at local, regional, national or international levels. There are PES programmes confined to a specific watershed, such as the Platanar Hydroelectric in Costa Rica, reaching a geographical scale of below 100,000 ha. National programmes, such as the Programme of Payment for Hydrological Environmental Services of Forests in Mexico, or sub-national programmes like the Bolsa Floresta in Manaus, Brazil, span more than 10 million hectares. International agreements promoting PES, such as Norway's International Climate and Forest Initiative in the Congo Basin, go beyond even that.

The scale of a PES scheme is typically determined by its *financing sources*. There are four main sources of financing: (1) government finance; (2) user finance; (3) market finance and (4) donations and grants. While very useful particularly during the inception of such programmes, the financing of PES schemes through government finance, donations and grants typically raises concerns regarding their sustainability, as their long term success much depends on continued governmental support, making them vulnerable to the economic and political climate at the time. When financed directly through user fees, PES programmes have a more direct and reliable source of revenue.

Two major models can be differentiated with regard to the *governance* of PES schemes: in public schemes, the government acts as an intermediary between the providers and the beneficiaries of the environmental service. As public schemes tend to pursue a political agenda and objectives other than environmental conservation, these tend to be less efficient (Wunder 2005). Private schemes, by contrast, are generally more focused locally, and often have a more specific and better defined objective. These schemes typically involve private entities as buyers and sellers of the services. Upon the basis of an agreed contract, the negotiated price or non-monetary compensation is given to the service providers directly or through an intermediary.

As a global overview of the status of different PES schemes, Ecosystem Marketplace conducts continuous research on the payments for watershed services, the voluntary carbon markets and the biodiversity markets. In 2008, Stanton et al. (2010) identified a total of 216 payments for watershed services (PWS), most of them as part of programmes developed in the last decade. Of these, Latin America was home to the highest number of identified programmes (101), with 36 active programmes contributing USD 31 million to watershed conservation measures and impacting 2.3 million hectares. The Asian region hosts 56 active programmes, 47 of which are in China and account for USD 7,800 million and 270 million hectares. These include China's national programme, under which water payments are bundled with other environmental services.

In 2010, land-based projects supplied the largest volume (28 MtCO2e, metric tonne carbon dioxide equivalent) of credits transacted in the voluntary carbon market (roughly 45 %), with a volume-weighted average price of USD 6/tCO2e. This represents a more than fivefold increase in volume from 2008 (see Fig. 9.1), when just 5.3 MtCO2e credits were transacted; accounting for a total area of 2.1 million hectares under projects influenced by forest carbon sequestration or avoided

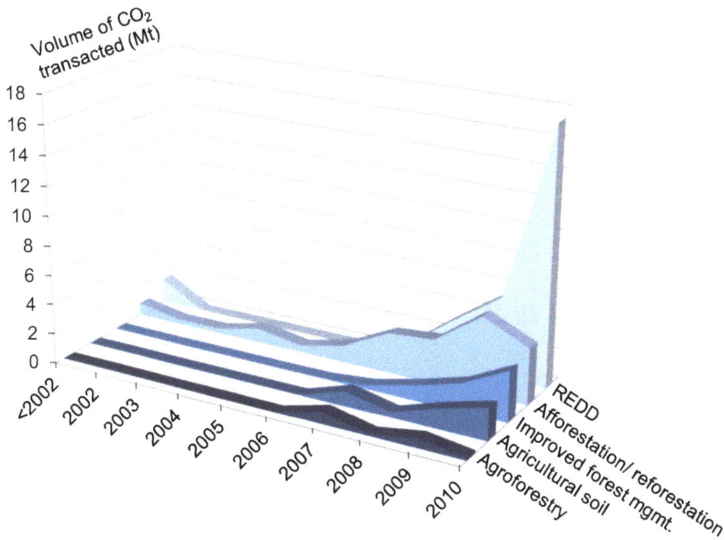

Fig. 9.1 Historic transaction volumes on the voluntary carbon market from forestry and other land use types (Source: After Peters-Stanley et al. (2011))

emission activities. The Global South led these trends as governments, foundations and buyers built local markets and highlighted forestry's potential as a large-scale climate solution. Additionally, conservation efforts and international politics spotlighted projects reducing emissions from deforestation and forest degradation (REDD), which generated 29 % of the transacted credits. Other land-based projects such as afforestation/reforestation (A/R) projects, improved forest management (IFM) projects, agricultural soil (Ag soil) and other forestry (in general) projects were also present with 6 %, 5 %, 3 % and 2 % of the credits transacted in the voluntary market, respectively (see Fig. 9.1). As in the PWS, Latin America dominated the market supply with the region's forests harbouring 81 % of all REDD credits and over half (15.7 MtCO2e) of all forestry credits transacted in 2010 (Peters-Stanley et al. 2011; Hamilton et al. 2010).

In 2010, 45 biodiversity markets were active and 27 were reported to be in development. The global annual market size falls between USD 2.4 and 4.0 billion with at least 187,000 ha of land restored or preserved each year (101,000 ha more than in 2009) (Madsen et al. 2011). By contrast to the programmes mentioned above, the biodiversity markets were dominated by the North American region with 15 existing compensatory mitigation programmes (followed by the Central and South American region with seven active markets).

Finally, Milder et al. (2010) estimated that by 2030, markets for biodiversity conservation in developing countries could benefit 10–15 million low income households, carbon markets 25–50 million, watershed protection markets 80–100 million, and markets for landscape beauty and recreation 5–8 million. Whether these estimates are feasible or not, very little is known about the internal distribution of the payments, the improvements to local sustainable development brought about by PES – including changing the attitudes and behaviour of providers and receivers of marketed environmental services – and the effectiveness of the country-wide policies on conservation and recuperation of natural resources.

9.4.2 Approaches to Determine the Value of Environmental Services

The lack of an effective attribution of monetary values to the environmental benefits provided by ecosystems has been cited as a major cause of their continuous degradation (MEA 2005). Through the valuation of environmental services it is possible to determine how environmental services benefit current and future generations, and how human land use choices and other activities impact the provision of these services. Quantifying the value of an environmental service constitutes an integral step in the design of a PES scheme.

The way ecosystems and biodiversity are valued is determined by how humans perceive nature (TEEB 2010). While some valuation approaches are rooted in a largely anthropocentric perspective, where the natural system and its interactions are analysed with an instrumental notion, other approaches originate from a more biocentric way of thinking, which also includes non-human and intrinsic values of natural capital stocks. The choice of valuation method will, therefore, via the output it produces, influence the future behaviour of the stakeholder groups vis-à-vis their environment. In order to obtain the most comprehensive assessment, the economic valuation must be complemented by analyses of non-monetary values, such as socio-cultural, religious and physical valuations. The concept of the total economic value (Fig. 9.2) is most commonly used to estimate these values in economic terms.

A variety of methods have been developed for the economic valuation of environmental services (see Table 9.1). However, most of these are characterised by specific limitations and constraints, and only a few of them have been employed in the tropics (e.g., Bernard et al. 2009; Kaiser and Roumasset 2002; Lele 2009; Pattanayak and Kramer 2001; van Beukering et al. 2003).

Economists often measure values based on the market behaviour of costumers. These are categorised as direct market valuation approaches. However, it is not always possible to infer the value of environmental services from the market. In such cases, indirect valuation methods might be used, including revealed preference approaches, where people's choices and behaviours related to environmental services are revealed on the market, and stated preference approaches, where hypothetical markets are established and the values of the environmental services are estimated with the help of survey methods.

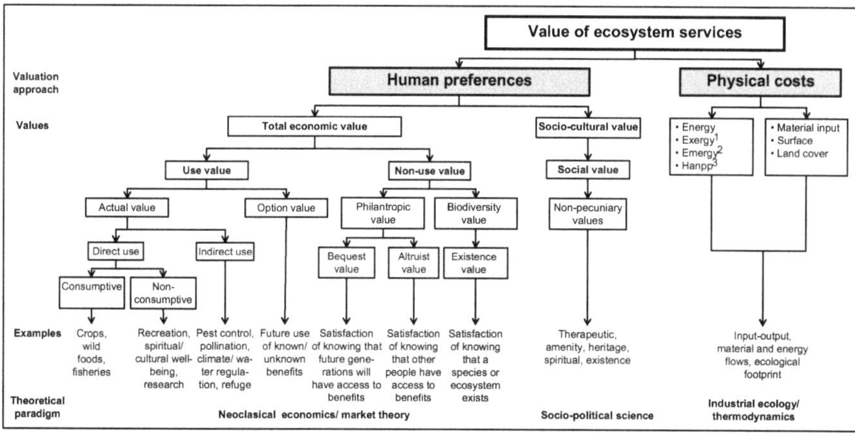

Fig. 9.2 Values of ecosystem services (Source: Adapted from Gomez-Baggethun et al. (2010), MEA (2005), TEEB (2010))
[1]Maximum amount of useful energy that can be extracted when matter is brought into equilibrium with its surroundings.
[2]Total amount of exergy of one kind that is directly or indirectly required to make a given product or to support a given flow.
[3]Human appropriation of net primary production (Hanpp): the difference between potential and actual (remaining) vegetation after harvest (for more details see Erb et al. 2009; Hau and Bakshi 2004; O'Neill et al. 2007)

Direct market valuation is often not the most suitable approach for environmental services. The main limitation is the absence of competitive markets, where sufficient supply and demand is available to correctly reflect the price of the services. Market valuation of environmental services is also complicated by the fact that there is only limited knowledge available with regard to the services being provided, the markets they refer to are partially distorted and the costs involved in restoring or technologically replacing these services are not entirely known. Therefore, the results of such valuations often represent only partial information and the true value of the services is frequently underestimated.

The revealed preference approach aims to determine the value of environmental services based on market data or the costs involved in their consumption. The analyses are often complicated by the fact that it is difficult to obtain sufficiently accurate and reliable data for these valuations. Marketable goods are generally poor proxies for these services as they usually exclude non-use values, and often the relationship between land use and the services is not well-defined.

While equally applicable to use and non-use values, the stated preference approach often delivers inconsistent and unreliable results given that stated preferences are very context-specific and variable, and depend on factors such as the

Table 9.1 Valuation approaches and methods

Approach		Method	Examples
Market valuation	Price-based [3+4]	Market prices	Mainly applicable to the 'goods' (e.g., fish), but also some cultural (e.g., recreation) and regulating services (e.g., pollination)
	Cost-based [3+4]	Avoided cost	The value of a flood control service can be derived from the estimated damage if flooding were to occur
		Replacement cost	The value of groundwater recharge can be estimated from the costs of obtaining water from another source (substitute costs)
		Mitigation/restoration cost	E.g., cost of preventive expenditures in the absence of wetland services (e.g., flood barriers) or relocation
	Production-based [4]	Production function Factor income	E.g., how soil fertility improves crop yield and, by extension, farmer incomes; how improved water quality increases commercial fisheries and so the incomes of fishermen
Revealed preference [3+(4)]		Travel cost method [3+(4)]	E.g., part of the recreational value of a site is reflected in the amount of time and money that people spend travelling to the site
		Hedonic pricing [3+4]	E.g., clean air, presence of water and aesthetic views will increase the price of surrounding real estate
Stated preference [1+2]		Contingent valuation	Respondents are asked their willingness to pay for improved environmental services. Often the only way to estimate non-use values
		Choice modelling/conjoint analysis	Applicable through different methods, e.g., choice experiments, contingent ranking, contingent rating and pair comparison
		Contingent ranking	Respondents are asked to rank a large number of alternatives with combinations of environmental goods and prices
		Deliberative group valuation	Allows shortcomings of revealed preference methods to be addressed, such as preference construction during the survey and lack of knowledge of respondents about how to allocate values

Approaches mainly applicable to the following ecosystem values: [1]use value, [2]non-use value, [3]direct use, [4]indirect use
Source: Adapted from TEEB (2010)

interviewer's experience, the questionnaire design, the interviewee's personal situation, and the potential discrepancy between a hypothetical willingness to pay and actual behaviour. The assumption that an individual can estimate accurately, in a common metric, the values of the environmental services provided is very naive.

A method increasingly being used in ecosystem services valuation is communal discussion or group valuation, which draws attention to complex services in group dynamics. The use of focus groups in this approach can help overcome the time and money constraints typical of individual surveys. However, this method also has implications with respect to the sample size and the expression of personal opinions.

It is essential to point out that the mere fact of assigning environmental services a monetary value does not reflect their full importance or holistic value. Non-monetary issues need to be taken into consideration when a more complete perspective is sought. The 'true' price of environmental services cannot be determined through valuation methods alone. There is no single most appropriate valuation method that suits every scenario. Depending on the specific socio-economic context and the conditions of the ecosystem, the most suitable methods must be selected. Social perceptions, political views and bargaining power are crucial in the complex negotiations between the stakeholders over the final price of the services (Smith et al. 2006). Therefore, valuation techniques, as tools for decision-making, need to be explicit about the methods they employ and decision-makers should appropriately acknowledge their limitations.

9.4.3 *Effectiveness, Efficiency and Equity of PES*

There is an increasing number of PES programmes being implemented worldwide, and investors, governments and the general public are interested in seeing the current and future programmes designed and implemented in the most effective and equitable manner, and this done as efficiently as possible. Consequently, PES is certain to become rooted as one of the most attractive mechanisms to provide environmental services, for both buyers and sellers, and for society as a growing number of funding sources and programmes seek to preserve the environment.

9.4.3.1 Effectiveness

In order for a programme to achieve its main goal, in this case the provision of environmental services, it must first improve conditions vis-à-vis the business as usual provision, and so be effective. This improvement, also termed additionality, can only be demonstrated where the original conditions (baseline) have been determined and the changes in the provision of the services, related to the influencing system or environment (i.e., land use change), can be measured. More often than not, programmes are implemented and run merely assuming these parameter settings, which can lead to a failure to meet their chief objective, rendering them ineffective. For example, if the minimum habitat requirement for a population of a particular species to persist is not clearly known and a smaller than required area is preserved for this purpose, land use change in the neighbouring area may be critical and reduce significantly the probability of successful preservation, resulting in a failure to achieve the intended objectives. This lack of 'effective additionality' is commonly observed in PES programmes focused on avoiding deforestation where contracts are made with service providers who would have kept the forest independently of their participation in the programme or where, as a consequence of protecting one forested area, a similar area is being deforested somewhere else (leakage).

Including participants enjoying higher profitability from conservation than from agriculture, or with very small areas of land, is not effective as they do not represent a major threat to the forests, and so no significant additionality is achieved. Wunder (2007) argued that the most promising scenario for PES schemes is where the margin of profitability is slightly lower under the desired land use (i.e., conservation) than under the undesirable uses, and that whenever the opportunity costs are too high to be covered by such mechanisms, other instruments such as command-and-control measures should be implemented.

Another way to bring a programme closer to achieving its objectives is through effective targeting (Pagiola 2008; Wünscher et al. 2008). This presupposes selection criteria where, depending on the objectives of the PES programme, specific areas or participants will be targeted and rated according to their deforestation risk (present and future), biodiversity values, carbon sequestration potential, water quality improvement and stream flow regulation potential, to name but a few, as well as additional objectives relating to the characteristics of the participants (i.e., poverty rates, access, land area), which may even be spatially correlated.

There are several additional issues described by Jack et al. (2008) that also affect effectiveness. Government subsidies, for example, often thwart PES objectives, like in the Brazilian cerrado where, stimulated by government subsidies, soy production expanded dramatically. Before beginning a PES programme in a particular area, it is necessary to understand the factors affecting the supply of the environmental services. In some cases, better outcomes can be attained without the need for new programmes, but merely by addressing these counterproductive policies.

Attention should be also paid to changes over time. The modification of environmental, socio-economic and political contexts as well as changes to opportunity costs can greatly influence the effectiveness of PES programmes. Therefore, it is advised that possible changes over time be envisaged in the design phase and that flexibility be allowed for over the lifespan of the programmes.

There have been very few studies focused on the effectiveness of PES programmes in comparison to business as usual conditions, for example, in similar areas that are not included in the programme (Pattanayak et al. 2010). Therefore, it is important that PES programmes encourage and make space for such evaluations in order to contribute to better outcomes and improved opportunities for current and future programmes.

9.4.3.2 Efficiency

Optimising the efficiency of PES mechanisms allows not only for the inclusion of greater areas in a selected programme with a defined budget but also makes these schemes more attractive for investors and donors, thus expanding PES programmes in size and number, consequently increasing the desired final outcomes.

Special attention must be given to the design, measurement and monitoring of the programme and the desired services so as to achieve efficiency. Through the process of monitoring and verification, non-compliant participants can be sanctioned or even expelled from the programme (conditionality). Consequently, the scarce resources can be allocated in more *productive* areas, increasing the efficiency and effectiveness of the programme. Nevertheless, such measures should include and analyse any increment in the transaction costs stemming from these activities, so that these do not outweigh the potential benefits. Conditioning these contracts entails a quid pro quo arrangement that encourages the permanence of the services provided. However, these decisions depend on the political agenda and even if non-compliance is detected, sanctions might not be applied (Bond et al. 2009).

Targeting can also improve cost-effectiveness, by boosting the performance of the mechanisms. In so doing, efficiency is increased as there are more benefits (greater provision of environmental services) for a given budget. It is important to keep in mind that targeting, even though it turns PES into more effective and efficient mechanisms, entails higher transaction costs that should be calculated by means of cost-benefit analyses and internalised so that the most advantageous level of targeting specificity can be reached.

The best means of achieving a high level of efficiency to date has been to include in the project design the calculation of the participation costs. These include the compliance costs, the transaction costs and the opportunity costs. The compliance costs are the costs borne by the participants with the purpose of undertaking the agreed activities for the service provision, such as forest rehabilitation efforts. Transaction costs, on the other hand, include costs associated with the design, negotiation, enforcement and monitoring of PES programmes. Transaction costs mainly depend on the number of participants, the size of the properties, information availability, access and institutional strength (including monitoring capacities). Hence, implementation of PES programmes in areas with limited access and with very little corresponding information available will incur high start-up costs, especially if transactions are carried out with smallholders. Finally, opportunity costs should also be included in order to estimate the potential service provider profiles (i.e., farmers who are less dependent on their forest land or who make little profit from their crops) and the potential amount of land (environmental service provision) that can be included in the programme for a given budget. Including participants with very high opportunity costs can be an effective measure as they are more likely to continue deforesting areas in order to expand their profitable activities (i.e., oil palm or soybean plantations). However, this measure is unlikely to prove efficient due to the excessive payments incurred in covering the participants' opportunity costs, which only permits the inclusion of a smaller area in the programme, with lower overall environmental benefits as a consequence.

Given that efficiency in PES schemes is achieved through the maximisation of the net benefits, taking into consideration the environmental benefits and the corresponding costs, and through the various participation costs and the potential provision of environmental services from each particular site, fixed payments are not necessarily the most efficient alternative for PES programmes. Differentiating the payments according to the aforementioned conditions means it is possible to avoid making payments that are either too low, which are unattractive (resulting in lower participation rates and/or lower compliance), or too high, both of which serve to reduce the efficiency of the programme.

Merely targeting effectiveness and efficiency in PES schemes, although often seen as being very attractive and even the 'best current option,' does not provide for the sustainability of programmes in the long term, and so poses a threat to the permanence of the provision of the environmental services.

9.4.3.3 Equity

An important concept influencing the long term viability of PES is equity. Corbera et al. (2007) stressed the importance of equity, stating that, when aiming to produce more efficient outcomes, PES programmes are likely to reinforce inequalities in access to resources (and existing power structures), triggering social tensions. Although there are many ways to define equity, they approached equity from three different perspectives, namely equity in access, in decision-making and in outcome.

The first refers to the probability of an individual participating in a particular PES programme. This is very much linked to the human and social capitals (i.e., access to information, knowledge and networks), natural capital (land and forest resources) and financial capital (Porras et al. 2011). Farmers without information about the programme, with a lack of knowledge regarding their options (and with insufficient capacity in enforcing these options) and with no lobbying capacity (and/or reduced network facilities) are at a disadvantage in terms of accessing these PES mechanisms. In addition, the transaction costs involved are, for many poor farmers, prohibitive, which is again an access limitation. To reduce these transaction costs, collective contracts (as applied in Mexico's and Costa Rica's national PES programmes) can be drafted, where individuals and communities are awarded a common contract reducing the individual participation costs (Bond et al. 2009). Therefore, if it is an intention of the PES programmes, and of the institutions behind them, to include poor farmers (pro-poor initiatives), transaction costs should be lowered and, where possible, covered by the service buyers or intermediary institutions, so that they do not create a barrier to people who cannot afford these start-up costs. These intermediary institutions can also encourage the participation of smallholders by supporting them in the design and negotiation phases. This corresponds with the premise that PES mechanisms should be flexible and explicitly designed for each specific set of circumstances.

In order for this to be effective, and to ensure positive outcomes among poor participants, these stakeholders should be included in the decision-making processes, so that they can also influence the conditions and outcomes of the programmes. Equity in decision-making is crucial to the success and sustainability of a programme. Providing equal opportunities to all participants in decision-making processes ensures greater social benefits and higher involvement in the provision of the environmental services through a sort of takeover of the PES programmes' objectives. It also undermines the differences in the bargaining power of wealthier stakeholders (i.e., large landowners) and poor smallholders, and reduces possible vulnerabilities in the programme.

Given their low opportunity costs, the characteristics of their location and the risks of deforestation on their land, theoretically, poor farmers should be more likely to become participants of PES programmes. Porras et al. (2011) also arrived at this conclusion but added that, in practice, the wealthier landowners are probably more likely to adopt a PES programme, not just because they have greater access to information and more connections to influential people but also because they are the ones who possess better and greater assets, while at the same time having a lesser dependence on these assets (higher availability of service provision). This issue describes the difficulties stemming from the allocation of outcomes, more specifically equity in outcomes.

The allocation of outcomes is very much dependent on the decision-making processes and on access to the programme. These will determine the areas where the project will be implemented (e.g., through some targeting methods) as well as the amount of payments and the conditions under which these will be delivered. Moreover, as equity can be viewed from different perspectives, it may be considered unfair whenever a participant receives the same payment for an area that provides more of the desired environmental services as that received by a participant with a less profitable area. On the other hand, a participant may deem it unfair that although he has set aside the same expanse of land as his neighbours he receives a lower return due to a system of differentiated payments. The same assumption can be made when considering the percentage of area set aside for service provision in relation to the entire property area.

All of this can lead to a conflict between the search for equity in the PES programme and the achievement of its environmental objective (main goal). Therefore, apart from the mechanism used, whether equal or differentiated payments, the nature of the service providers should be noted. Even if the integration of poor farmers into a programme as a poverty alleviation measure results in a less efficient mechanism for the provision of environmental services in the short term, this may prove in the long term to be a more effective and even more efficient mechanism. A programme seen to be positive for both the environment and for society, helping to undermine the differences between wealthier and poorer farmers rather than increasing them, has better chances of success than a programme that is more efficient environmentally (in the short run) but which serves to generate social conflict (because of targeting and selection criteria).

Concerns persist over how to estimate certain components such as the valuation and distribution of the opportunity costs between this and the coming generations, which must be assessed and studied.

9.4.4 Trade-offs Between the Three Dimensions

It is generally not possible to achieve efficiency, equity and effectiveness objectives to the same degree. In other words, the outcomes will not usually encompass the full attainment of these three principles simultaneously. Therefore, the interest group must go through a decision-making process, arriving at a trade-off between efficiency, equity and effectiveness. Trade-offs can be distinguished by their levels of scale, i.e., macro and micro levels. Here, stakeholders (as an individual or as a group) make decisions of national character (macro) or very specific (land use at the plot level) which may have, at the same time, a local or a more generalised impact.

In this section, the focus is on the trade-offs between macro levels, where the design (including selection criteria), planning and implementation of PES schemes are involved.

As Pascual et al. (2010) described, there is not a single equity or fairness criterion but rather several criteria. Depending on the perception, it can range from distributive equity (or even relative distributive equity) to equity related to the opportunity costs incurred. The diverse selection of criteria presented illustrates equity scenarios that have in turn different implications for the efficiency outcomes, thus locating the set of alternatives along a trade-off curve between those principles. Le Grand (1990) emphasised that equity and efficiency cannot be evaluated similarly because the latter is actually the way of achieving an objective and not an objective itself. In spite of the validity of this assertion, a decision – taking into account both principles – must be made. Yet, each alternative will affect the outcomes, for the participants and the project, and therefore efforts should be made to identify them and take them into consideration during the decision-making process.

As a result, information is a key element that should be provided by researchers and through the experiences gleaned from the monitoring processes of previous projects. Here, it is of great importance that all stakeholders have complete information – transparency – so as to avoid influencing the decisions of policy-makers and thus (deliberately) further confine the final outcomes of the programme to one or more of the three principles. Therefore, the higher the level of ignorance (lack of information) on the part of one or more of the stakeholders – including policy-makers – involved in the trade-off, the more unequal, inefficient and ineffective the PES programme will be.

If we underestimate the equity principle and favour efficiency criteria (i.e., by including poor participants only if they provide ES efficiently), the potential risk of civil unrest (whenever socio-economic inequalities increase), leakage issues etc. may threaten the permanence of the programme and, therefore, its effectiveness.

Information concerning the potential participants in the programme is needed in order to explore power relations between them and to assess the status of their assets and their limitations. This will have a positive influence mainly with respect to the equity principle.

Moreover, information about the baseline of the services provided is also essential so that additionality can be calculated so as to ensure the delivery of the services and thus the effectiveness and efficiency of the programme. Another factor to take into account is the cost of the provision of service, the participation costs including opportunity costs and transaction costs. This calculation is very challenging whenever other parameters are considered, such as costs (and benefits) related to the providing services for present and future generations, and the valuation of non-renewable services. Its difficulty also lays in the inclusion of moral values and equity issues in the calculation.

When calculating, costs involved in the trade-off decision should also be included in the transaction costs (Grimble and Wellard 1997). As discussed in the previous section, transaction costs will vary depending on the design and implementation of the programme. The same occurs depending on the condition of the country and/or implementation area, with variables such as access, maturity of markets and organisation.

Decision-makers should, therefore, decide from among the several options available, taking into account the repercussions in terms of the participation costs and acknowledging the further implications for the equity, efficiency and effectiveness outcomes. Even public schemes with reduced budgets should be aware and act in accordance with these principles during trade-off analysis. Possible final outcomes should be considered so as to avoid cutting the wrong corners.

Reducing PES expenditures at all costs, by reducing participation costs (i.e., no targeting, modest design, lowest opportunity costs), may seem an efficient option, but it is likely to be unprofitable in the long term given the low additionality, higher risks of non-compliance, equity issues likely to affect at some point the effectiveness of the programme, etc. By contrast, investing in participation costs may in some cases provide the best trade-off outcomes with respect to the three principles. Whenever there is certainty and determination, programmes can be designed to have site-specific arrangements and remain flexible. This may result in non-monetary benefits to the service providers with one of the five capitals (from the livelihood assets), other than the financial, which at most affects their well-being. Addressing local concerns can ensure higher compliance in the programmes and thus higher effectiveness. This may also result in a more efficient allocation of resources than with monetary incentives, having a greater and longer lasting provision of environmental services with lower net present costs.

Moreover, investing in physical, social and/or human capital (when a shortage is evident) can secure land tenure, reduce vulnerability, strengthen local institutions and capacity building, and, in many cases, be more equitable than money incentives (which may even have negative impacts).

An ideal win-win situation for providers and users in terms of efficiency and equity criteria will ensure more often than not the effectiveness of a PES programme. A PES scheme that does not satisfy one of the target groups cannot be certain of its success in the long term, reducing its effectiveness. For example, whenever the provision of an environmental service becomes too expensive for the user (in relation to other options), making it inefficient, it poses a threat to the permanence of the programme, even though it may be highly equitable.

Furthermore, a PES programme with a fixed budget and very high costs for the provision of services will not have the wherewithal to cover the same amount of services as a more efficient programme. Consequently, a smaller number of providers will benefit from the programme and, significantly, future generations will be affected adversely by a more limited provision of services, both of which increase inequality.

Alternatively, a programme that delivers benefits below the cost of the opportunities of the providers, but which increases inequity or benefits to just those providers who attempt the greatest use of their forests, will not be sufficiently equitable though it may be very efficient for all users. Any such programme is also likely to face problems of effectiveness whenever social conflicts arise.

All in all, trade-offs must be incorporated and assessed, and decision-makers have to weigh up the most beneficial long term outcome from the different possibilities, in keeping with the objectives of the programme. A good starting point is probably the setting of a minimum acceptable equity level that is to be achieved and to define from there the most efficient way to achieve it. The alternative selected should be equitable enough to be of advantage for the participants (ES providers), so that they are committed to being part of the PES programme. This should also occur in an efficient manner so that it attracts users and other (international) decision-makers to apply the scheme so that ultimately it is effective enough to ensure the benefits of the programme for this generation and the next.

9.5 Outlook

PES is an expanding and sound mechanism of environmental conservation that is becoming ever more relevant, especially when the progress of the voluntary carbon markets and the projections of the REDD + activities are taken into consideration. The latter has been of great concern in recent years and considerable effort is being devoted to accelerating and strengthening its development (mainly at national level) and to implementing pilot initiatives. Although international funding is very low in relation to the needs, it remains significant in comparison to earlier attempts.

Existing PES schemes are far from financially sustainable and, to date, the market mechanism does not appear to be the solution. Market measures alone cannot provide sufficient guarantees of environmental sustainability and of local development under conditions of equity and climate justice. To rely solely on tradable services with values dependent upon fickle and volatile international markets (i.e., commodity prices of oil, gold and rice) is overly optimistic, even naïve. For instance, the value of carbon sequestration, as a regulating service, on the exchange market will arguably be influenced by technological development. This may be achieved through the development of more cost effective solutions replacing this regulating service, or even through the widespread use of energy from renewable sources, reducing the demand for tradable carbon credits; the latter occurring over a longer timeframe. The long term success of these schemes will, therefore, depend on their ability to overcome these limitations. Market mechanisms still possess many gaps (some of which could be filled by means of regulation) and need to mature. Moreover, current schemes should be replaced by more flexible mechanisms and the inclusion of institutions, so that the sustainability of the system is ensured.

Another main issue affecting current schemes is the gap between the resource users and property rights. A common solution would appear to be out of the question in this respect as tenure systems differ greatly from country to country. Well defined and formalised tenure rights will certainly increase the costs of a scheme, especially when done individually. However, unclear tenure rights lead to conflict (particularly an increase in the value in forest land) and increasing inequity (where poor people's rights may be usurped). The securing of tenure rights will support people conserving forests and ensure higher and longer term engagement in PES programmes (contributing to greater efficiency and effectiveness).

Furthermore, when looking at the three criteria and the trade-off to be considered, decision-makers should design PES schemes maximising effectiveness and efficiency based on the foundation of a satisfactory and suitable equity level.

Throughout the process of selection of the expected outcomes for each of the three principles of the programme (and the further design and implementation processes), decision-makers and the stakeholders exerting influence over them are a central factor. Genuine efforts on the part of decision-makers and an open-door policy are needed to shape the desired outcomes and the success of PES programmes.

Finally, PES is an instrument of significant potential in certain particular contexts. It should not be held up as a uniform means to solve deforestation and water issues worldwide. It should simply be seen as a mechanism helping to achieve the objectives of ecosystem conservation and rural development (through poverty alleviation). It is by no means a final solution, merely part of it.

References

Bernard F, de Groot RS, Campos JJ (2009) Valuation of tropical forest services and mechanisms to finance their conservation and sustainable use: a case study of Tapanti National Park Costa Rica. For Policy Econ 11:174–183

Bond I, Grieg-Gran M, Wertz-Kanounnikoff S, Hazlewood P, Wunder S, Angelsen A (2009) Incentives to sustain forest ecosystem services: a review and lessons for REDD. Natural resource issues no 16. IIED/CIFOR/WRI, London/Bogor/Washington, DC

Broughton E, Pirard R (2011) What's in a name? Market-based instruments for biodiversity, Analyses 03/11, Institute for Sustainable Development and International Relations, Paris

Corbera E (2011) Payments for environmental services: are they a solution for conservation? European Forest Institute Winter School, Freiburg/Br, 24 March 2011

Corbera E, Kosoy N, Martínez Tuna M (2007) Equity implications of marketing ecosystem services in protected areas and rural communities: case studies form Meso-America. Glob Environ Chang 17(3–4):365–380

Cordero D, Moreno-Díaz A, Kosmus M (2008) Manual para el desarrollo de mecanismos de pago/compensaciones por servicios ambientales. GTZ/Inwent, Lima

Ehrlich PR, Ehrlich AH (1981) Extinction: the causes and consequences of the disappearance of species. Random House, New York

Engel S, Wünscher T, Wunder S (2009) Increasing the efficiency of forest conservation: the case of payments for environmental services in Costa Rica. In: Palmer C, Engel S (eds) Avoided deforestation: prospects for mitigating climate change. Routledge, Oxford

Erb K-H, Krausmann F, Lucht W, Haberl H (2009) Embodied HANPP: mapping the spatial disconnect between global biomass production and consumption. Ecol Econ 63:328–334

FAO (2004) Payment schemes for environmental services in watersheds. Regional forum, Arequipa, Peru. FAO Regional Office for Latin America and the Caribbean, Santiago de Chile, 9–12 June 2003

Gomez-Baggethun E, de Groot R, Lomas PL, Montes C (2010) The history of ecosystem services in economic theory and practice: from early notions to markets and payment schemes. Ecol Econ 69(6):1209–1218

Grimble R, Wellard K (1997) Stakeholder methodologies in natural resource management: a review of principles, contexts, experiences and opportunities. Agric Syst 55(2):173–193

Hamilton K, Chokkalingam U, Bendana M (2010) State of the forest carbon markets 2009: taking root and branching out. Forest Trends, Washington, DC

Hau JL, Bakshi BR (2004) Expanding exergy analysis to account for ecosystem products and services. Environ Sci Technol 38:3768–3777

Jack BK, Kousky C, Sims KRE (2008) Designing payments for ecosystem services: lessons from previous experience with incentives-based mechanisms. Proc Natl Acad Sci 105(28):9465–9470

Kaiser B, Roumasset J (2002) Valuing indirect ecosystem services: the case of tropical watersheds. Environ Dev Econ 7:701–714

Kosoy N, Corbera E (2010) Payments for ecosystem services as commodity fetishism. Ecol Econ 69(6):1228–1236

Landell-Mills N (2002) Developing markets for forest environmental services: an opportunity for promoting equity while securing efficiency? Philos Trans R Soc Lond A 360:1817–1825

Le Grand J (1990) Equity versus efficiency: the elusive trade-off. Ethics 100(4):554–568

Lele S (2009) Watershed services of tropical forests: from hydrology to economic valuation to integrated analysis. Curr Opin Environ Sust 1:148–155

Madsen B, Carroll N, Kandy D, Bennett G (2011) 2011 update: state of biodiversity markets. Forest Trends, Washington, DC

Mayrand K, Paquin M (2004) Payments for environmental services: a survey and assessment of current schemes. Report elaborated by UNISFERA for the Commission for Environmental Cooperation of North America

MEA (Millennium Ecosystems Assessment) (2005) Ecosystems and human well-being: current state and trends. Findings of the condition and trends working group of the Millennium Ecosystem Assessment, vol 2. Island Press, Washington, DC

Milder JC, Scherr SJ, Bracer C (2010) Trends and future potential of payment for ecosystem services to alleviate rural poverty in developing countries. Ecol Soc 15(2):4 [online] http://www.ecologyandsociety.org/vol15/iss2/art4/

Muradian R, Pascual U, Rodríguez LC (2008) Revisiting the relationship between equity and efficiency in PES. International Association of Ecological Economics, Nairobi

Muradian R, Corbera E, Pascual U, Kosoy N, May PM (2010) Reconciling theory and practice: an alternative conceptual framework for understanding payments for environmental services. Ecol Econ 69(6):1202–1208

O'Neill DW, Tyedmers PH, Beazley KF (2007) Human appropriation of net primary production (HANPP) in Nova Scotia, Canada. Reg Environ Change 7:1–14

Pagiola S (2008) Payments for environmental services in Costa Rica. Ecol Econ 65(4):712–724

Pagiola S, Arcenas A, Platais G (2005) Can payments for environmental services help reduce poverty? An exploration of the issues and the evidence to date from Latin America. World Dev 33(2):237–253
Pascual U, Muradian R, Rodríguez LC, Duraiappah A (2010) Exploring the links between equity and efficiency in payments for environmental services: a conceptual approach. Ecol Econ 69 (6):1237–1244
Pattanayak SK, Kramer RA (2001) Worth of watersheds: a producer surplus approach for valuing drought mitigation in Eastern Indonesia. Environ Dev Econ 6:123–146
Pattanayak SK, Wunder S, Ferraro PJ (2010) Show me the money: do payments supply ecosystem services in developing countries? Rev Environ Econ Policy 4(2):254–274
Peters-Stanley M, Hamilton K, Marcello T, Sjardin M (2011) Back to the future – state of the voluntary carbon markets 2011. Ecosystem Marketplace & Bloomberg New Energy Finance
Porras I, Chacón-Cascante A, Robalino J, Oosterhuis F (2011) PES and other economic beasts: assessing PES within a policy mix in conservation special session on 'Instrument Mixes for Biodiversity Policies,' ESEE 2011, Istanbul
Rapidel B, Le Coq JF, Beer J, DeClerck F (eds) (2011) Ecosystem services from agriculture and agroforestry: measurement and payment. Earthscan, London
Rico L, Ruiz M, Reyes F, Barrasa S, Contreras E (2011) Efficiency of payments for environmental services: equity and additionality in a case study from a biosphere reserve in Chiapas, Mexico. Ecol Econ 70(12):2361–2368
Rosa H, Kandel S, Dimas L (2004) Compensation for environmental services and rural communities: lessons from the Americas. Int For Rev 6:187–194
Smith M, de Groot D, Perrot-Maîte D, Bergkamp G (2006) Pay – establishing payments for watershed services. IUCN, Gland
Stanton T, Echavarria M, Hamilton K, Ott C (2010) State of watershed payments: an emerging marketplace. Ecosystem Marketplace, Washington, DC
Swallow B, Kallesoe M, Iftikhar U, van Noordwijk M, Bracer C, Scherr S, Raju KV, Poats S, Duraiappah A, Ochieng B, Mallee H, Rumley R (2009) Compensation and rewards for environmental services in the developing world: framing pan-tropical analysis and comparison. Ecol Soc 14(2):26 [online] http://www.ecologyandsociety.org/vol14/iss2/art26/
TEEB (The Economics of Ecosystems and Biodiversity) (2010) The economics of ecosystems and biodiversity for national and international policy makers. Earthscan, London
van Beukering PJH, Cesar HSJ, Janssen MA (2003) Economic valuation of the Leuser National Park on Sumatra, Indonesia. Ecol Econ 44:43–62
Wunder S (2005) Payments for environmental services: Some nuts and bolts. CIFOR Occasional Paper No. 42. CIFOR, Bogor, Indonesia. 24p
Wunder S (2007) The efficiency of payments for environmental services in tropical conservation. Conserv Biol 21(1):48–58
Wünscher T, Engel S, Wunder S (2008) Spatial targeting of payments for environmental services: a tool for boosting conservation benefits. Ecol Econ 65(4):822–833

Chapter 10
Financing Forests for Rural Development

Davide Pettenella and Mauro Masiero

Abstract Rural development strategies often neglected forests in the past, but in the last two decades the attitude of policy makers and the interests of public and private investors have changed considerably. Although the forestry sector has inherent characteristics that increase the complexity of investments, forest financing has grown in importance and became a focal point for the implementation of rural development policies. Finncial actors are conventionally distinguished as private, i.e. profit oriented investors, or public, i.e. development oriented investors. As regards finance instruments in use, they could be distinguished as conventional (i.e. also utilized outside forestry), or innovative ones (i.e. forestry-specific). According to UNEP figures, USD 64 billion have recently been invested in the forest sector every year, but global need for funding for sustainable forest management is estimated by the Collaborative Partnership on Forestry to be between USD 70 and USD 160 billion per year. While direct public sector investments in forestry played a decisive role in the past, recent figures show that they are generally stable or declining, especially with regard to bilateral flows. At the same time private sector direct investments are on the increase, but they remain concentrated in a few countries. In the near future private forest investments will probably be mainly in planted forests. A consequent global shift in the industrial wood supply from temperate to tropical zones and from the Northern to Southern hemisphere is predicted by several studies. Financial flows for investments in the timber sector will shift accordingly: this progress is not free of social and environmental risks. While the forestry sector presents new business opportunities for private operators, specific international and national policy tools for the governance of the changes are greatly needed.

Keywords Rural development • Forest financing • ODA • DDI • FDI • REDD • TIMO

D. Pettenella (✉) • M. Masiero
Department of Land, Environment, Agriculture and Forestry, University of Padova, Agripolis – Via dell'Università 16, 35020 Legnaro (Padova), Italy
e-mail: davide.pettenella@unipd.it; mauro.masiero@unipd.it

J. Pretzsch et al. (eds.), *Forests and Rural Development*, Tropical Forestry 9,
DOI 10.1007/978-3-642-41404-6_10,

10.1 Introduction

Rapid structural changes in society and the economy – such as demographic and consumption growth, globalization of the market and policy, improved communication, changes in trade patterns for commodities, and the emergence of important non-farm activities – have led to some evident threats, but also significant opportunities for rural development. It is widely recognized that traditional sectoral policies need to be upgraded and, in some cases, substituted with more appropriate instruments (OECD 2006). While rural development strategies often neglected forests in the past because they were mistakenly viewed as being outside the mainstream of agricultural development (World Bank 2008), in the last years the importance of forests in rural development has been increasingly recognized in both developed and developing countries. The issue of forest finance has thus become a focal point for the implementation of rural development policies. This is reflected in objectives and tools related to forest strategies defined by major international institutions:

- The United Nations Forum on Forests established Four Global Objectives on Forests (UNFF 2006) which member states agreed to achieve by 2015 and the fourth objective deals with forest finance: "Reverse the decline in official development assistance for sustainable forest management and mobilize significantly increased new and additional financial resources for the implementation of sustainable forest management";
- The Forest strategy of the World Bank (2004) is aimed to foster rural development providing "institutional and policy support for community joint forest management, governance and control of illegal activities, building markets, and financial instruments in support of private investment in sustainable forest conservation and management";
- The Food and Agriculture Organization's Strategy for forest and forestry (FAO 2010) recognized forest finance in two of its priorities: "the support of the development and implementation of national and international financial mechanisms to support sustainable forest management" and the "increased use of financial mechanisms to facilitate information sharing and database development, and to build capacity to strengthen forest management and to reduce deforestation and forest degradation";
- The United Nations Development Programme and the United Nations Environment Programme have based their pro-poor growth strategy on sustainable natural resources management. Forest products and services are recognized as key assets in the household economies of the poor. When properly financially supported, the productivity of these assets can be raised substantially, leading to increased income and livelihood opportunities (WRI 2005; UNDP 2010).
- In addition, public institutions have declared the need for investments in forest-based rural development as a strategy for accelerating progress toward the Millennium Development Goals (MDGs).

As a consequence of these official commitments the question of broadening and diversifying the financial basis for forest management has emerged as a key theme in the international forest policy dialogue (Savenije et al. 2008). However, looking at the concrete actions undertaken in the last years by many developing countries, it could be stated that the role of forests in rural development and poverty reduction has been marginalized (Simula 2008). A clear illustration of that is given by the Poverty Reduction Strategy Papers (PRSPs). As a requisite for receiving external development funds, the World Bank and International Monetary Fund required less developed countries to develop PRSPs. These are prepared independently by governments in order to define objectives and develop actions such as budget and policy priorities for development and poverty reduction. Contreras-Hermosilla and Simula (2007) reviewed PRSPs of 43 countries in order to determine how they address various forest issues. Although two-thirds of countries (28) had significant representations of forest links to poverty reduction and development, the remaining third of countries that did not discuss forests in relation to poverty reduction and developments include some countries where forests represent a significant source of income, and countries with substantial forest cover such as Indonesia and Vietnam. Countries that set forest-specific priorities in their PRSP and other national development strategies are less likely to attract funding for forest-based rural development.

This might explain why, notwithstanding the formal commitments by many intergovernmental and national institutions, in recent years a general trend towards a stagnation of public financing for sustainable forest management has been observed, including investments from international aid agencies (Gutman 2003). In the meanwhile, integration of the forest sector into the capital market is still limited as well as access to private capital, although private financing already exceeds that of the public sector (Canby and Raditz 2005) and interest from private financial institutions in forestry investments is growing.

10.2 Some Definitions

Except in the sections that describe the more conventional understanding of these concepts, the following connotations will be used throughout this book:

Domestic Direct Investment (DDI) is a category of investment that reflects the objective of establishing a lasting interest by an enterprise (direct investor) in an enterprise (direct investment enterprise) that is resident in the same economy (i.e. country) than the direct investor. Direct investment involves both the initial transaction between the two entities and all subsequent capital transactions between them and among affiliated enterprises, both incorporated and unincorporated (OECD and IMF 2004).

Foreign Direct Investment (FDI) is a category of investment that reflects the objective of establishing a lasting interest by a foreign enterprise (direct investor) in an enterprise (direct investment enterprise) that is resident in an economy other than that of the direct investor. The lasting interest implies the existence of a long-term relationship between the direct investor and the direct investment enterprise and a significant degree of influence on the management of the enterprise. The direct or indirect ownership of 10 % or more of the voting power of an enterprise resident in one economy by a foreign investor is evidence of such a relationship (OECD 2008b).

Microfinance is a specific branch of finance that involves the delivery of traditional and new financial services, such as loans, savings, insurance, transfer services and other products, targeting them at low-income clients. Pioneered by Mohammad Yunus who founded the Grameen Bank in Bangladesh in the 1970s, microfinance institutions today are spread all over the world (including in developed countries such as the United States) and count millions of the world's poor among their clients. This segment of the population has often not had access to traditional banks (Agrawala and Carraro 2010).

OECD Development Assistance Committee (DAC) is an international forum of many of the largest funders of aid, including 24 members. The World Bank, International Monetary Fund and United Nations Development Programme participate as observers. The DAC has the mandate to promote development co-operation and other policies so as to contribute to sustainable development, including pro-poor economic growth, poverty reduction, improvement of living standards in developing countries, and a future in which no country will depend on aid (OECD 2010a).

Official Development Assistance (ODA) includes flows of official financing administered with the promotion of the economic development and welfare of developing countries as the main objective, and which are concessional in character with a grant element of at least 25 % (using a fixed 10 % discount rate). By convention, ODA flows comprise contributions of donor government agencies, at all levels, to developing countries ("bilateral ODA") and to multilateral institutions. ODA receipts comprise disbursements by bilateral donors and multilateral institutions (IMF 2003).

10.3 Old Constraints and New Driving Forces

The forestry sector has inherent characteristics that increase the complexity of financing in addition to conventional challenges of rural development investments. Problems related to investments in the forestry sector include undervaluation of the multifunctionality of forests, perceived high risks of forest investments, the long-term nature of the forestry cycle – which introduces extra risk to the investment – and the uneven distribution of costs and revenues over time (Simula 2008). As regards multi-functionality their translation into financing mechanisms is not easy and may not be possible in many cases. The driving forces behind forestland misuse are subsistence and direct financial revenues and they cannot be curtailed by mere theoretical calculations of forest values (Lammerts van Bueren 2002): this is why undervaluation of forests multifunctionality is one of the reasons for the continuation of unsustainable practices leading to deforestation and forest degradation (Savenije et al. 2008; Indufor 2010). On the other hand, if a more multifunctionally-oriented forestry is desirable and can contribute to diversification of revenues, this can also lead to conflicts arising amongst different objectives and stakeholders (Solberg and Miina 1997; Hellström 2001; Niemelä et al. 2005; Janse and Ottitsch 2005).

Problematic aspects related to forest investments may also include advance payments of relatively high amounts, as in the case of forest plantations. Since opportunity costs of such capital may be very high, tree planting has been strongly subsidized by governments and development agencies in the past. For example, in many parts of Latin America, Oceania and Asia, plantation programs paid more than 75 % of the establishment costs with additional allowances made for land, maintenance and many others costs (Brown 2000). According to one estimate, around two billion USD were granted each year in subsidies to industrial forest plantations: this sum is four times greater than the annual development assistance given to forest conservation (White et al. 2006; UNEP 2009). In the last years the role of subsidies has been integrated by direct investments from large industrial and financial institutions attracted by new business opportunities in the forest sector.

Increasing demand for wood and wood-based products, especially in fast growing nations like Brazil, China and India, have caused a relevant growth in forest plantations investments. Planted forests constitute about 7 % of the global forest area, covering around 264 million hectares, with a steady increase in all regions and sub-regions since 1990 (FAO 2010). Plantation products play an increasingly important role in the secondary product industry (especially for pulp, paper and reconstituted panels) and they constitute the majority of the value of the aggregate tropical timber trade (Scherr et al. 2004).

Critics have observed that both public policies and private financial institutions have favored big companies and encouraged large-scale planting of monocultures, with a limited involvement of any stakeholders other than industry and government (Bass et al. 1996). As a matter of fact small and medium forest enterprises have normally been excluded from access to financial support; thus, when external finance is available, it is normally from informal or non-institutional credit suppliers, many enterprises being too small to benefit from significant access to capital and other resources, even if under the same rules as the large companies in delivering their products to the market. However, with the increased role of microfinance, small-scale investments are developing also in the forest sector, benefiting forest-based small scale enterprises and communities, especially in those cases where basic requirements related to food security are satisfied: most small scale enterprises operate their forest-based activities jointly with other processing, service or agricultural activities, so they seldom occur as separate enterprises (FAO 2005a). Microfinance is partly linked to another fast developing financial sector, represented by the so called ethical finance, basically intended as finance in which returns are in accordance with social and environmental concerns. Many private financial operators are aware of the risks that investments in unsustainable forest management can pose on financial and reputational liabilities. At the same time they widely acknowledge new business opportunities for forest investments such as green and low carbon solutions (PwC and WBCSD 2010).

In accordance with Corporate Social Responsibility principles and tools, producing timber and timber-based products is not the only reason for investing in forest management. Traditional wood production still remains important, but is complemented (and sometimes surpassed) by a wider recognition of the economic role of Non-Timber Forest Products (NTFPs) and of the functions based on ecosystem services like biodiversity, landscape, recreation and, above all, carbon sequestration. Carbon offsetting forest projects have received keen global attention with the establishment of the Clean Development Mechanism and Joint Implementation projects within the framework of the Kyoto Protocol. Moreover, the development of Reduced Emission from Deforestation and Forest Degradation (REDD) projects will undoubtedly increase the interest of potential investors in the forestry sector.

The enhanced attractiveness of forests as investment opportunities is encouraging for the whole sector. The parallel growth of environmental and social concerns and investing opportunities feeds a real need for tools allowing reliable and transparent assessment of financing channels as well as proper rating systems of different investments options on the basis of environmental and social performances of all actors involved in the process.

Finance instruments utilized for forest-based rural development could be distinguished as conventional finance instruments utilized in numerous other types of investments besides forestry, or innovative finance instruments that are specifically designed for forest finance. Both traditional and innovative finance instruments can be used in traditional and innovative markets for forest products (Fig. 10.1).

Markets for timber and to a certain extent NTFPs are well known, and are utilized (or could be better utilized) to sustain the forestry sector. Market mechanisms that encourage the user of other, currently not marketed, environmental goods and services to pay (to the producer/resource manager) are less known but are worth exploring for their potential to contribute to the sustainable financing of national forest programs.

10.4 Funding for Forest-Based Rural Development

10.4.1 A General Classification of Forest Finance Resources

According to FAO (2005b), more than 600 large financial actors invest in forest-based activities worldwide. Investors are conventionally distinguished as private, i.e. profit oriented investors, and public, i.e. development oriented investors (Holopainen and Wit 2008; Simula 2008).

When considering existing financial flows, a distinction between public and private investments is again normally made. Public funds include domestic public investments and Official Development Assistance (ODA). Private funds include Foreign Direct Investment (FDI) and Domestic Direct Investments (DDI), even under the form of private investments by institutional investors, commercial banks and export credit agencies.

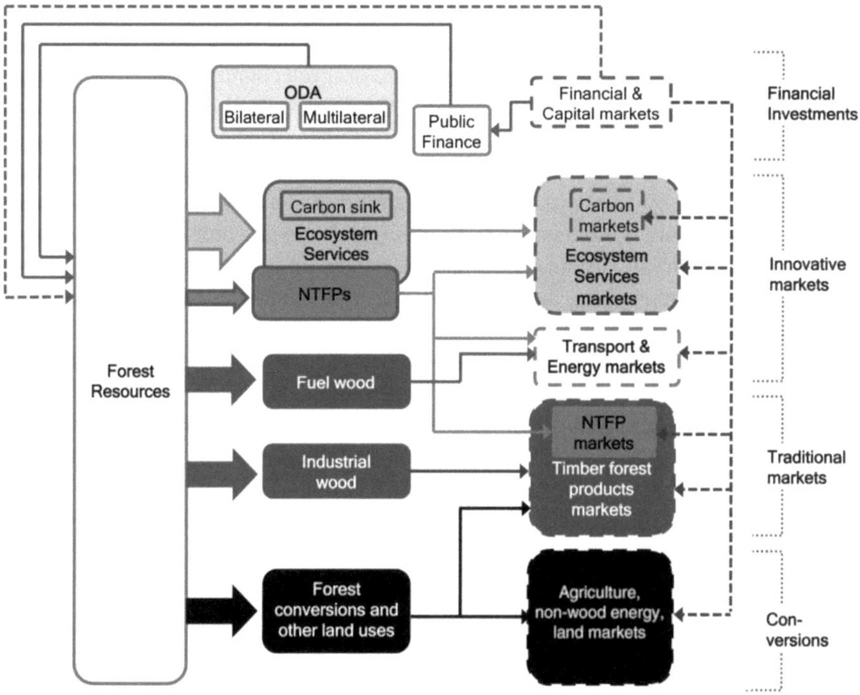

Fig. 10.1 Forest resources, markets and financial institutions (Source: Adapted from UNEP (2011))

Forest financing also includes contribution from NGOs and philanthropy, as well as other funding forms that may be classified as mixed, public and private, since they are characterized by the co-existence of public and private sources of funds.

Forms of financing are dynamic over time and new types of investment partnerships are being set up: between governments, international donor agencies, civil society, local community and the private sector forest investors. These partnerships have significant potential to enable forest-based rural development, because they include a wide array of stakeholders. Due to the emergence of innovative instruments of forest finance based on payments for environmental services, some forest investments that were financed mainly by public institutions are becoming private sector investments. The REDD projects are probably the most remarkable example of this trend. At the moment they are financed by bilateral ODA and multilateral ODA, however, this may partly change in the future: the interest of profit-oriented (private) investors will increase when REDD projects become eligible for credits for international carbon trading.

10.4.1.1 Public and Predominantly Public Funding

ODA funds are the main source for rural development funding, a flow of official financing administered to promote economic development and welfare of developing countries (IMF 2003). ODA can be bilateral when finance is disbursed by a government development agency of one country, and multilateral, when finance is disbursed by government development agencies from two or more countries, or by multilateral institutions such as the World Bank[1] and the United Nations. There are many multilateral ODA initiatives focused on forests and rural development, such as the Program on Forests (PROFOR) at the World Bank, or the Collaborative Partnership on Forests developed by 14 international organizations and secretariats with substantial programs on forests. ODA supports institutional capacity building, forest policy reform, promotion of sustainable forest management and rural development in national development plans, and creation and implementation of economic, social and environmental incentives for sustainable forest management.

Public funding includes regional development banks, which are particularly focused on rural development and poverty alleviation. In the 2002–2006 period the International Bank for Reconstruction and Development and International Development Association invested USD 517 million in "stand-alone forest projects" and "projects with dominant forest components" (Contreras-Hermosilla and Simula 2007). Two World Bank bodies, the International Finance Corporation (IFC) and the Multilateral Investment Guarantee Agency (MIGA), finance forestry through predominantly public investments in the private sector. Particularly, MIGA is involved in FDI in developing countries, by providing risk insurance (guarantees) against country, political and other non-commercial risks. The main criticism of FDI in relation to rural development regard the fact they are typically channeled into infrastructure and larger-scale investments, rather than small or medium-scale enterprises that might to a larger extent benefit the poor. The number of forest related projects financed by MIGA is still negligible. However this might change with the establishment of agribusiness, manufacturing and services sector, and of programs such as micro-financing and small investment program (World Bank 2010a, b).

10.4.1.2 Private and Predominantly Private Finance

The most significant example of private forest finance is the activity of timberland funds. The United States has the most developed finance mechanism specialized in timberland investments: Timberland Investment Management Organizations (TIMOs), and Timberland Real Estate Investment Trusts (T-REITS). These are private companies who buy, manage and sell timberland on behalf of institutional investors such as pension funds, insurance funds, endowments, conventional banks,

[1] There are five World Bank agencies: the International Bank for Reconstruction and Development (IBRD) provides debt financing on the basis of sovereign guarantees; the International Finance Corporation (IFC) provides funds and advise primarily to the private sector; the International Development Association (IDA) provides interest-free loans or grants; the International Centre for Settlement of Investment Disputes (ICSID) works with governments to reduce investment risk; and the Multilateral Investment Guarantee Agency (MIGA) provides insurance against certain types of risk, including political risk, primarily to the private sector.

etc. (see Box 10.1). They mainly operate in the USA, but also in Canada, New Zealand, Australia, South Africa and many countries in South America (FAO 2007).

The institutional investors tend to include timber in their investment portfolio because timber has low-to-negative correlation with other asset classes such as stocks and bonds. This is because the majority of returns come from biological growth, which is not correlated with any economic factor and is, in normal investment conditions, positive. Trees grow regardless of the economic and capital market conditions. Therefore, the institutional investors tend to increase returns and lower risks by adding timberland management activities to their investment portfolio (Healey et al. 2005).

Pension funds are one of the main institutional investors in the forest sector. Usually, pension funds dedicate 1–3 % of their financial assets to timberland investments. This is still a large amount as for example, PGGM pension fund from the Netherlands has an overall financial asset over USD 100 billion, and the typical amount dedicated to forest investments is between USD 75 and USD 150 million (Plomp 2010).

Box 10.1 TIMOs and T-REITs

Timber Investment Management Organizations (TIMOs) are private companies acting as investment managers with the primary aim to maximize the growth in the value of timberland assets. Timberlands are owned as illiquid direct investments or partnership shares, generally in separate accounts, but frequently in pooled funds. Investors in TIMOs are interested in total return and capital appreciation. Two main types of investment models for TIMOs exist. The first one is named "separate accounts model": an investor buys and manages timberland for returns over an indefinite term. The second one is called "closed-end funds": multiple investors purchase timberland jointly and intend to hold it for a set period, such as 10–15 years, before selling it.

The TIMO vehicle suits many institutional investors, who may not want to directly buy and manage planted forests, or may want to put only small amounts of their funds into investment, so need to have a structure which can accumulate funds from many organizations to provide large funds.

Timberland Real Estate Investment Trusts (T-REITs) are private or public companies owning and operating income-producing real estate. Historically REITs manage apartment buildings, shopping centers, offices, hotels or warehouses, but they recently started to pay growing attention to timberland assets. Their primary business is managing groups of properties to produce income, and they are required to distribute most of their profits as dividends. They have high degree of liquidity through the public trading of shares on a stock exchange.

(continued)

Box 10.1 (continued)

The number of TIMOs has grown significantly in the last 30 years: from only 2–3 in the early 1980s to more than 25 in 2007. Today the largest private planted forest owner in the world (Plum Creek, USA) is a T-REIT. It has been estimated that the total value of such tools grew from USD 2.0 billion in 1990 to USD 50 billion in 2007.

Source: FAO (2007), Fernholz et al. (2007)

10.4.2 Finance Instruments for Forest-Based Rural Development

Conventional finance instruments include grants, loans, credits, equity investments, co-funding, guarantees, insurances and savings. Moreover new instruments like microfinance, leasing, remittances, and payments for environmental services (PES) are emerging.

Grants are the most common finance instrument disbursed by ODA. Unlike loans or credits, grants do not usually have to be repaid. *Loans* can be obtained from most banks, but development banks usually provide privileged rates of interest, with an initial interest free period, repayable over the long term. To justify a loan, the forest project has to be developed in a strong business case. Loans can be distributed to a variety of beneficiaries, including governments, projects, forest companies, individual forest farmers etc. under several forms of funds that cover a wide range of mechanisms to raise money through national/international transfer payments. National Environmental Funds, for example, use collected revenues as disbursements for environmental and conservation purposes. They can have a public or private nature (or even mixed) and can be site/issue specific or multi-issue. In the second case they address environmental management and sustainable development in a broader sense. Different kind of funds exist (NFP Facility 2011):

- Endowment funds that use exclusively revenues from capital investments to finance activities;
- Sinking funds that gradually disburse their capital and investment income over a fixed period of time, taking the revenues into account;
- Revolving funds that receive regular – and in ideal cases fixed – income, e.g. from taxes and levies, which complements and replenishes the original capital, to provide a source of finance for specific activities.

Equity investments enable persons and institutions to invest in shareholding of a company managing or implementing a sustainable forest management project. The investment may make an enterprise viable or enable it to expand, while the new shareholder will benefit through shareholder voting rights and dividends on profits. Several examples may be mentioned. Precious Woods Ltd. is a Swiss company with

FSC certified forest operations in Brazil, Costa Rica and Gabon, whose main shareholders are banks, insurance companies and Swiss pension funds (Precious Woods 2011). The Terra Capital Fund is a USD 15 million private equity fund that invests in private enterprises generating conservation benefits through sustainable use of biodiversity in Latin American countries, which have ratified the Convention on Biological Diversity (Argentina, Bolivia, Brazil, Chile, Ecuador and Paraguay). The Fund is a 10 years venture capital fund launched with a USD 5 million Global Environmental Facility (GEF) grant and participated in by several actors including the International Finance Corporation, the Swiss Government, Triodos Bank, and others. Investment fields cover organic agriculture, native species aquaculture, native species reforestation, NTFPs and nature tourism (Terra Capital 2000). Another relevant example is the Lignum Fund in Chile (see Box 10.2).

Guarantees are obligations of one party (the guarantor) to assume responsibility for the debt obligation of a borrower if that borrower defaults, as in the case of MIGA investments where the World Bank provides country and political guarantee for private investors. A special form of guarantee is represented by bonds, such as the so-called Environmental Performance Bonds, i.e. payments made by an operator prior to the commencement of project activities. Such bonds are then returned to the operator at the end of the project if certain predetermined environmental performance standards are met. If not, the money is to be used to support mitigation measures to restore appropriate environmental conditions. Environmental Performance Bonds are normally held on deposit by government bodies.

Similarly to the previously described items, *insurances* are also a form of financial risk management, in which a company can insure itself against uncertain financial or other losses by buying an insurance based on a specific investment risk assessment. Insurance products for investors in the forest sector commonly cover aspects like losses of growing trees, fruits and yield due – for example – to fire and allied perils or wind storms, business interruption due to increased costs of working, etc.

Savings represent the income not spent, or deferred consumption. They mainly refers to personal finance, but besides benefiting poor people, they are important for small-scale enterprises as a strategy to mitigate income fluctuations, overcome unexpected expenditures and emergencies, and similar.

Leasing is a transaction in which the owner (the lessor) of a productive asset allows another party (the lessee) to use an asset for a predefined period against a rent (lease payment). The lessee becomes responsible for all operational costs including maintenance and repairs of the asset. Examples include the leasing of degraded public forestlands to private companies in India, and to poor households in Nepal. Land tenure through leasing can facilitate access to microfinance services and credit, thus improving benefits for lessees.

Remittances are transfers of money back home from seasonal and long-term migrants. For many developing countries migrants' remittances represent a major source of income for local households. IFAD (2009) estimates a global value of remittances over USD 300 billion a year, which surpass FDI and ODA combined. It is also estimated that one third of remittances go directly to rural areas and,

Box 10.2 Lignum Fund in Chile

In 2003 Fundación Chile joined the investment bank Asset and the real estate investment fund Independencia to create an investment vehicle that would allow financial investors to invest in the Chilean forestry sector. The result of such co-operation was the Lignum Fund, a publicly listed fund whose sole purpose is to invest in the forestry sector in Chile. Being the first forestry fund set up in Latin America, Lignum Fund became effective after the approval of the Internal Regulations by the Chilean Securities and Insurance Commission, which took place on January 3rd, 2006 and is renewable for an additional 4-years period. Asset raised a total of USD 39 million for the Lignum Fund, and negotiated forestry management contracts and off take with the two large Chilean forestry companies, CMPC and Arauco. Fund investors include the Chilean pension funds, Chilean life insurance companies and family offices. The Lignum Fund, managed by Foresta AFI, a joint venture between Asset and Independencia, is fully invested and is now in its divestment process.

The Lignum Fund acquired approximately 12,000 ha of immature pine and eucalyptus forests, and planted approximately 15,000 ha of land with pine and eucalyptus under long-term land-use right agreements with small and medium landowners. These forestry assets will be managed, harvested, and commercialized under long-term agreements signed with selected Chilean forest companies. On the basis of these contracts, the Lignum Fund intends to issue a securitized financial instrument backed entirely by net cash flows generated from the harvest and commercialization of its forestry assets. Fund investors will receive cash proceeds as both dividends and a return of capital. Upon the final liquidation of the Lignum Fund, the fund investors will also receive an in-kind distribution of the subordinated tranche of the securitized bonds.

Source: Asset (2011), Independencia (2011)

although they are mainly used to cover household expenses, they can also be used as start-up capital for small livelihood activities such as forest-based enterprises.

Some of the conventional finance instruments can be designed to target especially low-income people and small enterprises; this is the case of *microfinance*, an innovative set of instruments that is particularly significant for forest-based rural development (FAO 2005a), see also Box 10.3.

Innovative finance instruments that are specifically designed for forest investments are mainly oriented towards PES. Carbon sequestration is the environmental service that currently receives the majority of public and private funds. Macqueen (2010) studied the 16 most substantial funds for forest climate change mitigation, mainly focused on REDD projects. The study estimates that these funds attracted pledges of USD 21.78 billion, mainly derived from bilateral and multilateral ODA, with the high predominance of the Hatoyama Initiative of the Government of Japan. However, forest rural people, and indigenous groups still do not have enough decision-making power and control over these funds, as the funding is received mainly by governments. REDD projects could have a greater role in forest-based rural development if rural people and indigenous groups' access to these funds is improved, together with their property rights, rights to free, prior and informed consent, recognition of traditional knowledge and the forest stewardship role (UNFCCC 2010).

Box 10.3 Microfinance in Forestry

Microfinance has been successfully implemented in the forestry sector both with regard to traditional activities related to timber production, as well as to NTFPs and services. Examples include:

- *Nepal:* NTFPs are a basic resource for livelihoods in Nepal, especially for people in mountain areas. In Parbat district (western Nepal) 91 % of population is dependent on agriculture and forestry, while NTFPs provide more than 12 % of local GDP. Although several banks, microfinance institutions, NGOs, savings and credit groups have been operating in Parbat for a long time, only a few of them provide credit to forest-based enterprises. It has been estimated that the total amount of microcredit investment needed in forest-based enterprises in the district would be around USD 325,000. In order to improve microfinance in forestry two dedicated programs were launched between 1998 and 2001: Micro-Enterprise Development Program (MEDEP) – funded by the Nepalese Government and UNDP – and Livelihoods and Forestry Programme (LFP) – funded by the Nepalese Government and DFID. MEDEP and LFP played a major role in promoting forest-based enterprises in Parbat district. With a total microcredit of about USD 60,000 nearly 280 new forest enterprises were created. An average net increase of 380 % in the income of the entrepreneurs from the 1.5 year or older enterprises was observed. As a result, the average net income per day became significantly high in comparison to other districts in west Nepal.
- *Senegal*: the Project de Gestion Intégré des Ecosystems Senegalaises (PGIES) was launched in 2002 with funding from GEF, UNDP and the Government of Senegal. The project aims to promote community based integrated ecosystem management in the four most important eco-regions within Senegal, to reduce deforestation, forest degradation and biodiversity loss. PGIES involves the establishment of a network of villages responsible for the management of a community nature reserve. Upon transfer of these forest reserves from the local government to the community, a management plan for the reserve was developed, which includes activities such as the creation of fire breaks, surveillance by specially appointed eco-guards and active fire prevention by the community. Livelihood activities were also

(continued)

Box 10.3 (continued)

supported, including vegetable gardening, animal husbandry and fruit tree planting. Pressure on the forest was reduced, thereby avoiding emissions from further deforestation and capturing carbon by allowing the forests to regenerate. PGIES has established microfinance institutions to support communities' members. Such institutions provide low interest rate loans (10 %) for eligible "environmental activities", as determined by the microfinance institutions governing body. Five percent of all interest payments are used to finance an Environment Fund which in turn is used to support the protection and rehabilitation activities undertaken within the reserve.

Source: Binayee et al. (2004), FAO (2005a), UNEP (2011)

10.4.3 Forest Finance for Rural Development: Where Are We Going?

The issue of the quantification of resources needed for sustainable forest management is controversial and has been part of the international agenda since the United Nations Conference on Environment and Development (UNCED) held in Rio de Janeiro in 1992. UNCED (1992) estimated that the annual flows to the forestry sector required for the 1993–2000 period were approximately USD 31.2 billion, a figure revised upwards to USD 33 billion per year in 1996 (Chandrasekharan 1996).

According to more up-to-date and detailed estimates by UNEP (2011), USD 64 billion have recently been invested in the forest sector every year: USD 18 billion for forest management and USD 46 billion in forest product processing and trade. The majority of this is domestic investment (90 %) mainly concentrated in developed countries and associated with plantations and processing facilities for pulp, paper and biofuel. ODA accounts for only around 7 % of the total investment in forestry worldwide (Tomaselli 2006). Overall, the picture for developing countries is that, apart from official development assistance, financing is primarily domestic and relies heavily on internal cash flows, since lending and equity capitals are, as already noted, difficult to access (Streck et al. 2010).

While direct public sector investments in forestry played a decisive role in the past, recent figures show that they are generally stable or on a declining trend in most countries, although still important – especially when considering the ODA contribution – for low-income ones. In particular, with few exceptions, bilateral flows remained more or less steady, while an increasing engagement of multilateral sources has been observed. At the same time private sector direct investments, both domestic and foreign, are on the increase, but they remain concentrated in few countries (Tomaselli 2006). By 2005, about a third of the stock of global foreign direct investments had gone to developing countries (Borregaard et al. 2008), but those countries that hold relevant forest resources and have good potential for forest industry did not manage to attract any substantial amounts of these investments. There are several reasons for this, one of the main being that the majority of investments are cross-border mergers and acquisitions geared towards augmenting technological assets or accessing new markets by purchasing existing firms (Laaksonen-Craig 2004). The introduction of new investment opportunities, even outside timber production, could probably invert this trend. As commented by UNCTAD (2010), the relative weight of developing and transition economies as both destinations and sources of global FDI is expected to keep increasing: these economies, which absorbed almost half of FDI inflows in 2009, are leading the FDI recovery after the global economic crisis.

Compiling a quantitative historical assessment of the external sources of forest funding and investments worldwide is not an easy task: difficulties involving data collection, compilation and analysis of information are widely recognized. This is due to several reasons, mainly including reliability and completeness of available data. Reliability is strongly related to effectiveness and transparency of official reporting systems, while completeness refers to the existence of gaps both in chronological terms and in sectoral ones. For example, with the exception of FDI, little information exists about investment by the private sector. Problems may also exist in terms of considerable annual variations in the financing flows: this is particularly evident in the case of sources that record commitments rather than real disbursements (e.g. OECD) because decision making in the case of large projects can create wide variations in the data.

A tentative overview of trends in forest financing is presented in the following pages, basically distinguishing between public investments (bilateral and multilateral ODA) and private ones.

10.4.3.1 Forest Sector ODA

The relevance of forest sector ODA reduced in the last years: this is witnessed by Resolution 2006/49 of the Economic and Social Council of the United Nations that, as an outcome of the sixth session of the United Nations Forum on Forests, called on governments to reverse the decline in ODA for sustainable forest management. While there are no precise data on forest sector ODA, annual estimates range from USD 0.5 to USD 1.7 billion (OECD and World Bank quoted in Savcor Indufor 2006). According to OECD (2008b) public official aid to forestry has remained stable in real terms over the past decade, with an average value of about USD 500 million between 1995 and 2008 (Fig. 10.2). On the other hand, since total ODA has risen sharply, the share of aid to forestry has declined: it represented 0.8 % of the development assistance funds in the mid-2000s compared to 1.1 % at the end of the 1990s, when figures were probably higher due to the effect of the 1992 UNCED Summit in Rio de Janeiro. However, with reference to bilateral investments, Simula (2008) observed that data made available by OECD through its Development Assistance Committee (DAC) Credit Reporting System may be incomplete, providing just a partial view because of the weaknesses in DAC members' reporting

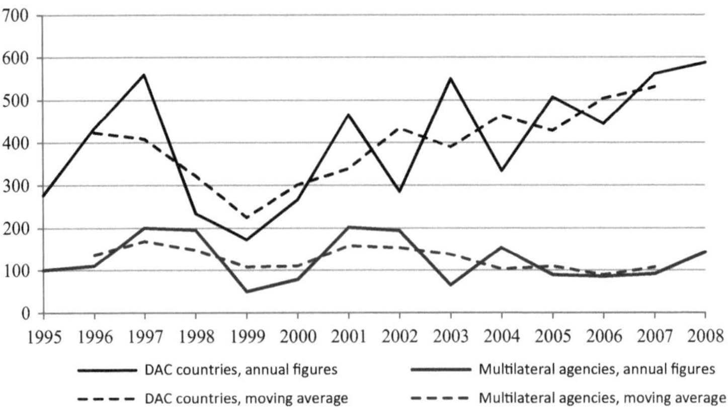

Fig. 10.2 Annual ODA in the forest sector (1995–2008), million USD (Source: Our elaboration from OECD (2008c, 2010b))
Note: 3-year moving averages and annual figures in current prices. DAC countries include EU institutions

systems and the presence of several gaps in the past data. As commented directly by OECD (2000), annual DAC data on aid to forestry are available from 1995 onwards.[2] Prior to that year, forestry activities are included, but are not separately identifiable within the wider "Agriculture, forestry and fishing" sector. Furthermore the definition of aid to forestry excludes aid to forest industries, which is coded elsewhere in the classification (industrial sector) but not identifiable (OECD 2008d). Finally, although specific figures for the forest sector are reported[3] by the DAC Credit Reporting System, forest components in multisector projects and programs primarily aiming at rural development, natural resource management, biodiversity or environmental management are not recorded separately and this

[2] The Credit Reporting System permits aid to forestry to be examined over a longer period as specific forestry sector codes have existed from the outset of the system. Analyzing the evolution in aid to forestry in 1973–1998, OECD (2000) observed it was similar to that of ODA in general. A constant growth in the 1970s and 1980s could be observed as aid to forestry increased from a few tens of millions of US dollars a year to over half a billion a year. When converting data to constant dollars, however, it can be observed there was a real growth only up to the early 1980s, after which flows remained broadly stable. DAC countries' commitments of bilateral aid to forestry over the whole period amounted to a total of US$ 5 billion (current), and ODA lending to forestry by the multilateral development banks to US$ 3 billion.

[3] According to OECD DAC Credit Reporting System classification, aid to forestry is divided in six sub-sectors: (sub-code 31210) Forestry policy and administrative management (including: forestry sector policy, planning and programs; institution capacity building and advice; forest surveys; unspecified forestry and agro-forestry activities); (31220) Forestry development (including: afforestation for industrial and rural consumption; exploitation and utilization; erosion control, desertification control; integrated forestry projects); (31261) Fuelwood/charcoal (including: forestry development whose primary purpose is production of fuelwood and charcoal); (31281) Forestry education/training; (31282) Forest research (including artificial regeneration, genetic improvement, production methods, fertilizer, harvesting); (31291) Forestry services.

may lead to additional underestimation of global invested values. As regards biodiversity, it has been estimated that for the 2003–2006 period forest-related DAC bilateral aid covered about 13 % of the overall biodiversity-related aid, i.e. about USD 313 million per year on a grand total of about USD 2.7 billion per year (OECD 2008d).

A survey carried out by PROFOR (2003) showed DAC-reported bilateral donors' contribution record was just about half of the total funding volume. Based on this, PROFOR estimated that during the 1986–1997 period, bilateral and multilateral ODA resources dedicated to the forest sector grew from USD 784 million in 1986 to USD 1,270 million in 1997 (Tomaselli 2006). A study by El Lakany et al. (2007) calculated the annual volume of ODA financing at a level of about USD 1.5 billion, while in 2008 Simula estimated an amount of about USD 1.9 billion. It should be remembered that in its estimations on the annual flows of funds to the forest sector for the years 1993–2000 UNCED calculated that 18 % of the overall value – i.e. USD 5.7 billion – were to be covered through concessionary funding from ODA budgets.

10.4.3.2 Bilateral ODA

Bilateral ODA to forests comes from a limited number of sources. About 90 % originates from a short list of donors, including European Union (EU) institutions, Finland, France, Germany, Japan, the Netherlands, Switzerland, the United Kingdom, and the United States. Although flows are quite irregular, the overall trend is positive with the total contribution from DAC countries doubling between 1995 and 2008. Japan plays a major role: in the same period it contributed to more than 45 % of the global DAC members' ODA value on average, reaching nearly 75 % in 2003. Not considering the Japanese contribution, the global bilateral ODA calculated in 2008 would be more or less identical to that calculated in 1998. Increases can be observed even with reference to other countries, but only EU institutions and Finland show relevant volumes in absolute terms, and sudden growth can be observed in 2008 for Australia and Norway. In many other cases forest ODA declined, especially in the cases of the Netherlands, Sweden and the United Kingdom. Several reasons can explain this decline (El Lakany et al. 2007; Simula 2008): on the one hand there is a widely reduced allocation to project and program funding with, on the other, a shift from sectoral approaches to budgetary support and broader development strategies that respond to the Millennium Development Goals, the sectoral allocation of which is done by the recipient country. A general trend to no longer consider forests as a self-standing priority, but as part of the climate change and other environmental agendas is consolidating. There is an increasing use of multilateral agencies as funding channels because they have a competitive advantage in those recipient countries where bilateral donors cannot effectively operate because of governance constraints.

When considering recipients of forestry bilateral ODA, Asia plays a prevalent role, receiving more than 60 % of total aid to the sector, with a 80 % peak in 2003.

India has been the major recipient in the last years (31 % of total ODA to the forestry sector), followed by China (13 %) and Vietnam (12 %). Africa receives about 20 % and Central-Latin America about 11 %. In terms of sectoral funding, the thematic area "forestry development" (OECD code 31220) in the last years received almost two-thirds (63 %) of the total bilateral ODA, followed by "policy and administrative management" (33 %), while remaining activities only obtained marginal contributions (OECD 2008c).

In addition to traditional grant financing for targeted projects and programs, bilateral donors have introduced new instruments such as sector-wide approaches, program support, budgetary support, debt-for-nature swaps, etc. (Simula 2008). It can be expected that bilateral ODA will not increase or will even decrease in the next years, and a higher fraction of the ODA will probably be channeled through multilateral institutions. In the meanwhile investments will focus on new forest-related instruments of the climate change initiatives.

10.4.3.3 Multilateral ODA

Commitments to forestry among the multilateral agencies are mainly provided by the World Bank – represented primarily by IFC, IDA and IBRD – with a total of nearly USD 200 million in average over the 1996–2004 period (Tomaselli 2006). In the period 2002–2006 the IBRD and IDA invested USD 517 million in "stand-alone forest projects" and "projects with dominant forest components" (Contreras-Hermosilla and Simula 2007). Simula (2008) estimated that between 2000 and 2006 development banks invested USD 457 million in forests, of which the African Development Bank (AFDB) invested USD 352, followed by the Asian Development Bank with USD 65.6 million. Although there are no official statistics for forest-based rural development finance, according to AFDB (2010), USD 31.6 million were invested in 2009 for 14 environment sector projects that are directly (stand-alone forest project) or indirectly (projects with forest components) related to forest-based rural development. Because of their structure multilateral sources have a much broader geographic scope than bilateral agencies and when reporting their funding allocated to forest-related projects do not indicate whether or not some of the amounts are duplicated across agencies. Moreover, OECD seeks to collect data on aid activities by multilateral organizations on the same basis that for bilateral donors, but at present, sufficient data are received only from the World Bank group, the regional development banks and IFAD, which together account for approximately 40 % of multilateral ODA. Sectoral data for the European Commission and United Nations, each of which represents some 30 % of multilateral ODA, are largely incomplete (OECD 2000, 2008a). Multilateral ODA registered by OECD Credit Reporting System showed a quite irregular pattern between 1995 and 2008, with a couple of peaks in 1997–1998 and 2001–2002 and an average value around USD 125 million (OECD 2010b).

As for other sources, online information on forestry financing by regional development banks indicate their combined funding volume in 2000–2006 as USD 457 million, i.e. about USD 65 million per year (Simula 2008). Another relevant component is that of the Global Environment Facility (GEF). Since 1991 the GEF has allocated approximately USD 1.5 billion to forest initiatives, supplemented by more than USD 4.5 billion in co-financing (GEF 2009). GEF is planning to expand its support to actions reducing deforestation and provide up to USD 1 billion for the implementation of a dedicated REDD Program throughout the period 2010–2014 (GEF 2010).

Important contributions have also been provided by the International Tropical Timber Organization (ITTO), that since 1987 has made USD 314 million available to finance around 800 projects and activities through the Administrative Account, based on contributions by ITTO members, and the Bali Partnership Fund, again mainly from voluntary contributions. From 2000, the yearly allocations range between USD 15 million and USD 20 million, the three main contributors being Japan, Switzerland and the United States, which have collectively accounted for 90 % of the cumulative voluntary contributions since 1987. Their role as contributors has decreased, but this has been offset by contributions from other donors. The number of recipients of ITTO contributions increased over time, up to the inclusion of 33 producer members and 3 developing consumer members. Specific figures suggest a high degree of concentration because more than 50 % of the total investments are absorbed by eight countries – Indonesia, Malaysia, Ghana, the Philippines, Brazil, China – and the Republic of Congo – while the share of 12 developing member countries has been 1 % or less of the total for each (Hardcastle and Umali 2007).

As for the FAO National Forest Program (NFP) Facility, by mid-2006 nine donors had committed USD 15.5 million, while for the 2007–2012 phase the proposed budget is USD 33.7 million (Savcor Indufor 2006), with USD 2 million extra requested in 2010 to continue operations as planned through mid-2012 (NFP Facility 2010).

10.4.3.4 Private Sector FDI in Forestry

Although there is no systematic information available on domestic or foreign direct private investments in the forest sector, many authors agree on the fact that they play and will play a key role. Since the late 1980s, the world has experienced a strong expansion of direct investments flows mainly under the form of FDI as a result of the globalization process, associated to the financial markets liberalization. In the second half of the 1990s, the direct investments intensified even more, due to different driving factors such as the extension of bilateral and multilateral commercial treaties, the intensification of privatization processes and the growing trend of fusions and trans-frontier acquisitions, mainly in the USA, EU and Japan (IADB 2004). This trend was maintained until the end of the 1990s, then inverted until 2003 when a new increasing period started, reaching a peak in 2007. FDI then experienced a drastic decline in 2008 and 2009 in line with global economic crisis: global FDI then inflows fell a further 37 %, while outflows fell some 43 % (UNCTAD 2010). It can also be noticed that in the last few years FDI declined in value and share, and Domestic Direct Investments (DDI) became even more important.

Forest industries may not have been the forerunners in this field, but during the last 30 years FDI shown a steep increase in the forest sector (Laaksonen-Craig 2004; El Lakany et al. 2007; Simula 2008) even with reference to developing countries (Savcor Indufor 2006; Borregaard et al. 2008). FDI considerably exceed ODA contributions and remain important for foreign exchange earnings, skills and technology transfer. According to IADB estimations (2004) the global amount of direct private investments in the forestry sector (forest, industry and trade) exceeds USD 60 billion a year, representing about 1 % of total direct investments worldwide, while – on the basis of UNCTAD figures – Savcor Indufor (2006) calculated the forest and wood products sector accounts for about 2.6 % of total FDI stock in developing countries, especially in Asia and the Latin America/Caribbean region. The World Bank estimates that direct investments for developing countries may range between USD 8 and USD 15 billion (PROFOR 2003). It should be underlined, however, that in the case of developing countries most of the increase in forest sector FDI concentrated in the plantations, and pulp and paper sectors, rather than in natural forests. According to El Lakany et al. (2007) the present growth rates for forest plantations results in investment requirements of about USD 4 billion per year, while Spek (2006) commented that in the pulp and paper sector developing countries account for a small proportion of total capacity, but they dominate capacity growth, also with a view to meeting rising domestic demand as in the case of China and India. Whereas in industrialized countries forestry FDI are dominated by the manufacturing and processing sectors, in developing countries the emphasis is on primary sector activities. Private foreign direct investment to forest industries in developing countries has grown at a rapid rate since 1990. It should be mentioned that UNCTAD figures for FDI in the forest sector are aggregated with those of agriculture and fisheries; moreover, forestry investments may only be a segment of a large multilevel manufacturer or conglomerate and will therefore not be listed under the forestry sector. Finally, FDI figures track only equity investment with at least 10 % control of the voting stock in the company, and ignore domestic finance, minority participation, loans and other non-equity cross-border flows. The figures also ignore foreign remittances to the families of expatriate workers, which represent a value almost double the flows of ODA and might support investments in informal or small-scale operations (Canby and Raditz 2005).

10.4.3.5 Private Sector DDI in Forestry

The bulk of private forest investments remains domestic (PROFOR 2003; Canby and Raditz 2005; Tomaselli 2006) but information on these is even less clear than that related to FDI. According to Tomaselli (2006) domestic investment constitutes over 90 % of private sector flows to the forest sector. Domestic private investment can be divided between company and community or farmer investment. Company investment can also be divided between larger scale industrial forestry enterprises (whether for natural forest concessions or plantations) and small and medium forest enterprises (Mayers 2006). According to Molnar et al. (2004) community investment in their own resources may amount to USD 1.3–2.6 billion per year, equivalent to annual ODA flows to forestry.

10.5 Outlook

As observed by UNEP (2011), the forest sector will be able to face the future increasing demand for products and services through a combination of changes in the external driving forces and internal efficiency gains. In particular through:

- Reducing the external pressure to forest conversion (e.g. higher efficiency in the use of existing farmland);
- Enhancing efficiency in the production of traditional forest products (timber and NTFPs), also with new plantations;
- Establishing new markets for forest-based environmental services.

Private investors and financial institutions are all involved in such development. As a matter of fact in the last years private funding initiatives have already grown to a large extent bringing new opportunities. Private funds can fill the gap left by traditional public funding mechanisms that are on the decrease, providing a relevant support to rural development strategies.

Quite a lot of forecasting exercises are underlining the increasing focus that will be placed on planted forests by future private forest investments. Studies and forecasts may differ in specific figures, but they all agree on the developing role of planted forests in wood supply. Production could increase to 44 % of the overall wood production by 2020 (Carle et al. 2002) and 75 % by 2050 (Sohngen et al. 1999, 2001). Dick (quoted by Evans and Turnbull 2004) estimated that forest plantations would take over the whole industrial timber supply by 2050. More recently, Carle and Holmgren (2008) calculated that plantations will be able to provide from 66 % to 80 % of the world's industrial wood supply by 2030. Almost all the previously mentioned studies predicted a global shift in the industrial wood supply from temperate to tropical zones and from the Northern to Southern hemisphere. Financial flows for investments in the timber sector will shift accordingly.

This development is not free of social and environmental risks. Some profit-oriented initiatives based on monocultures are already giving no or limited attention to the protection of local communities and natural resources (Fernholz et al. 2007). While the approach of integrating rural development in forest finance is widely present among public institutions, the majority of private sector investors are still reluctant to acknowledge this concept.

On the other hand, natural forests will most probably enter new market opportunities linked to PES. While methodologies for estimating the economic value of environmental services have still to be refined, and mechanisms to reward them are largely perfectible, the opportunity to market a broader range of products and services could have a remarkable impact on forests (see also Chap. 9). The growth of the "green economy" and the demand for "green" products – for the sake of profit, reputation needs or legal constraints – will probably attract an increasing amount of capital and investments. New actors and operators will also enter the forestry sector.

While the forestry sector presents new business opportunities for private operators, specific international and national policy tools for the governance of the changes are greatly needed. With proper policies, based on a mix of command and control instruments and provision of financial and technical support, it is possible to ensure that forests and their services are attributed their real values. Only in this way they can directly and increasingly contribute to rural development.

Acknowledgments The authors wish to express their gratitude to Dragana Stojkovic who has provided useful insight on the paper during her stage at the University of Padova.

References

AFDB (2010) Annual report 2009. African Development Bank Group, Tunis

Agrawala S, Carraro M (2010) Assessing the role of microfinance in fostering adaptation to climate change. Organisation for Economic Co-operation and Development, Paris

Asset (2011) Lignum Forestry Investment Fund. http://www.assetchile.com/descripcion-de-transacciones/el-fondo-de-inversion-forestal-lignum/. Accessed Jan 2013

Bass S, Nussbaum R, Speechly H, Morrison E (1996) Paper farming: the role of plantations in the sustainable paper cycle. The International Institute for Environment and Development (IIED), London

Binayee SB, Subedi BP, Sapkota IB, Pun L (2004) Microfinance for small scale tree and forest products enterprises: opportunities and challenges for the local producers in forestry sector. Nepal microfinance case study. ANSAB, Kathmandu

Borregaard N, Dufey A, Winchester L (2008) Effects of foreign investment versus domestic investment on the forestry sector in Latin America (Chile and Brazil). Demystifying FDI effects related to the Environment, vol 15, Discussion paper number. Working Group on Development and Environment in the Americas, Tufts

Brown C (2000) The global outlook for future wood supply from forest plantations, Global forest products outlook study working paper series. Food and Agriculture Organization of the United Nations, Rome

Canby K, Raditz C (2005) Opportunities and constraints to investment. Natural tropical forest industries. Forest Trends, Washington, DC

Carle J, Holmgren P (2008) Wood from planted forests. A global outlook 2005–2030. For Prod J 58(12):6–18

Carle J, Vuorinen P, Del Lungo A (2002) Status and trends in global forest plantation development. For Prod J 52(7–8):12–23

Chandrasekharan C (1996) Status of financing for sustainable forestry management programmes. In: Proceedings of the UNDP pretoria workshop on financial mechanisms and sources of finance for sustainable forestry, UNDP, Pretoria, Jun 1996

Contretras-Hermosilla A, Simula M (2007) The World Bank forest strategy: review of implementation. World Bank, Washington, DC 845.

El Lakany H, Jenkins M, Richards M (2007) Background paper on means of implementation (Contribution by PROFOR to discussions at UNFF-7, April, 2007). The Program on Forests (PROFOR) at the World Bank, Washington, DC

Evans J, Turnbull JW (2004) Plantation forestry in the tropics. Role of silviculture, and use of planted forests for industrial, social, environmental and agroforestry purposes. Oxford University Press, Oxford
FAO (2005a) Microfinance and forest-based small-scale enterprises. Food and Agriculture Organization of the United Nations, Rome
FAO (2005b) Sources of funding to support sustainable forest management. Food and Agriculture Organization of the United Nations, Rome
FAO (2007) Corporate private sector dimensions in planted forest investments. Food and Agriculture Organization of the United Nations, Rome
FAO (2010) FAO strategy for forests and forestry. Food and Agriculture Organization of the United Nations, Rome
Fernholz K, Bowyer J, Howe J (2007) TIMOS and REITS: what, why, & how they might impact sustainable forestry. Dovetail Partners, Minneapolis
GEF (2009) New climate for forests. GEF action on sustainable forest management. The Global Environment Facility, Washington, DC
GEF (2010) The GEF incentive mechanism for forests. A new REDD + multilateral finance program. The Global Environment Facility, Washington, DC
Gutman P (ed) (2003) From goodwill to payments for environmental services. A survey of financing options for sustainable natural resource management in developing countries. WWF-Macroeconomics Program Office, Washington, DC
Hardcastle P, Umali R (2007) Review of the implementation of the Yokohama Action Plan. International Tropical Timber Organisation (ITTO), Yokohama
Healey T, Corriero T, Rozenov R (2005) Timber as an institutional investment. J Altern Invest 8 (3):60–74
Hellström E (2001) Conflict cultures – qualitative comparative analysis of environmental conflicts in forestry, vol 2, Silva Fennica monographs. The Finnish Forest Research Institute & The Finnish Society of Forest Science, Helsinki, p 109
Holopainen J, Wit M (eds) (2008) Financing sustainable forest management. Tropenbos International, Wageningen, pp vi–xvi
IADB (2004) Estudio sobre inversión directa en negocios forestales sostenibles. Inter-American Development Bank, Washington, DC
IFAD (2009) Remittances: sending money home. The International Fund for Agricultural Development (IFAD), Rome
IMF (2003) External debt statistics: guide for compilers and users – Appendix III, Glossary. International Monetary Fund, Washington, DC
Independencia SA (2011) Lignum Fund profile. Available at www.independencia-sa.cl/cgi-bin/procesa.pl?plantilla=/v2/eng/fondo.html&id_cat=5. Accessed Jan 2013
Indufor (2010) Financing forests and sustainable forest management in Small Island Developing States. Second macro-level paper. Prepared on September 2010 for the United Nations Forum on Forests. Economic and Social Council of the United Nations, New York
Janse G, Ottitsch A (2005) Factors influencing the role of non-wood forest products and services. For Policy Econ 7(3):309–319
Laaksonen-Craig S (2004) Foreign direct investments in the forest sector: implications for sustainable forest management in developed and developing countries. For Policy Econ 6(3): 359–370
Lammerts van Bueren E (2002) Preface. European tropical forest research network news 35:1. European Tropical Forest Research Network, Wageningen
Macqueen D (2010) Review of funds which aim to protect tropical forests. The International Institute for Environment and Development (IIED), London
Mayers J (2006) Small-Medium Forestry Enterprises. The 'best-bet' for reducing poverty and sustaining forests? Paper presented at international tropical forest investment forum: issues and opportunities for investment in natural tropical forests, Cancun, 26–27 Apr 2006

Molnar A, Scherr S, Khare A (2004) Who conserves the world's forests? Community-driven strategies to protect forests and respect rights. Forest Trends, Washington, DC
NFP Facility (2010) Minutes of the Donor Support Group meeting friends of the facility 9 February 2010 Rome. Food and Agriculture Organization of the United Nations, Rome
NFP Facility (2011) Capacity development for national forest programmes. Overview of financing mechanisms. Food and Agriculture Organization of the United Nations, Rome. http://portals.wi.wur.nl/nfp4all/?Overview%20of%20financing%20mechanisms. Accessed Jan 2013
Niemelä J, Young J, Alard D, Askasibar M, Henle K, Johnson R, Kurttila M, Larsson TB, Matouch S, Nowicki P, Paiva R, Portoghesi L, Smulders R, Stevenson A, Tartes U, Watt A (2005) Identifying, managing and monitoring conflicts between forest biodiversity conservation and other human interests in Europe. For Policy Econ 7(6):877–890
OECD (2000) Official development assistance to forestry 1973–98. Organisation for Economic Co-operation and Development, Paris
OECD and IMF (2004) Glossary of foreign direct investment terms and definitions. Organisation for Economic Co-operation and Development/International Monetary Fund, Paris/Washington, DC
OECD (2006) The new rural paradigm: policies and governance. Organisation for Economic Co-operation and Development, Paris
OECD (2008a) OECD benchmark definition of foreign direct investment, 4th edn. Organisation for Economic Co-operation and Development, Paris
OECD (2008b) Is it ODA? Fact sheet. Organisation for Economic Co-operation and Development, Paris
OECD (2008c) Official development assistance to forestry 1995–2006. Unpublished report. Organisation for Economic Co-operation and Development, Paris
OECD (2008d) Statistics on biodiversity-related aid. Organisation for Economic Co-operation and Development, Paris
OECD (2010a) DAC Mandate 2011–2015. Organisation for Economic Co-operation and Development, Paris
OECD (2010b) Creditor reporting system aid activities database. Organisation for Economic Co-operation and Development, Paris. http://www.oecd.org. Accessed Jan 2013
Plomp R (2010) Forestry investments at PGGM. Presentation at Congress "Investment in Sustainable Forestry", Leusden, 4 Nov 2010
Precious Woods (2011) Shareholder structure. Shareholders holding more than 3 % of the share capital of Precious Woods Holding Ltd. http://www.preciouswoods.com/index.php?option=com_content&task=view&id=119&Itemid=231. Accessed Jan 2013
PROFOR (2003) Report of the forest investment forum. Investment opportunities and constraints. In: Proceedings. The World Bank, Washington, DC, 22–23 Oct 2003
PwC and WBCSD (2010). Sustainable forest finance toolkit. Pricewaterhouse Coopers LLP and the World Business Council for Sustainable Development (WBCSD). Available at www.pwc.co.uk/pdf/forest_finance_toolkit.pdf. Accessed Jan 2013
Savcor Indufor (2006) Study on international financing mechanisms for sustainable forest management. Report for ministry for foreign affairs of Finland. Savcor Indufor Oy, Helsinki
Savenije H, Holopainen J, van Dijk K, Barua SK, Oksanen T (2008) Financing for sustainable forest management: an overview. In: Holopainen J, Wit M (eds) Financing sustainable forest management. Tropenbos International, Wageningen, pp vi–xvi
Scherr S, White A, Khare A (2004) For rendered services: the current status and future potential of markets for the ecosystem services provided by tropical forests. International Tropical Timber Organization (ITTO), Yokohama
Simula M (2008) Financing flows and needs to implement the non-legally binding instrument on all types of forests. Prepared for The Advisory Group on Finance of The Collaborative Partnership on Forests. The Program on Forests (PROFOR) at the World Bank, Washington, DC
Sohngen B, Mendelsohn R, Sedjo RA (1999) Forest management, conservation, and global timber markets. Am J Agric Econ 81(1):1–13

Sohngen B, Mendelsohn R, Sedjo RA (2001) A global model of climate change impacts on timber markets. Am J Agric Econ 26(2):326–343
Solberg B, Miina S (1997) Conflicts management and public participation in land management. In: EFI proceedings, Joensuu
Spek M (2006) Financing pulp mills. An appraisal of risk assessment and safeguard procedures. CIFOR, Bogor
Streck C, Lehmann A, Rau A, Coren M (2010) The challenge of urgency: incentivizing private sector early action in REDD+. http://www.climatefocus.com/documents/the_challenge_of_urgency_incentivizing_private_sector_early_action_in_redd. Accessed Jan 2013
Terra Capital (2000) The Terra Capital Fund. Presentation at IUCN Congress "Developing & Investing in Biodiversity Business", Amman, 7 Oct 2000
Tomaselli I (2006) Brief study on funding and finance for forestry and forest-based sector. Report prepared for UNFF7. STCP, Curitiba
UNCED (1992) Agenda 21. United Nations Conference on Environment and Development, Rio de Janeiro
UNCTAD (2010) World investment report 2010. Investing in a low carbon economy. United Nations Conference on Trade and Development, New York/Geneva
UNDP (2010) Accelerating progress towards millennium development goals. UNDP's work in Environment and Sustainable Development. United Nations Development Programme, New York
UNEP (2009) Vital forest graphics. FAO/UNFF and United Nations Environmental Programme, Nairobi
UNEP (2011) REDDy set grow. Part 1 – a briefing for financial institutions. Opportunities and roles for financial institutions in forest carbon markets. United Nations Environmental Programme (UNEP) Finance Initiative, Genève
UNFCCC (2010) Statement on behalf of the International Indigenous Peoples' Forum on Climate Change. http://unfccc.int/files/meetings/cop_16/statements/application/pdf/101210_cop16_hls_iipfcc.pdf. Accessed Jan 2013
UNFF (2006) Report of the sixth session of the United Nations Forum on Forests. United Nations General Assembly. The United Nations Forum on Forests, New York
White A, Bull GQ, Maginnis S (2006) Subsidies for industrial plantations: turning controversy into opportunity. Arbor Vitæ, IUCN/WWF For Conserv Newsl 31:15
World Bank (2004) Sustaining forests. A development strategy. World Bank, Washington, DC
World Bank (2008) Forests sourcebook: practical guidance for sustaining forests in development cooperation. World Bank, Washington, DC
World Bank (2010a) MIGA: cultivating agribusiness growth. World Bank, Washington, DC
World Bank (2010b) MIGA: microfinance – critical to development. World Bank, Washington, DC
WRI (2005) World resources 2005: the wealth of the poor – managing ecosystems to fight poverty. World Resource Institute (WRI)/United Nations Development Programme/United Nations Environment Programme/World Bank, Washington, DC

Chapter 11
Land Use Planning for Sustainable Forestry

Holm Uibrig, Andre Hilbrich, and Gérard Hutter

Abstract The planning and overall management of rural land use are widely recognized as being complex and complicated processes. The state and the dynamics of land use derive from both the natural endowment of the area under consideration and related attributes of the human society. Forest land use has not long been a particular focus of land use planning and rural development. However, increasing human populations with growing needs locally and in a global context, the transformation of forests to other land use types such as agriculture and pasture, but also for purposes of settlement, mining, technical infrastructure, etc., have resulted in efforts to forecast the sustainability of land use based on historical development, the current state and potentials. Land use practices producing results other than those expected have led to the development and implementation of various participatory approaches ahead of exclusively technocratic means of planning. Accordingly, contemporary land use planning is characterized by argumentation stemming from a combination of top-down and bottom-up procedures. Forests continue to play a secondary role relative to other rural land uses, especially agriculture and grassland. Nevertheless, a recognition of the multiple production, protection and service functions of the large proportion of forests worldwide increasingly justifies and impels the adoption of innovative concepts such as adaptive strategy development and strategic spatial planning approaches to ensure an appropriate integration of forests and their management in rural development at local, landscape and regional level.

H. Uibrig (✉)
Institute of International Forestry and Forest Products, Technische Universität Dresden, Postfach 1117, 01735 Tharandt, Germany
e-mail: holm.uibrig@forst.tu-dresden.de

A. Hilbrich • G. Hutter
Leibniz Institute of Ecological and Regional Development, Weberplatz 1, 01217 Dresden, Germany
e-mail: a.hilbrich@ioer.de; G.Hutter@ioer.de

J. Pretzsch et al. (eds.), *Forests and Rural Development*, Tropical Forestry 9,
DOI 10.1007/978-3-642-41404-6_11,

Keywords Rural regions • Land resources • Land degradation • Land use types • Land classifications • Public planning • Participation • Integrated land use planning • Strategic spatial planning • Adaptive management

11.1 Introduction

The relationship between land, population and land management embodies basic evidence of the dependence of humans on the production of food and other goods, on the provision of services necessary to make a living and for continued socio-economic development. Monitoring of the utilization of land over recent decades has shown that changes to land cover and land use have occurred at an unprecedented rate and spatial extent (Rindfuss et al. 2003). Apart from the effect of the reduction of the area of land allocated to the production of food and fodder and/or reduced productivity, biodiversity has also declined drastically as a consequence (Bridges and Oldeman 2001; Tengberg and Stocking 2001). The establishment and expansion of industry, mining, settlements and various infrastructural facilities have frequently led to the loss of productive agricultural and forest land, and of the natural vegetation (Wild 2003). Another trend that can be related to the ongoing process of forest degradation and the transformation of forest to other land use types is climate change, which impacts upon all spatial scales, from the global to the micro-level (Cooper and Arblaster 2007; IPCC 2007).

Land use planning and management are generally considered a means to take advantage of the natural endowment of the geographical space in order to optimize the utilization of land and stabilize socio-economic systems for the purposes of continued economic and social development. A distinction is often made between land use planning for urban and for rural development (e.g., Randolph 2004; Wehrmann 2011).

Amler (1992, p. 215) structured framework conditions for space-related planning, referring to a large number of projects in numerous developing countries. Some of the factor groups identified included the centralization of planning and decision making, the lack of or an incomplete legal framework, competition between governmental institutions and overlapping mandates, inadequate setting of objectives, low level sector planning, inadequate planning methods, poor qualification and/or motivation of the planners, unsatisfactory data and unreliable statistical data collection and the dynamics of the subject of the planning process. Smoke and Romeo (1997, cited by Dalal-Clayton et al. 2003, pp. 156–159) listed factors constraining rural planning from the institutional perspective: a complex and poorly coordinated institutional framework, incentives to maintain confusion, poor public accountability, flawed performance incentives, unsatisfactory donor strategies and low absorption capacity.

The constraining factors have, to differing degrees, changed for the better through recent developments in the economy and in human society. Certainly, the processes of decentralization and devolution occurring in numerous countries have highlighted not only responsibilities but also brought about commitments regarding socio-economic development at lower administrative levels. Principles of governance increasingly focus on local populations and their empowerment, on public-private partnerships and so on. These changes also impact upon the prevailing systems of planning. An appropriate level of involvement of land users and other relevant stakeholders has been identified as being key to more successful strategies targeting the sustainable development of land use and of management (Dalal-Clayton et al. 2003; Wehrmann 2011). The alternative of either a top-down or a bottom-up approach associated with technocratic versus participatory planning has been modified through the adoption approaches that integrate the assessment of permanent land attributes with needs, culture and other factors characterizing the particular human society. This is expected to contribute to resolving issues stemming from the conflicting needs and interests of land users and other relevant stakeholders in adopting sustainability-oriented land use and regional development. Forests increasingly play an essential part in this given their production, environmental and social service functions.

Presented in this chapter are a few concepts pertaining to space-related/regional planning. The particular type of land use planning employed in rural regions is characterized. The historical roots and different fields of application of land use planning within the planning system are explained, and the methodological essentials relevant to the varying planning concepts and their refinement over time are highlighted. The shift from technocratic approaches to participation and integration in spatial planning is outlined. The main emphasis is on forest land as an essential component of the geographic space. Forest land and its use is a particular focus in the context of contemporary strategy development and strategic spatial planning approaches.

11.2 Definition of Concepts of Space-Related Planning

There is no standardized definition of fundamental terms in the context of land use, its planning and management worldwide. For the purposes of this chapter a definition of terms as given in FAO (1999) after FAO/UNEP (1997), FAO (1995) and Choudhury and Jansen (1998), and widely employed, is used.

Land and **land resources** refer to a delineable area of the earth's terrestrial surface, encompassing all attributes of the biosphere immediately above or below the surface, including those of the near-surface climate, the soil and terrain forms, the surface hydrology (including shallow lakes, rivers, marshes and swamps), the near-surface sedimentary layers and associated groundwater and geo-hydrological reserve, the plant and animal populations, the human settlement pattern and physical results of past and present human activity (terracing, water storage or drainage structures, roads, buildings, etc.).

Land use is characterized by the arrangements, activities and inputs by people to produce, change or maintain a certain land cover type (FAO 2007 after Di Gregorio and Jansen 1998). Land use defined in this way establishes a direct link between land cover and the actions of people in their environment.

Land cover is the observed (bio-)physical cover on the earth's surface (FAO 2007 after Di Gregorio and Jansen 1998). It does not matter whether the land cover is used by people in the way defined above or not.

Land use planning is the systematic assessment of land and water potential, alternative patterns of land use and other physical, social and economic conditions, for the purpose of selecting and adopting land use options that are most beneficial to land users without degrading the resources or the environment, together with the selection of measures most likely to encourage such uses.

Land utilization type (LUT) is a use of land in terms of products and services, the inputs and operations required to produce these products and the provision of these services, and the socio-economic setting in which these processes are carried out. In the strict meaning of the term, it describes a synthetic, simplified, representative land use type for the purpose of land suitability evaluation, such as a plantation of a particular forest tree species producing industrial timber under a specific technical and economic setting. It is necessary to distinguish between the LUT, as described above, and the actual, or real land use observed and described in the field (Choudhury and Jansen 1998).

Strategic planning is a disciplined effort to produce fundamental decisions and actions that shape and guide what forest management is, what forest managers do, and why they do it (Bryson 2004). Strategic planning can be fashioned in different ways. Given the possibility of change and the existence of uncertainties with regard to the context conditions, it is useful to distinguish between different modes of strategic planning such as programming and scenario planning.

Adaptive management (AM) is a structured, iterative process of decision making in the face of uncertainty. Key aspects of AM include accepting uncertainty and change as unavoidable, and treating learning as an explicit goal and management policies as deliberately monitored and evaluated experiments. Essential requirements include the possibility of necessary adjustments, a strong interaction between science and management and a systematic planning of all elements and steps of the adaptive management cycle (AMC) (Fig. 11.6) from the outset (Allan and Stankey 2009; Holling 1978).

11.3 The History of Land Use Planning at a Glance

Planning the use of land has always meant dealing with a great variety of subjects, by representatives of various disciplines and over time (an extensive history was provided in Amler 1992, pp. 25–37). The real beginning of the discipline land use planning remains arbitrary. In the case of the silvicultural and management planning of forests in the European context, Olschowy (1978) claimed that a well developed standard was already in place in the seventeenth century. In the German state Saxony, the regulation of forest utilization – 'Forst- und Holzordnung' – by Kurfürst (elector) August dates as far back as September 8th, 1560. There are also reports of examples of space-related administrative work, such as construction planning for towns and Napoleon's (eighteenth century) national road network planning.

Thünen's (1826) theory with regard to agricultural location marks another milestone in terms of the agricultural utilization of land. The distance between the location of the residence and the site of production was defined the most essential economic parameter. The differentiation of the capability of the land, referring to site quality and climate, was not taken into consideration. The quality of soils and the favorability of the climate were assumed to be identical within a socio-economic unit.

The opposite of this theory sees land use based upon the natural differentiation of the geographic space. The corresponding theory of land use in accordance with the geographic space was supported by methodological approaches for the allocation of industrial locations at the beginning of the twentieth century. Weber (1909) availed of 'site factors'. Christaller (1933) developed the conceptual framework of central locations for regional planning, as is employed in regional planning to this day. Further methodological development was characterized by sector approaches, for example, a mathematical-spatial system of the relationship between the economy and a honeycomb structure of the landscape.

Weber's ideas had a fundamental impact on the development of an 'epoch of planning'. The self-administration of municipalities in Germany, planning in overcrowded regions and the Housing and Town Planning Act (1909) in Great Britain were further examples of space-related planning.

The GOERLRO Plan developed by the State Commission for Electrification in Russia in 1920 is deemed to be the first regional plan developed under a socialist system.

In the USA, the State of Wisconsin issued the first Law on Land-Use in 1929, with the first plan following for the district of Oneida, Wisconsin, in 1933 (McAllister 1973). The plan defined areas to be used for forestry, recreation, agriculture and areas without any restrictions in terms of further utilization. The background to this kind of planning was extensive forest exploitation leading to widespread forest destruction, uncontrolled forest fire, soil erosion following extensive cultivation and also delayed and unpaid tax contributions by landholders. Further regulations pertaining to land use planning were prepared by the National Resource Planning Board. However, a strong anti-planning fraction within the parliament successfully stalled these efforts. Only during the 1960s were the activities of the board re-activated.

Sector-specific planning emerged to fill the gap resulting from limited engagement in planning processes. Jacks (1946) published a first monograph entitled 'Land Classification for Land-Use Planning', linking the assessment of the capacity of land with land use planning.

From 1967 onwards the European Commission on Agriculture (ECA), Geneva, in cooperation with the FAO, organized several conferences on rural development and land use planning. Planning for rural regions was deemed synonymous with land use planning. The objective was to formulate a standardized approach to this kind of planning.

For more than three decades, there has been a strengthening of the understanding and recognition of the need for the planning of land use. In the year 1981 the

Table 11.1 Historical development of space-related planning

Time	Milestone	Focus of planning
Pre-seventeenth century	Forest land management standards	Forest utilization, hunting
Seventeenth, eighteenth century	Establishment of towns	Construction, road networks
Late eighteenth, early nineteenth century	Forestry planning	Sustainable forest production
Early nineteenth century	Industrial development, focus on distance between production and consumption	Agriculture, husbandry, forestry
Nineteenth, twentieth century (epoch of planning)	Decentralized administration, overcrowded areas	Sector planning
Late nineteenth, early twentieth century	Natural differentiation of geographic space	Allocation of industrial locations
1920s	GEORLRO plan in the Soviet Union	National electrification
1920, 1930s	Land use planning in the USA	Rural land use
Since 1933	Central locations, honeycomb structure of economy, mathematical models	Various sectors, infrastructure
1940s	Land capability classification	Rural land use with emphasis on agriculture
1960s	Focus on rural development	Rural land use
Since 1980s	Land suitability classification	Sustainable production in agriculture (protection of soil from erosion)
Since 1980s	Multifunctional forestry	Sustainable forestry (economic, social, environmental)

Sources: Amler (1992), Jacks (1946), Bridges et al. (2001)

twenty-first General Assembly of the FAO passed the World Soil Charta in response to the increasing loss and degradation of soils worldwide. Environmentally sound and sustainable production in agriculture was placed at the center of efforts to plan the future use of land.

In forestry, medium-term planning based on site quality has developed and been refined over the course of the last two centuries. However, application has been limited to intensively managed forests and land uses other than forestry have been left largely unconsidered. Continuous land degradation, and strategies to control this process (Bridges et al. 2001), as well as biodiversity loss and the declining availability of freshwater, will be of paramount importance in the context of land use for many years to come. An overview of the development of the main focus areas in spatial planning is provided in Table 11.1.

11.4 Forest Land Use Planning for Rural Development

11.4.1 Integration of Land Use Planning Within the Public Planning System

Three levels of administration can be distinguished for public space-related planning: (1) development planning at national level, (2) sector planning and regional planning at sub-national level and (3) area utilization planning at local level. The purpose of regional planning is to specify the general objectives of the national development planning according to the prevailing conditions within a defined area. Regional plans are obligatory for governmental institutions. They serve as an orientation for the local population and refer to its needs. Area utilization plans in effect at subordinate administrative levels represent a specification of the regional plan for smaller territories, applying at a higher mapping scale and with a higher level of detail with regard to the utilization of land. Work plans comprise programs of management that are not usually a matter of land use planning (Young 1993).

Winckler (1999) highlighted a series of topics related to planning at the regional level, namely regional rural development, environmental action planning, management of water catchment areas, development of mountainous regions, coastal zone management, buffer zone management, desertification control, etc. Of the different rural development strategies tailored to the prevailing natural, socio-economic and socio-cultural conditions, and adhered to over time (Hammer 1999, p. 291), land use planning and resource management have come to prominence since the mid-1980s. These aspects have become central facets of rural development (Van Lier et al. 1994; Wehrmann 2011).

Land use planning currently focuses on the management of resources, dealing mainly with infrastructure, water, energy, agriculture and forestry, but also environmental protection and ecological development. It is generally understood not as a short term project but as a long term process, to be harmonized with the relevant interests and demands.

In most cases, land use planning is a means to prevent and to resolve land use conflicts. Such conflicts may arise where the resource land is consumed over the course of activities not providing for the maximum, sustainable production of goods and services, or where resource use violates governmental land use policies (Vergara 1996, p. 92). On the whole, the number of competing users of land is increasing steadily and the resource is becoming increasingly scarce. The status of tenure is frequently also linked to land use conflicts (Vergara 1996, pp. 93–94, Wehrmann 2011). Levels in the administrative hierarchy, the actors and the objectives of land use planning vary in accordance with the state of establishment and functioning of national planning systems, the problem specifications and conflict situations, and the expected outcomes of the land use planning procedure.

11.4.2 Methodological Essentials of Land Use Planning

11.4.2.1 Land and Geographic Information Systems

Data on land is a basic requirement for land use planning and management, irrespective of the planning approach followed. The value of the data and the effectiveness of a planning process are closely related to the quality of the data and the manner in which it is made available. A diverse group of professionals and institutions/organizations, both governmental and private, provide the necessary data. These include surveyors, cartographers, geographers, foresters, farmers, computer scientists, land use planners, etc. (Dale and McLaughlin 1988; De Graf et al. 1999; FAO 1998).

An information system may be formally defined as a combination of human and technical resources, together with a set of organizing procedures that produces information in support of managerial requirements (Hill and Aspinall 2000). The raw alphanumeric, graphic and/or digital data must be processed so that they are transformed into useful information that can be understood by a decision maker (Dale and McLaughlin 1988).

The management of a land information system includes the acquisition and assembly of data; data processing, storage and maintenance; data retrieval, analysis, and dissemination. Spatial and non-spatial information can be distinguished. Resource information is an essential subgroup of spatial information. Dale and McLaughlin (1988) subdivided resource information into geographic information systems (GIS) and land information systems (LIS). The critical factors for this subdivision are the scale of representation and further use. While GIS typically focus on the biophysical and socio-economic attributes of land, large scale land information is typical for parcel-based systems serving juridical and fiscal purposes. Parcel-based information is preferably administered in governmental cadastral offices. This type of information system is not yet implemented in many countries.

The land information system can be subdivided into groups, providing four specific kinds of information, as indicated in Fig. 11.1. Cadastral information is preferably parcel-based, whereas environment and infrastructure information is polygon referenced.

Land information systems may present the data in the form of hard results, such as maps or land title certificates, or as soft results such as information or professional advice. A land information system can supply attribute data that may be presented in verbal or numerical form, spatial data that may be shown on maps, and temporal data indicating how current the data is. The attribute or text data may describe phenomena within a place, such as the characteristics of the soil and the activities taking place on the land, or between places, as in networks and activity flows (Dale and McLaughin 1988; Hill and Aspinall 2000).

For centuries in forestry, the long production periods required for trees providing timber have led to the creation and management of information related to forest land, in order to secure property rights and sustainable utilization. The increasing management intensity of forests, and management of forests for purposes other than

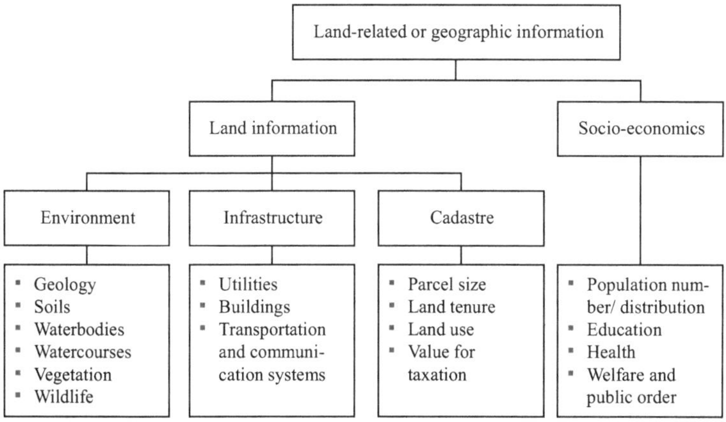

Fig. 11.1 Structure of land-related information (Source: Dale and McLaughlin (1988, p. 11, modified))

wood production, have resulted in a wide variety of attributes and details within the respective planning systems.

Enterprises engaged in agriculture and forestry increasingly use mobile GIS/LIS devices, which support resource management planning, monitoring and control at the micro-level. This also applies in the context of ecological development. The application of global positioning systems (GPS) facilitates field work.

In further developing land information systems, Grove et al. (2002) integrated land-related data into a human-environment database consisting of space, time and human decision making dimensions. The purpose was to understand the past better, to monitor the present and to predict and plan for the future. The model is applicable at any scale, from individual to global (Grove et al. 2002). With the further advancement of information and communication technology, new stimuli for the development of integrated rural planning applications can be expected.

11.4.2.2 Classification/Typification of Land Use and Forest Functions

Land use may change over time and may also drive environmental change. Both are critical to sustainable development. The definition of categories of land use (classes, groups, types, kinds) is to take stock of (1) the various forms of land use (genesis or development of the earth's surface), to identify (2) land use structure (pattern: location within the geographic space) and to identify (3) land use within a given territory (dynamics: interactions between the various types). Land use classification requires the generalization of an almost infinite variety of land uses, and their representation in maps of various scales and in other file types.

The definition of categories of land use constitutes a specific classification. In geography terms it is referred to as typification (Neef 1967, p. 72). Types of land use can be defined by employing only a few (agriculture, forestry) or numerous characteristics (plantation of a specific tree species, e.g., *Acacia* sp., for the production of a certain assortment of wood, such as industrial roundwood). The sequence of broad types plus more narrowly defined sub-types is called a row of types (e.g., agriculture – irrigated tillage – rice paddy; forestry – planted forest – teak plantation). The typology provides the means to select the attributes relevant to the objectives of inventory and planning. The types identified form the physical and spatial basis for further analysis, for planning and for management.

The numerous land use classifications, from global to local, result from the many technologies available for the identification of land use classes. In addition to solely terrestrial approaches, there exist a variety of remote sensing techniques and combined methods. Even census data may be incorporated, especially at the macro-level. The classifications correspond to different scales of representation on maps (hard copy or digital) and to the degree of generalization. A report produced by a Sri Lankan/Swiss remote sensing project (Dept. of Geography et al. 1988) provided a comprehensive review of the history, technical aspects and results of one such project, and of the land use classification in particular.

On maps and in GIS layers, land use classes are indicated by words, symbols, colors or combinations thereof. For example, the classification and representation on maps, as developed for the land use survey in Great Britain carried out from 1930 to 1954 based on field survey using ordinance maps (Stamp 1950; Board 1968), and the classification proposed for a world land use survey (Van Valkenburg 1950) have had a substantial influence on a large number of land use classifications used today (Table 11.2).

Classifications of land cover/land rarely use rows of various specified land uses at comparable levels of detail. Therefore, a further subdivision of forest land represents a separate task to be carried out by the corresponding experts. In contrast to this, in the case of a MRC/GTZ forest cover monitoring project conducted in the Lower Mekong Basin during the 1990s, where the main focus was on changes to the forest cover, other land uses were identified by general classes only (Box 11.1).

In the year 1995, the FAO published a land use classification based on the degree of modification of the ecosystem. Under this classification population is understood to be a dynamic object, and also the acting subject, attributed with experience, ideas and activities, inciting interactions between land and land use in a certain economic and social context. Following assessment of the degree of ecosystem modification, functional and biophysical land use classes are distinguished (Table 11.3).

With respect to the harmonization of class-sets of recurrent inventories, two parameters should suffice to describe a land use, namely function and activity. The availability of high quality data facilitates the harmonization of data sets collated at different times. This refers to positional and thematic accuracy, which is still subject to research (Jansen 2005, pp. 6–7).

Table 11.2 Classification of land uses applied in field surveys

No.	Description	Letter marking	Color marking
1	*Forest* and woodland to be marked	F	Dark green
2	*Meadowland* and permanent grassland	M	Light green
3	*Arable* or tilled land, fallow, temporary grassland and market gardens	A	Brown
4	*Heathland*, moorland, commons and rough hill pasture	H	Yellow
5	*Gardens*, allotments, orchards, nurseries, etc.	G	Purple
6	*Land* that is agriculturally unproductive, e.g., buildings, yards, mines, cemeteries, etc.	W	Red
7	*Ponds*, lakes, reservoirs, ditches, dykes, streams and anything containing water	P	Blue

Source: Stamp (1950, pp. 22, 24)
Ordinance survey map, scale 6 in.: 1 mile (1: 10,560); quarter sheets 2 by 3 miles each

Box 11.1 Classification of Land Cover and Land Use Within the MRC/GTZ Forest Cover Monitoring Project 'Assessment and Monitoring of the Lower Mekong Basin Forest Cover'

The analysis of the changes to forest cover was based on different remote sensing data and ground checks. The classification of the land cover and land use was assessed as a key for the visual interpretation of satellite images and aerial photographs. The mixture of forest at different stages of exploitation, agricultural land and forest regrowth proved to be a challenge in the definition of classes of land cover/use.

Classes and codes of the Lower Mekong forest cover map

1. Forest	Crown cover >20 % and forest percentage >40 %		Code
Evergreen and deciduous formations			
1 **Evergreen and semi-evergreen forests**	Evergreen	Continuous cover: dense/closed	11
	Evergreen	Continuous cover: open/disturbed	12
	Evergreen	Mosaic of forest blocks	13
	Semi-evergreen	Continuous cover: dense/closed	17
	Semi-evergreen	Continuous cover: open/disturbed/degraded	18
	Semi-evergreen	Mosaic of forest blocks	19
2 **Deciduous forests**			20
		Continuous cover	21[a]
		Mosaic of forest blocks	22[a]

(continued)

Box 11.1 (continued)

3	**Coniferous forests**	Coniferous pure	31
		Coniferous mixed	32
4	**Regrowth of secondary forest**		40
	Specific forest types		
5a	**Bamboo forest**		51
5b	**Swamp forest**		52
5c	**Mangrove forest**		53
5d	**Forest plantations**		54
5e	**Other forests**		55
2. Non-forest		Crown cover <20 % and forest percentage <40 %	
	Woodland and grassland		
6	**Wood- and shrubland**		61
7	**Grassland**		62
	Mixed pattern of shifting cultivation areas		
8a	Mosaic of regrowth, shrub, and recent fields of shifting cultivation	Percentage of fields under cropping < 30 %	81
8b	Mosaic of regrowth, shrub, and recent fields of shifting cultivation	Percentage of fields under cropping > 30 %	82
	Other classes		
9a	**Agriculture**		91
9b	**Barren land**		92
9c	**Rock area**		93
9d	**Urban area**		94
9e	**Water**		95
9f	**Other land**		96
9g	**Clouds**		99

[a]Only possible for wet season images

The analysis of two test areas within the province Vientiane provided evidence of the ongoing decline in the area of natural forest cover (code 11) by about 60 % between 1981 and 1995. Agriculture (code 91) doubled during the same period, from 52 to 101.2 ha. Due to a continued shifting cultivation mosaic of regrowth, shrub and recent fields, cropping > 30 % of the area (code 82) underwent a fivefold increase. A mixture of forest of various conditions and other land uses such as shifting cultivation indicates one of the critical limitations in the classification of forest land use.

Source: Stibig (1996), Sysavath (1997), Saylath (1998)

Table 11.3 An initial approach to an international framework for the classification of land use

Level I Degree of ecosystem modification	Level II Functional land use	Level III Biophysical land use
Uses based on natural ecosystems	Not used	
	Conservation: total; partial	
	Collection	Plant products
		Animal products
		Plant and animal products
Uses based on mixed natural and managed ecosystems	Agro-silvopastoralism	Forest products, cropping, livestock and aquaculture on same holding
Uses based on managed ecosystems	Production forestry	Management of natural forests
		Management of planted forests
	Livestock	Nomadic grazing
		Non-intensive grazing
		Intensive livestock production
		Confined livestock production
	Arable crops	Shifting cultivation
		Sedentary cultivation
		Temporary cropping
		Permanent cropping
		Wetland cultivation
		Covered crop production
	Mixed livestock and crop production	
	Fisheries	Fishing
		Aquaculture
Settlement and related uses	Recreation	
	Mineral extraction	Mining
		Quarrying
	Settlement	Residential
		Commercial
		Industrial
		Infrastructure
	Uses restricted by security	

Source: FAO (1995, p. 22), slightly modified
Note: Land use phases on irrigated land, type and sequence of crops, intensity of inputs, etc.

11.4.2.3 Land Capability and Land Suitability Classification

Land Capability Classification

Land capability classification is a technique to determine the most suitable use for any area of land. Land capability classification refers to the assessment of the capacity for producing crops, or combinations of crops, under optimum management; or its suitability for non-agricultural types of land use (e.g., Jacks 1946; Davis

1976). During the initial stages of development of this approach there was no sharp distinction made between land capability classification and land use planning (Jacks 1946, p. 10; Klingebiel and Montgomery 1961; King 1966). Nowadays, any such land classification may be deemed a starting point of land use planning.

It is common practice in the qualitative classification of land capability to distinguish between the physical (inherent) and the economic and social attributes of land (e.g., Obeng and Smith 1963; Young and Goldsmith 1977; Carpenter 1981). The physical characteristics include geology, climate and topography. They are permanent in the sense that they cannot be profoundly altered by humans. Geology, climate and topography are combined in the factor soil. Though modifiable by humans and other living organisms to some extent, soil is the main factor in a physical classification. A physical land classification can serve as a basis upon which to superimpose economic and social classifications, modified with changes to economic and social conditions as well as changes to the objective of the plan.

Included amongst the essential economic data are uses of land, ownership types, land values, transportation facilities and markets for current and potential products. Social data address the distribution and population types, economic status of the (rural) population, etc.

A land capability classification with long term validity should be open for future developments in terms of techniques and science. It is, in fact, impossible to forecast with any degree of accuracy either economic or scientific developments; and a practical land classification, which takes these into account, must necessarily be vague.

The result of land capability classification is an indication of trends in land use within defined areas. In Table 11.4 classes of the widely employed capability classification employed by the United States Agriculture Department (USDA) are listed.

The classes can be subdivided into subclasses, specifying the limitations, such as risk of erosion, excess water, rooting zone limitations or climatic limitations. A possible third level provides more specific and detailed information at farm field scale (Kang and Tripathi 1992).

Reference can be made to case studies in numerous countries, including the USA, Canada, Ethiopia, Malawi, Zambia, India and China. The space-related analysis carried out as part of a field study conducted by Darr (2003) drew on a comparison of land capability classes of farm forests and competing land use types in three project areas in Laos to balance the degree of sustainability associated with the competing land use types (see Box 11.2).

Based on the visual interpretation of aerial photographs, terrestrial land survey and land use inventory, Badege Bishaw (1985) structured the area of a peasants' association in the eastern Ethiopian highlands into land capability classes, indicating where to intensify agriculture and where the land should be left for permanent vegetation. The physical attributes included in the assessment were altitude, parent material, slope, soil type, natural vegetation and current utilization of the land. The results were revealed in a map at a scale of 1:5,000, and used as a basis for the expansion of irrigated agriculture and for afforestation. Over the course of one

Table 11.4 Land capability classification of the USDA

Class	Characterization
Suitable for cultivation, with	
I	Ordinary practices to maintain productivity
II	More than ordinary practices
III	Considerable management inputs
IV	Very careful management
Not suitable for cultivation, but suitable for permanent vegetation, with	
V	No special restrictions or practices
VI	Severe restrictions in use
VII	Severe restrictions in use, unsuited to cultivation
Not suitable for cultivation, grazing or forestry	
VIII	Limitations that preclude use for commercial plant production

Source: Norton (1939), Hockensmith and Steele (1943), Kang and Tripathi (1992)

decade only, the area of planted forest increased from almost nothing to over 10 % of the association's entire land area of about 600 ha (Badege Bishaw and Uibrig 1989; Uibrig 1991). Recommendations for projects of this type were provided by Dent (1991).

Land capability classification has been criticized by a number of authors, especially from the forestry perspective. From examination of the literature, and as is reflected by the classes, it is apparent that agriculture is the main focus of the classification, leaving poorer quality land for permanent vegetation, forest, scrub, etc.

Box 11.2 Extent of Selected Land Use Types Over Land Capability Classes: Example from Laos

The competing land use and land cover types identified were agriculture, pasture, and planted and natural farm forests. The land attributes involved followed categories recommended for land use at the national scale in Laos and the referenced assessment of fertility of soil types. Soil fertility was found to be the most influential factor in local farmers' decision making with regard to land use, of a total of 14 factors listed in exploratory household interviews. Land capability maps were derived from remote sensed and GIS data. The results were verified using spot-checks on the ground.

Three village areas characterized by different biophysical and socio-economic key criteria, namely potential for permanent agriculture, the abundance of natural forest and individual land tenure situation, were identified for the study. Each village was located within the area of operation of an internationally supported natural resource management project. The projects supported the establishment of farm forests and provided technical advice, seedlings and other materials.

(continued)

Box 11.2 (continued)

The village *Nasom* was characterized by poor soil fertility and a mountainous climate, both features limiting the potential for agriculture. Subsistence tillage and livestock rearing represented the common livelihood strategies. The remote location and poor access to roads restricted market access. Property rights pertaining to tilled land and homesteads had largely been legitimized. *Napo* village enjoyed favorable access to markets and opportunities for off-farm employment due to the proximity to the capital. Located in the Mekong plain, the village was characterized by favorable conditions for agriculture and an extensive land reserve. Farming was based mainly around rice paddies, with the considerable production surpluses sold. Customary claims to homesteads, pasture and agricultural land, including fallow, were officially acknowledged. The third village, *Houamo-tai*, was situated along a national road and was the most accessible of the three study sites. A basic infrastructure, such as an electricity supply, was installed and created diverse opportunities for off-farm employment. Due to a ban on slash-and-burn, customary property claims to upland plots and fallow land had not been legalized, but were respected amongst the villagers.

The arable areas in all three villages corresponded mainly to capability class II (99 %, 86 % and 70 %). Pasture land ranged between classes II and VIII, indicating the traditional status and limited importance of animal rearing for livelihoods overall. The areas of natural forest were but remnants of the once abundant natural forests, predominantly left on the steeper slopes. In Napo village the presence of natural forest on plains land belonging to class II (74 %) indicated that the villagers derived their livelihoods from other income generating activities. Additionally, this forest land was considered to be an extensive land reserve for the expansion of permanent and temporary tillage. Farm forests had mainly been established on abandoned arable/pasture land of all capability classes.

Extent of the land use types within the land capability classes (in percent).

(continued)

Box 11.2 (continued)

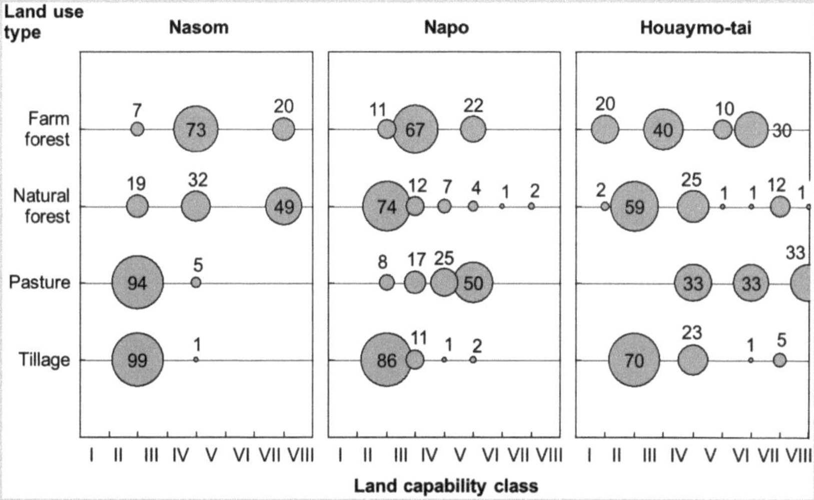

Tillage comprised home gardens, permanent and upland paddies.

Houaymo-tai was the only village where farm forests have been established on sites of capability class I, mostly on the sites of previously permanent paddy fields (20 %). This may have been linked to households that possessed abandoned agricultural land and derived their livelihoods entirely from alternative activities, such as trade and other off-farm employment. The differences between the land use types were statistically significant.

Source: Darr (2003), Darr and Uibrig (2004)

Land Suitability Classification

The objective of the land suitability classification is the comparative assessment of land performance when used for specific purposes. It involves the execution and interpretation of basic surveys of climate, soils, vegetation and other aspects of land in terms of the requirements of alternative types of land use or – applying the term often used in the context of land suitability classification – land utilization. Only the range of land uses applicable within the physical, economic and social context of the area in question are relevant for planning (cf. FAO 1976; Davis 1976). The following principles are fundamental to the approach and the methods employed (FAO 1984; FAO 1985):

- Assessment of land suitability and classification with respect to specified kinds of utilization;
- Evaluation of benefits obtained and inputs needed on different types of land;

- Adoption of a multidisciplinary approach;
- Comparison of more than one type of utilization during the evaluation;
- Incorporation in the evaluation of the physical, economic and social context of the area of interest.

From a forestry perspective, land suitability classification essentially deals with decision making between forestry and other major types of land use. Subsequently, the suitability assessment relates to the kinds of forestry that are relevant (Young 1993).

Land not yet assessed and classified has become increasingly rare. Underlying any recommendation for a change to the major land use as a result of a suitability classification are considerations that stem from various policy fields, preferences at different levels, customary and formal rights, etc. This leads to a multidisciplinary approach.

Land suitability classification in forestry concerns the functions of the forests and forest management. Forestry has very different features when compared with agricultural or grazing uses, with important consequences for the land suitability classification. These are (FAO 1984, pp. 2–3):

- *Long time scale:* the period from planting to tree harvest is rarely less than 6 years, frequently at least 15 and sometimes over 100 years. This means that forestry-related decisions, once taken, are difficult (or expensive) to change.
- *Multiple functions:* a particular forest land area usually provides more than just one specific good or service, including wood products, non-timber forest products, water, biological conservation, recreation and tourism, benefits to other land areas. Multiple-use forestry is designed to maximize the absolute benefits in goods and in services.
- *Levels of management intensity:* there is a wide range in the levels of management intensity in forests, from managed 'wild' lands with more or less unmodified natural vegetation to more intensively managed forests comprising selected species and involving measures (investments) to improve growth and form. Forest can be natural, modified to various degrees or planted. Silvicultural practices are manifold, including selective logging, natural regeneration, enrichment planting, pruning, intermediate cutting, planting of indigenous and exotic species in monocultures or in mixtures, site amelioration, etc. As management intensity increases, so also does the capital invested and the value of the product per unit area.
- *Conservation*: a conservation function or service is almost always present in forestry. Where the objective is wood production, conservation is still an essential component. In other types of forestry the protection of biodiversity, soil or water may be a primary objective.

Production forestry targets the growing of wood or timber of various qualities, as well as non-timber forest products (NTFP). Non-timber products are either raw materials for local use (fodder, litter for compost, food, etc.) or goods for sale to generate a cash income (resin, gums, rattan, etc.), and often a combination of both (medicinal plants, bamboo, fruits, honey, etc.). For all forest products, the objective should be the achievement of a sustained yield.

Conservation forestry includes watershed protection, comprising the control of run-off, soil erosion, earth movements, the regulation of river flow; the reclamation or improvement of degraded land through the planting of woody species, assisted or protected regeneration; the protection of the life of indigenous forest peoples; the conservation of natural ecosystems or selected species of flora and fauna; the reduction of emissions from deforestation and forest degradation (REDD), etc.

Forestry for recreation and tourism in tropical regions is often focused in national parks, with wildlife viewing an important component. Ecotourism, camping, etc. all depend on the availability of particular natural and/or technical facilities.

The purpose of multiple-use forestry is to meet more than only one major objective simultaneously. This refers to the production of any kind of product or service, such as the conservation of a selected component of a forest ecosystem plus recreation or tourism.

Agroforestry describes land use practices combining woody plants with agricultural crops, pasture or livestock overlapping on a specific area at one time or in sequence. Both ecological and economic interactions between the component elements play a part in characterizing the general practices and particular systems (Nair 1989). However, a dispute over whether agroforestry should be defined as land utilization under the general heading of either forestry or agriculture remains (Torquebiau 2000).

The need to plan in space and time: one of the main planning objectives in forestry is a sustained flow of outputs. This must be achieved on land exhibiting a wide amplitude of characteristics resulting in a highly variable potential for tree growth. The achievement of sustained production requires estimates of growth rates on the different sites, coupled with a scheme for planting and/or cutting cycles.

The classification sticks on land use systems. The systems represent particular parts of the land area which are characterized by qualities of the land and are compared with requirements of the land use types taken into account. Consequently, land utilization types must be defined. Particular crops grown in the area in question, e.g., wheat, sorghum, maize or potatoes, represent land utilization types in agriculture. Types of forest land utilization derive from the respective time scale, functions, management intensity, combination of space and time. Various major forestry land utilization types were described by Young (1993, pp. 822–825), among others.

Suitability refers to the fitness of a given unit of land for a specific use. Four levels are recognized, from general to detail. There are two orders, namely S (suitable) and N (not suitable). The order N is subdivided into the classes N1 (currently unsuitable) and N2 (permanently unsuitable). Three classes have been defined for the order S, reflecting varying degrees of suitability (S1 highly suitable, S2 moderately suitable, S3 marginally suitable). The classes can be subdivided into subclasses indicating particular limiting factors or the kinds of improvement measures required, such as e (erosion) and w (water logged), and further specified into units reflecting minor differences in the management required within the subclasses (e.g., e-1) (FAO 1984, pp. 16–18; Weigel 1986, pp. 21–22). Symbols for subclasses and units are identified based on individual projects.

Land suitability classification outlines alternative options for the use and management of land in the future and presents the consequences of these alternatives for the area in question. The results provide the following kinds of information:

- Description of land utilization types: basic descriptive data and management practices;
- Suitability maps: suitability of the land units for each of the relevant land utilization types;
- Estimates of the consequences of applying each relevant land utilization type to each land unit: inputs (material, labor, capital), outputs (material and intangibles, e.g. timber, tourist potential), environmental impact (positive and negative), social consequences (favorable – meeting local needs, adverse – displacement of population), economic analysis (based on pricing of benefits and costs);
- There may also be data from basic surveys and specialist studies (soil survey, forest inventory, economic and social data).

Commonly, a comparison between land utilization and a land unit requires an economic basis, both to compare the different kinds of output and to compare the production costs with the benefits.

Manithaphone Mahaxay (1996) developed a method of land suitability classification aimed at sustainable land use within a test area of about 3,600 ha in Laos (Box 11.3).

Land suitability classification has been incorporated into the FAO guidelines for land use planning (FAO 1993). This approach is characterized by ten consecutive steps, from baseline appraisal to implementation, monitoring and revision of the plan. However, in the event of limitations such as a lack of the necessary skills, a shortage of manpower, etc., the standard procedure may be modified (Dent 1991).

Box 11.3 Land Suitability Classification for Forest Land Use Planning: Sangthong Forest Training Site, Laos

The study area was situated in the Sangthong district of Laos, in the Vientiane municipality. Once almost entirely natural forest, the area had been exploited since the late 1970s, and especially intensively during the 1980s. During that time, the number of local residents present in two villages increased to over 400. This served to further reduce the natural forest cover to just a few remnant areas. Consequently, upland arable fields, abandoned agricultural land in various stages of succession and some secondary forest characterized the land use pattern.

The site was selected for the development of a land suitability classification method and the generation of a land use model striving for sustainability

(continued)

Box 11.3 (continued)

involving environmental and socio-economic criteria. The land area available for rehabilitation by means of reforestation, for forest conservation (protection), for agriculture (rice paddies, upland crops) and for settlement constituted the main land use types to be identified.

The evaluation was based on physical characteristics (landform and soil), biological attributes (vegetation and land cover) and socio-economic data. Particular parameters were employed for different types of land utilization.

Parameters employed for the different land utilization types

	Types of land utilization				
Parameter	Rice paddy	Upland crops	Reforestation	Protection forest	Settlement area
Slope	[a]	[a]	[a]	[a]	[a]
Soil depth	[a]	[a]	[a]		
Soil texture	[a]	[a]			
Soil pH	[a]	[a]			
Soil fertility	[a]	[a]			
Soil drainage	[a]	[a]			
Current land use	[a]	[a]	[a]	[a]	[a]
Accessibility	[a]	[a]	[a]	[a]	[a]
Access to water	[a]				[a]
Headwater of river				[a]	[a]

[a]Parameters considered

Primary data were collected by means of the visual interpretation of aerial photographs, ground checks and managed using GIS (ARC/INFO). Socio-economic data collected during interviews with members of local farm households using a prepared questionnaire were input into spreadsheets in Microsoft Excel. A topographic and recent land use map was integrated within the system.

Land suitability was assessed according to the classes I (highly), II (moderately) and III (marginally suitable), NS (not suitable) and NR (not relevant). The last class covers land area not assessed for the relevant land utilization types due to exclusion from the evaluation, including existing rice paddies, settled areas and Ray.

The land requirements for reforestation were fixed at a slope of >28 %, soil depth <30 cm, on degraded open land previously forest and accessible for seedling transportation. In the case of protection forest, adverse environmental effects with and without forest cover were considered.

(continued)

Box 11.3 (continued)

Land qualities in identical technical terms as requirements were also specified for the land utilization types taken into account. In the case of reforestation, the following were included: erosion hazard, rooting condition, land cover, accessibility, etc. The individual land quality classes, ranked from best to worst, were converted to scores ranging from 1 to 4. A weighting of the various parameters from 1 to 3 was used to rank the parameters with respect to the land utilization type under evaluation. The lowest figure from a combined score and weight indicated the highest suitability. The results of the land classification were documented in thematic maps.

The potential area for each land utilization type was derived from the most suitable category for the respective type taking into consideration the current land use and the general socio-economic conditions. The prioritization of land utilization was conducted by overlapping the various suitability maps and the future needs of the villagers, the college (now faculty) of forestry and the administrative bodies. Customary law was employed to allocate land to the individual households.

Finally, a future land use pattern within the project area was proposed. Almost half of the entire area of about 3,600 ha was classified as currently unstocked forest land with reforestation proposed. Some 120 ha should be dedicated protection forest. Current fallow land was split into future reforestation, upland crops and rice paddy.

Source: Manithaphone Mahaxay (1996)

11.4.3 Towards Participation in Land Use Planning

Land capability and land suitability classifications conducted by groups of experts have been carried out as a key activity for future land use development at regional and local scale. However, a lack of substantial improvements in land management and a continued degradation of land resources, including forests, prompted analysis of the driving factors behind this development (Dalal-Clayton et al. 2003). Unexpected and unplanned outcomes of land use planning and implementation were found to be dependent on institutional rather than technical problems, competing sector programs, the failure to develop integrated planning systems, hierarchical institutional structures and divided responsibilities, poor communication and collaboration between various disciplines, a failure to involve land users and to address all relevant factors, and an inability to integrate dissimilar factors (social, economic, ecological, political) (FAO 1995). As indicated in Chap. 10 of Agenda 21 (UNCED 1992), there is a need to strengthen institutions and co-ordination mechanisms, and for the creation of mechanisms to facilitate the active involvement and participation of communities and people at local level.

This highlights the shift away from the technocratic, top-down approach to planning and implementation in favor of a political process of participation (Simon 1990; Meliczek 1993; Amler et al. 1995). Determinants of this planning process are political mechanisms, the resolution of conflict situations and consensus. The constellation of power is most important for the result (participative approach) (e.g., WCED 1987; De Graf et al. 1999). Dalal-Clayton et al. (2003) developed a sustainability framework combining various techniques of analysis and planning. It is structured following the 'ladder of participation' (Arnstein 1969), from pure information through consultation to joint analysis with consultation, selective participation and finally participation in decisions. It is held that land use planning necessitates the participation of stakeholders.

The process of decision making should be structured according to (1) the identification of the perceived needs or objectives of various stakeholders and land users, (2) a review of the available natural and human resources, (3) identification of possible land use options (land use types), and finally (4) negotiation and selection of the combination of options to optimize the achievement of objectives. This process renders it necessary to combine both a top-down and a bottom-up approach in the form of interactive development (cf. WCED 1987; De Graf et al. 1999).

In line with the process of allocation of land to individuals and households in northern Vietnam, Kraienhorst et al. (1996) developed a micro land use planning approach, targeting the rehabilitation of degraded land through afforestation and the management of the resultant forests. The most suitable sites for afforestation should be determined at maximum possible probability to keep the forest once established maintained. Local people participated in the planning process, especially through the medium of meetings held at community and village levels. The approach is structured according to six steps: (1) *introduction* (background, approach, classification system, presentation and documentation, time requirements), (2) *preparation* (district level, community level – before, during and after first community meeting, village level – before first village meeting), (3) *implementation* (village level – during and after first village meeting, community level – before, during and after final community meeting), (4) *approval* (district level), (5) *dissemination* (district, community, village level – follow up) and (6) *calculation of costs*. The approach was further developed and refined over a series of afforestation projects closely linked to individual forest land allocation (Kirchhoff and Nguyen Khac Ninh 2006).

Loikkanen et al. (1999) published a guidebook dedicated to participation in natural resource management. The mission is to come up with a separate participation plan to support planning for the management of natural resources. The ideal is that common problems are solved through cooperation and interaction between stakeholders: internal problems within an organization, problems between planners, and problems concerning the public and interest groups. The goal is a widely accepted end result, justified from different perspectives.

Table 11.5 Interactive rural development

Level \ Linkage	Policies	Institutions	Information
International	Conventions	International task forces (e.g., FAO)	Perspective studies
⇕	⇕	⇕	⇕
National	Land use policy	National task forces (e.g., board, committee) Decentralization Awareness raising Strengthening of institutions	Respondents to national demands (economy, politics, environment, etc.) Decision making
⇕	⇕	⇕	⇕
Meso (e.g., district)	District land use plan	District committees Ownership Responsibility Stewardship	Demands; technical knowledge and assistance; decision making
⇕	⇕	⇕	⇕
Local (e.g., town)	Village land use plan	Land resource management group	Indigenous knowledge; decision making; demands; needs

Source: FAO (1999, p. 10)

11.4.4 Contemporary Spatial Planning Approaches

11.4.4.1 Integrated Land Use Planning

Participative planning for the pursuit of the sustainable management of land resources has been further developed, towards so-called integrated planning (FAO 1999). Integrated planning is considered to be an objective-oriented decision support mechanism and an interactive partnership between government and civil society to address their common concerns in relation to managing land resources sustainably. Interaction between land users, decision makers and professional and technical staff is more important than the documentation of the plan resulting from the exercise.

Integration in this context is characterized as follows (FAO 1999):

- It combines elements of both the top-down aspects of land resource assessment and the bottom-up approach based on grass root participation;

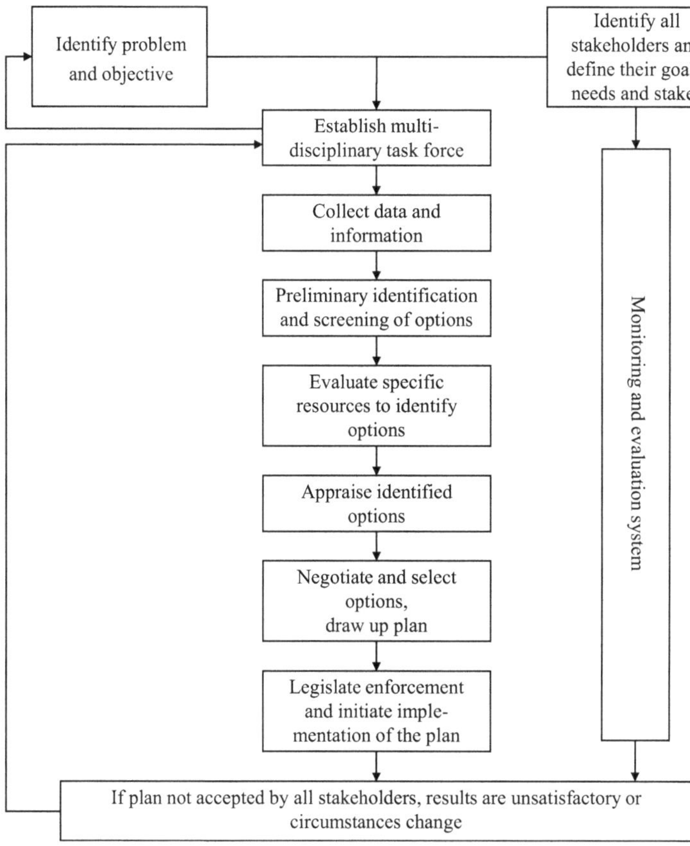

Fig. 11.2 The planning method (Source: FAO (1999, p. 14))

- It takes into account the complex biophysical and socio-economic variables that determine the land use system;
- It considers legal and institutional aspects facilitating the implementation of the plan.

Interactive refers to (FAO 1999):

- A negotiation process, in which land users engage with one another and with specialists;
- The involvement of different levels (e.g., national, sub-national) in the planning process.

Integrated planning for the sustainable management of land resources is demand-driven. Demand may be triggered by a development opportunity or problem perceived at either the local or the regional scale, or a concern of the government. Table 11.5 illustrates the exchanges and flows of knowledge, links and

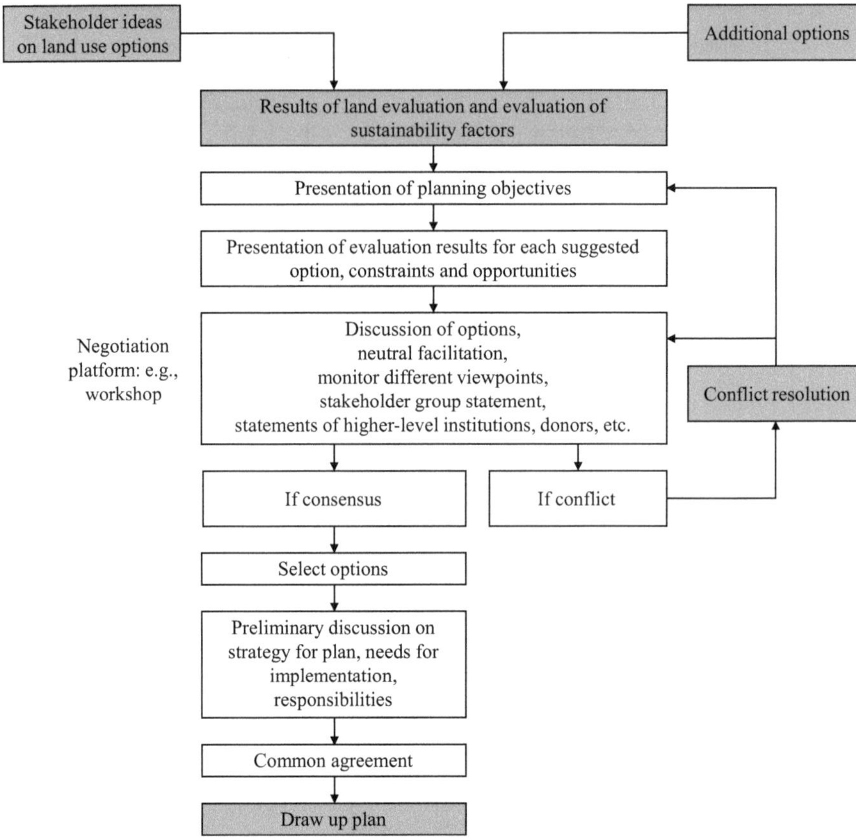

Fig. 11.3 The negotiation process in integrated land use planning (Source: FAO (1999, p. 31))

actions in interactive development. The four-level structure strives towards an optimization of interacting bottom-up and top-down approaches to development.

Seven key factors are associated with successful integrated planning for the sustainable management of land resources. These are: (1) clearly formulated objectives and problems to be solved; (2) the recognition of stakeholders and their differing objectives; (3) an enabling environment and regulatory policy; (4) effective institutions at local, sub-national and national level which are linked; (5) a platform for negotiation; (6) an accessible and efficient knowledge base; and (7) a set of planning procedures (FAO 1999, p. 11).

Figure 11.2 depicts an iterative and cyclical process with nine essential steps in integrated planning for the sustainable management of land resources. This method is independent of scale and level of detail.

A detailed characterization of the consecutive steps is presented in FAO (1999). The selection of the best option or the best range of options is determined by weighing the alternatives against the goals of the various groups of stakeholders.

In most cases it is hoped that land use will be decided by negotiation leading to trade-offs and consensus. The task force is responsible for arranging the forum in which negotiation can take place and providing mediation in order to reach consensus and resolve conflicts. The negotiation process is illustrated in Fig. 11.3.

In this context land conflict is to be accepted as a natural phenomenon. It refers to the various interests and activities and so impacts upon the environment as a consequence of the different objectives followed by the groups and individuals involved and those affected by the use of land (FAO 1998). Conflict resolution is a process whereby two or more parties improve their situation by cooperative action (informal or formal discussions, court) based on a mutual compromise.

Implementation of the plan requires a successful negotiation process. Furthermore, balanced incentives and regulations play an essential part in achieving the sustainable use of land.

From the unexpected results arising from programs and projects concerned with reforestation, sustainable management of forests, forest degradation, deforestation, soil erosion, land degradation, etc., it can be concluded that the various types of forestry land use have up to now been underestimated in terms of their potential contribution to sustainable land use and rural development. Furthermore, concerns linked to forestry land use and management may not have been sufficiently integrated in real world framework conditions set by competing policy fields and their specific objectives in the rural development context. The long time horizon of tree growth compared to other forms of primary production, as well as planning in space and time (see Chap. 4.2.3 – suitability classification), are particular limiting factors for a common understanding and acceptance of forestry-related arguments. This prompted the analysis of one such program and a search for a location-specific adaptive strategy in land use planning, and especially in relation to forestry land use (Box 11.4).

The results of this field study, as well as numerous practical projects, provided evidence of the necessity of participation by local populations, incorporating their experience, knowledge and attitudes in the planning of land use and so also reforestation. This provides a reason to further improve participatory planning approaches focusing on the utilization of natural resources.

Box 11.4 Human Ecological Analysis of Land and Forest Use by Hmong People Conducted for the Purpose of Harmonizing Land Use with the Governmental Reforestation Program in Vietnam

An explorative social research study, following principles of the human ecological model (HEM) proposed by Machlis et al. (1997), was conducted to assess the performance of reforestation projects in three villages predominately populated by members of the Hmong ethnic group in Lao Cai province, Vietnam. The search for reasons for deviations from planned reforestation goals was complemented by the formulation of an integrated

(continued)

Box 11.4 (continued)

planning model, in the form of a descriptive scenario for future reforestation. The principle purpose of the HEM was to identify *critical resources* as they are variables of open land (physical attributes, land use) and of forests (physical attributes, forest uses), socio-economic resources (village characteristics) and religious and cultural resources. The aspects of the *social system* studied at the village level were the social order (norms, hierarchy), social cycles (cropping season, collection seasons), social institutions (informal institutions) and their interactions. Higher social system characteristics were also addressed at community, district, provincial and national level, including formal institutions.

The results provided evidence of subsistence-oriented land management. The land use pattern was influenced by the topography and the specific location of a village on the slopes. Small patches of agricultural land were allocated to individual households following customary law. Cropping was done manually and followed experience accumulated over generations. Work in the forest was carried out using simple hand tools. The people's strong belief in gods supported the maintenance of sacred sites in the forests, which also served to prevent soil erosion. Particular forest and religious events helped to strengthen the social order and increase solidarity amongst the villagers. The village elders and other institutions played an irreplaceable role in resource mobilization, communication and also in the formulation of village rules, including the use of the forests.

Forest use was based on customary tenure, comprising community-owned (village), individually-owned and in one case a clan-owned forest. Unallocated (open access) forest was rare. The type and intensity of utilization underlay customary rules including fines. Household, village and clan forest was customarily governed by particular institutions such as the head of the household, the village head, the village elders, wardens and police. Of these, the village head and police were state initiated and locally respected.

Based on their limited shared knowledge, the Hmong designed an integrated planning approach. It followed the principle steps outlined in FAO (1999). An independent mediator was involved from the very outset of the planning process (diagnosis and design). Another peculiarity of the case study area was the independent design of future land and especially forestry use by both the official agency representatives assigned and the villagers. The mediated negotiation was expected to contribute to mutual understanding, the resolution of arguments and finally harmonization of the draft government plan and the initial local plan.

Source: Nguyen Tien Hai (2009), Nguyen Tien Hai and Uibrig (2011)

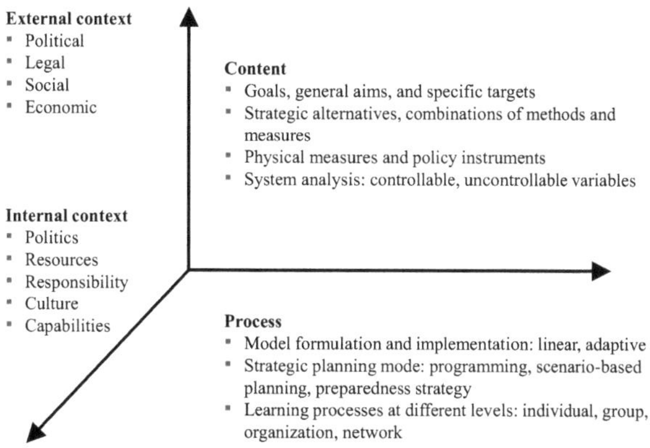

Fig. 11.4 Three dimensions of strategy development: content, process and context (Source: Hutter (2006))

11.4.4.2 Strategy Development and Strategic Spatial Planning

Strategic decisions are fundamental decisions influencing the overall welfare of an actor or community. Once a strategic decision has been made, the remaining options for other fundamental decisions decrease. Context can change, however. Therefore, conditions must be monitored to ensure that they still support the chosen strategy. Trade-offs between (a) decisions to determine future land deployment and (b) preserving flexibility for future options cannot be avoided. These must be handled within the process of strategy development (Volberda 1998; Sutcliffe and Vogus 2003).

Strategy development is, therefore, a multidimensional phenomenon that encompasses the dimensions content ('deciding what to do'), process ('deciding how to do it') and context ('aligning strategic decisions with context'). It requires time, resources, changing power structures and a capacity to learn (see Fig. 11.4).

It is necessary to recognize that spatial planning and plans are elements of strategy development. The *context of strategy development* comprises variables that cannot be altered within the chosen time perspective. Context reduces the complexity of possible decisions and actions to a manageable level by assuming many relevant variables as given. External context conditions are static boundary conditions by which to analyze the content and processes of spatial planning. For strategy development within rural regions, this comprises political, legal, economic and social conditions (e.g., policy guidelines of federal and state government, national spatial planning system, funding conditions determined by national government). Internal conditions influence urban and regional development. Political factors can determine the extent and duration of support for spatial planning and the manner of dealing with conflicts between public and private interests. Internal financial resources have an important impact on strategy development. A high level of resources currently not needed to accomplish predefined tasks ('organizational slack') facilitates experimentation with new solutions because strategic projects can be implemented without significantly changing existing priorities.

Table 11.6 Typical strategy development process models

	Linear model of strategy development	Adaptive model of strategy development
Process	Sequential process planning, programming, and implementation	Continuous alignment of content and process with context
	Top-down strategy making	Combination of bottom-up initiatives and top-down strategic decisions
Content	System of aims, targets and strategic alternatives	System of strategic alternatives, aims and targets
	Integrated set of strategic, operative and resource plans	Flexible configuration of resources
Context	Stable	Unstable
	Predictable	Limited predictability

Source: Based on Chaffee (1985) and Volberda (1998)

When looking at the real world conditions affecting strategy development, it is important to consider *two* process models: the linear and the adaptive model (Table 11.6). The linear model corresponds with a top-down view of strategy development. It works on the assumption that a single decision maker, or elite decision makers, can design an explicit, 'grand' strategy, based on a highly top-down, deliberate and analytical process (Volberda 1998, p. 38). The adaptive model assumes that strategy making requires both strong bottom-up and strong top-down forces (Burgelman 2002).

According to the linear model, the process of strategy development consists of a well-structured sequence encompassing strategy formulation and implementation:

1. Analyzing internal and external context conditions and cause-effect relationships; i.e., the inquiry into objective relationships largely seen as being independent of political and cultural context conditions and societal processes;
2. Formulating aims and targets on the basis of values and key principles of society in order to establish the evaluation basis for decisions and actions before action is taken;
3. Deciding on strategic alternatives after comparing and evaluating the effects of possible bundles of measures;
4. Implementing the chosen strategic alternative by adopting measures to achieve the desired effects of the program;
5. Continued controlling and learning, as context conditions can change and expected effects might not occur, while the unexpected might.

The linear process model of strategy making has its merits. Aims are formulated on the basis of a thorough understanding of context conditions. The evaluation of a broad range of strategic alternatives is undertaken to avoid restricting strategy development to current practices. Decision makers are encouraged to look for new solutions. The model represents a disciplined effort to produce fundamental decisions about medium to long term challenges for an organization or a set of organizations (Bryson 2004).

The hallmark of the adaptive process model is that strategy development does not move forward in a direct way through easily identifiable sequential phases. The process pattern is continuous, iterative and uncertain. An adaptive process is characterized by parallel processes of formulating and implementing strategic alternatives. Hence, formulation and implementation are more difficult to differentiate. Decisions with respect to formulating aims and targets, the analysis of the internal and external context, and combining measures are continuously aligned with the changing societal context. Political context conditions especially are taken into account as enabling and inhibiting factors. These not only affect the implementation of strategic alternatives. They also influence analysis and evaluation because political conditions restrict what actors notice and influence what they ignore.

Empirical work has revealed very different roles adopted by managers within such adaptive strategy processes. For instance:

- *Operative forest managers close to stakeholders and operational decisions* develop bottom-up initiatives to improve current forest management;
- *Strategic forest managers* connect these initiatives with the highly political process at the top level of decision making;
- *Decision makers at the top levels of forest management* are responsible for challenging and questioning the status quo, setting broad performance aims, coordinating, evaluating and legitimizing bottom-up initiatives.

Often, the origins of strategies lie at the lower levels of decision making. However, decision makers at the top select the initiatives leading to enduring resource commitments. Therefore, adaptive strategy can be the joint outcome of content-based, bottom-up initiatives and more formal top-down influences (Burgelman 2002).

The linear and adaptive models of strategy and planning may be complementary. Under specific conditions, it may be appropriate to use the linear model of strategy development. The linear model assumes a high degree of stability of the context conditions. Stability can be achieved, for instance, through a dominant coalition of political and administrative decision makers at regional level in favor of analyzing comprehensively the context conditions with respect to spatial planning and the implementation of strategic alternatives. The question whether the linear or the adaptive model is most appropriate for strategy development should, therefore, be answered on the basis of empirical findings.

Both the linear and the adaptive process model focus on a single organization. In the 'real world', land use planning for sustainable forestry will often also refer to inter-organizational relations. In this case, strategy development is based on interactions between multiple organizations, sometimes from one societal sphere (e.g., political and administrative institutions), sometimes from different societal spheres (e.g., political institutions and administrative units, business organizations, and initiatives from the so-called 'civil society'). Issues of network governance, participation and inclusion become important (e.g., Klijn 2008; Provan and Kenis 2007; Carlsson and Sandström 2008) because organizations experience difficulties formulating and implementing goals at network level ('goal-directed networks', Provan and Kenis 2007). Klijn (2008) stressed that concepts and guidelines with high applicability in single organizations do not apply as easily in networks of organizations. He argued that goal-setting and planning should not be taken for granted but carefully aligned to the specific circumstances at the organizational and network level of strategy development. This calls for strategic planning.

Strategic planning may be defined as a "disciplined effort to produce fundamental decisions and actions" (Bryson 2004, p. 6) that shape and guide what forest management is, what forest managers do, and why they do it. Strategic planning is an elusive phenomenon (Hutter and Wiechmann 2010). It is a set of regulations, procedures and process patterns that are partly applicable to the non-profit and public sector. Given changing and uncertain context conditions, it is useful to distinguish between three strategic planning modes (cf. Volberda 1998):

- *Programming* is appropriate under highly predictable conditions;
- *Scenario-based planning* considers different plausible futures; it is appropriate in the event of complex and dynamic context conditions that can be predicted with a sophisticated set of routines and information systems;
- *Preparedness strategies* increase organizational activities for coping with strategic surprise; a minimum of planning devices gives the strategy process a broad, overall direction in which actions and decisions can emerge adaptively.

In the academic literature on spatial planning, the renaissance of *strategic spatial planning* is unquestionably within the context of regional development (e.g., Healey 1997; Salet and Faludi 2000; Wiechmann 2008). In many respects, current approaches take part in a general shift within the planning system, from physical land use planning through extensive strategic planning to the articulation of a more coherent spatial logic for land use regulation, resource protection and investments in regeneration and infrastructure (Albrechts et al. 2003, p. 113). The question is, whether these concepts are different in nature from the comprehensive development plans of the 1970s and the project-focused planning approach of the 1980s. According to Brake (2000, p. 284), strategic spatial planning concepts share the following characteristics:

- All concepts indicate an atmosphere of awakening and assume the necessity of a regional repositioning in the national and international context;
- All concepts show a clear goal orientation (e.g., in the form of a vision);
- All concepts scale back the urban complexity to measure proposals to ensure implementation;
- The description of these measures is rarely detailed in terms of required resources, actors and milestones.

These new approaches adopt the idea of sustainability as the determining principle and basic message. Empirical research proves the popular belief that the pursuit of competitiveness in a globalizing economy has been the driving force for strategic spatial planning initiatives is unfounded (Albrechts et al. 2003, p. 126). Instead strategic spatial planning focuses on territorially integrated policy approaches and long range planning to improve the quality of life, to strengthen regional identity and to develop new forms of regional collaboration. It is an integrated yet focused approach to better utilize the forests' potentials in an increasingly complex and uncertain world. Against this background, strategic spatial planning can be understood to be a

> ... social process through which a range of people in diverse institutional relations and positions come together to design a plan-making process and develop contents and strategies for the management of spatial change. This process generates not merely formal outputs in terms of policy and project proposals, but a decision framework that may influence relevant parties in their future investment and regulatory activities. It may also generate ways of understanding, ways of building agreement, of organizing and of mobilizing to influence in political arenas. (Healey 1997, p. 5)

Strategic spatial planning is in most instances conceived of as a government-led strategic intervention at the urban and regional level (Albrechts et al. 2003, p. 114) and a deliberate effort to improve collective action. The underlying proposition is that regional governance has the capacity to, at least partly, manage or even control spatial development by influencing the political agenda and by defining common goals. In this regard, strategic spatial plan-making involves not only technical analysis but the development of spatial logics and metaphors that carry persuasive power as well as the construction of new institutional arenas at local and regional levels (Albrechts et al. 2003).

11.4.4.3 Adaptive Management

In the future – even more so than today – forest management will have to deal with uncertainty and change (e.g., global change including climate change, see also Chap. 3 on contemporary challenges). Complexity may also increase as a consequence of the incorporation of supplementary goals in forest management, leading to a more sustainable, multi-value forestry. Consequently, the basic idea is to develop a management system capable not only of coping with uncertainty, complexity and change, but also to use it as a means to learn about and improve current management (e.g., Heller and Zavaleta 2009; Millar et al. 2007; Walters and Holling 1990). This management should be designed in a systematic manner – based on the assumption that the preconditions for successful 'evolutionary' learning (trial and error) are progressively declining due to increasing uncertainty (Holling 1978).

One promising concept in this context is adaptive management (AM). AM is a conceptual framework to deal with change, complexity and uncertainty by planning and performing a comprehensive iterative and cyclical process, the adaptive management cycle (AMC) (Fig. 11.5).

Most contributions to AM can be classed as either 'passive AM' ('AM in its broadest sense', Millar et al. 2007) or 'active AM' ('AM in a strict sense'). Both are designed to systematically adjust and refine management by accomplishing the AMC depicted, including a targeted evaluation of the measures performed (Walters and Holling 1990). Passive AM is the basic application of this AMC,

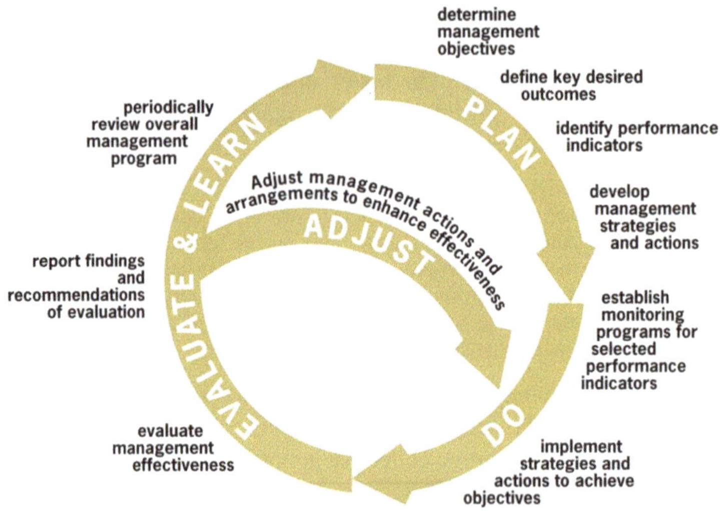

Fig. 11.5 Example of a general adaptive management cycle (Source: Jones (2005, 2009), reprinted with kind permission of the author)

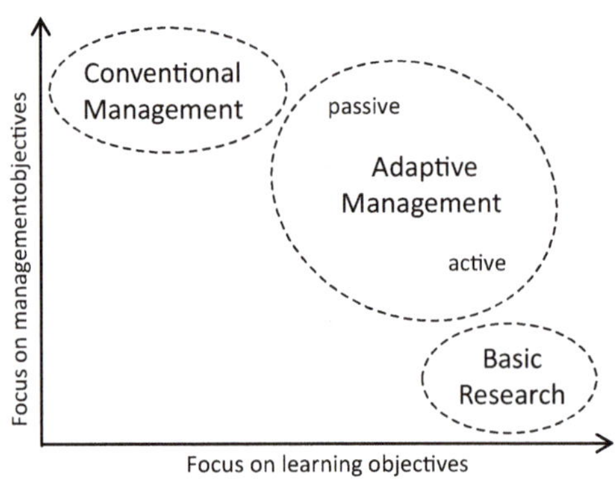

Fig. 11.6 Adaptive management in relation to management and research (Source: Marmorek et al. (2006), modified)

performing one 'best' measure, with the primary goal being to enhance the adaptability of the management. Going beyond this basic approach, an active AM performs (and compares) a range of alternative measures (chosen on the basis of certain criteria relative to the intended goals) simultaneously, which can encompass more ambitious goals such as spreading risk (cf. Millar et al. 2007) and systematic ('quasi-experimental', cf. Lee 1999; Margoluis et al. 2009) learning (Fig. 11.6). In some cases it may also be feasible to embed small scale and strongly focused active AMs into a well-monitored, large scale passive AM (Marmorek et al. 2006).

Originally developed to facilitate systematic acting and learning under uncertain conditions based on the complexity of ecosystems (e.g., Holling 1978; Lee 1993;

Walters 1986), AM is successfully employed in relation to a variety of problems (e.g., Allan and Stankey 2009; Marmorek et al. 2006). For instance, an increasing number of authors recommend AM as useful concept to enable sustainable forestry under changing and uncertain conditions (e.g., Allan and Stankey 2009; Kellomäki et al. 2008; Lasch et al. 2002; MacIver and Wheaton 2005). However, many discussions on these issues are ongoing (e.g., Allan and Gunderson 2011; Cash and Moser 2000; Ibisch et al. 2009; MacIver and Dallmeier 2000; Marmorek et al. 2006; Millar et al. 2007; Smith 2009). "What is needed, in addition to better communication and education and improved evaluation methods, is a set of conceptually sound yet practical criteria to help decision makers make thoughtful choices when it comes to the selection of problems that either are, or are not, appropriate for the application of AM techniques" (Gregory et al. 2006, p. 2414).

Some of the first applications of AM have been performed in the forestry sector, primarily in North America and Australia (e.g., Allan and Stankey 2009; Holling 1978; Marmorek et al. 2006; Peterson and Maguire 2005). "Adaptive approaches offer hope for the successful management of natural-origin and planted forests for their multiplicity of values, but driven and defined at the local levels within a global climate change envelope" (MacIver and Wheaton 2005). Such multiple values of forests, and the multiple objectives of forestry, may include issues such as silviculture, biodiversity, fisheries, recreation, soils and/or water (Allan and Stankey 2009). Examples of possible experimental treatments in forestry include variations of mechanical operations (thinning, retention and regeneration), rotation lengths, nutrient-inputs and water supply (Bunnel 2005; Seymour 2005; von Gadow and Kleinn 2005), tree species and provenances used (Kellomäki et al. 2008) and prescribed burning (Ritchie 2005) – usually designed within a set of treatment zones including controls and replications. For further examples see Allan and Stankey (2009) or Peterson and Maguire (2005).

To incorporate issues of climate change into forest management decisions, various authors use (or recommend the use of) predictive models to pro-actively compare the possible effects of alternative treatments, related to selected objectives (e.g., Holling 1978; Kellomäki et al. 2008; Millar et al. 2007). Integrated in an appropriately designed AMC, with due consideration given to possible changes to environmental conditions in particular, this approach can enhance adaptive capacity and reduce the vulnerability of forestry to global climate change (cf. MacIver and Wheaton 2005). Combined with other tools, such as multi-criteria analysis, this approach may generally lead to more sustainable and adaptive forest planning and management (cf. Fürstenau et al. 2006).

Key aspects of AM, and the main differences with respect to conventional management approaches include the acceptance of uncertainty and change as unavoidable. Learning is treated as an explicit goal; management policies as deliberately monitored and evaluated experiments, including the possibility of necessary adjustments. Consequently, policy options for AM have to be flexible and reversible (Holling 1978; Millar et al. 2007). Furthermore, there is a need for a strong interaction between science and management and a systematic planning of all elements and steps in the adaptive cycle from the outset (Allan and Stankey 2009; Holling 1978).

Analyzing case studies, Allan and Stankey (2009) identified a number of criteria for the success of AM projects, including "the ability to answer the specific questions of the project" and more general aspects such as 'learning' and 'building relationships'. Other criteria were 'effective communication and leadership' and the securing of 'long term funding and support (e.g., by building effective partnerships)'. These findings led to (at least) two insights: (1) conducting AM properly can contribute to organizing management planning in a participatory rather than a technocratic way. (2) There is a strong and bilateral dependency of aspects, generating a high potential for synergies between strategic (spatial) planning and AM, reinforced to some extent against the background of change, especially climate change.

11.5 Outlook

Land classifications and land use planning approaches in rural regions have been referred to over a longer period. Agriculture has been the major issue to date. Forested lands have often been deemed a source of capital readily converted to cash and a land reserve for the expansion of tillage, animal rearing and other uses. Recognition of the scarcity and even loss of forest-based products and services from various perspectives has contributed to an emphasis on forests as natural land cover, a productive land use option and also a provider of multiple services. Top-down and technocratic approaches in rural land use planning have to a large degree been supplemented and replaced by bottom-up and participatory approaches. The focus in the past on specific economic sectors has evolved to intersectoral and integrated approaches combining technical methods and participatory procedures best suited to the particular problem situation.

Increasing requirements in terms of the number, intensity and combination of land use types, growing numbers of stakeholders and rising populations, as well as risks affecting the maintenance and improvement of the land's capacity have demanded the creation of comprehensive science-based approaches to future land use development. Contemporary research and development deem *strategic planning* and *adaptive management* to best serve long term development. Both approaches have been tested and applied in urban centers. These experiences should be drawn upon in order to adopt and to adapt these techniques to rural regions; especially as rural regions are increasingly linked to peri-urban and urban centers by various functional relationships.

References

Albrechts L, Healey P, Kunzmann K (2003) Strategic spatial planning and regional governance in Europe. J Am Plan Assoc 69:113–129

Allan CR, Gunderson LH (2011) Pathology and failure in the design and implementation of adaptive management. J Environ Manage 92:1379–1384

Allan C, Stankey GH (eds) (2009) Adaptive environmental management: a practitioner's guide. Springer/CSIRO, Dordrecht/Collingwood

Amler B (1992) Landnutzungsplanung für Entwicklungsländer – Methoden der Standorteignungsbewertung und Landnutzungsplanung für den ländlichen Raum in Entwicklungsländern, vol 85, Landschaftsentwicklung und Umweltforschung, Schriftenreihe des Fachbereichs Landschaftsentwicklung der TU Berlin. Technische Universität Berlin, Berlin

Amler B, Betke D, Eger H, Ehrich C, Kohler A, von Lossau A, Müller U, Seidemann S, Steurer R, Zimmermann W (1995) Landnutzungsplanung – Strategien, Instrumente, Methoden. Gesellschaft für Technische Zusammenarbeit (GTZ), Eschborn

Arnstein SR (1969) A ladder of citizen participation. J Am Inst Plan 35(4):216–224

Badege Bishaw (1985) Beiträge der Forstwirtschaft zur ländlichen Entwicklung im Hararghe Hochland Äthiopiens am Beispiel des Projektes Legeambo. Dipl.-Arbeit, Technische Universität Dresden, Tharandt

Bishaw B, Uibrig H (1989) Management of community woodlots in Hararghe highlands, eastern Ethiopia. Paper presented to the 1st natural resource improvement conference, Addis Ababa, 7–8 Feb 1989

Board C (1968) The importance of the land-utilization survey of Britain: towards an assessment twenty years after. In: International conference on land-use maps, Budapest, pp 165–173

Brake K (2000) Strategische Entwicklungskonzepte für Großstädte – mehr als nur eine Renaissance der 'Stadtentwicklungspläne'? Arch Kommunalwiss 2:269–288

Bridges EM, Oldeman LR (2001) Food production and environmental degradation. In: Bridges EM, Hannan ID, Oldeman LR, de Vries FWTP, Scherr SJ, Sombatpanit S (eds) Response to land degradation. Science Publishers, Enfield/Plymouth, pp 36–43

Bridges EM, Hannan ID, Oldeman LR, de Vries FWTP, Scherr SJ, Sombatpanit S (eds) (2001) Response to land degradation. Science Publishers, Enfield/Plymouth

Bryson JM (2004) Strategic planning for public and nonprofit organizations. A guide to strengthening and sustaining organizational achievement. Jossey-Bass, San Francisco

Bunnel FL (2005) Adaptive management for biodiversity in forests – it can be done! In: Peterson CE, Maguire DA (eds) Balancing ecosystem values: innovative experiments for sustainable forestry, vol 635, Proceedings of a conference general technical report PNW-GTR. U.S. Department of Agriculture, Forest Service, Pacific Northwest Research Station, Portland, p 389

Burgelman RA (2002) Strategy is destiny. How strategy-making shapes a company's future. The Free Press, New York

Carlsson L, Sandström A (2008) Network governance of the commons. Int J Common 2:33–54

Carpenter RA (1981) Assessing tropical forest lands: their suitability for sustainable uses, vol 3, Natural resources and environment series. Tycooly International, Dublin

Cash DW, Moser SC (2000) Linking global and local scales: designing dynamic assessment and management processes. Global Environ Change 10:109–120

Chaffee EE (1985) Three models of strategy. Acad Manage Rev 10:89–98

Choudhury K, Jansen LJM (1998) Terminology for integrated resources planning and management. http://www:mpl.ird.fr/crea/taller-colombia/FAO/. . ./landglos.pdf. Checked 10 Jan 2012

Christaller W (1933) Die zentralen Orte in Süddeutschland. Gustav Fischer, Jena

Cooper T, Arblaster K (2007) Climate change and the rural environment in a European context: implications for land use and land use policy. A background paper for the LUPG & BfN conference 'Future Policies for Rural Europe 2013 and Beyond – Delivering Sustainable Rural Land Management in a Changing Europe', Brussels, 19–20 Sept

Dalal-Clayton B, Dent D, Dubois O (2003) Rural planning in developing countries supporting natural resource management and sustainable livelihoods. Earthscan, London

Dale PF, McLaughlin JD (1988) Land information management. An introduction with special reference to cadastral problems in Third World countries. Clarendon Press, Oxford

Darr D (2003) Farm forestry in semi-subsistence and its inter-dependency with the land tenure system – case studies from central Laos. M.Sc. thesis, Technische Universität Dresden

Darr D, Uibrig H (2004) Promotion of farm forestry in Laos enhances creation of individual land property. Asia Pac J Rural Dev 2:39–51
Davis KF (1976) Land use, McGraw-Hill series in forest resources. McGraw-Hill, London/New York
De Graf HJ, Musters CJM, Ter Keurs WJ (1999) Regional opportunities for sustainable development – theory, methods and applications. Kluwer Academic, Dordrecht
Dent D (ed) (1991) Land use planning applications: proceedings of the FAO expert consultation 1990, Rome, 10–14 December 1990, vol 68, World soil resources report. FAO, Rome
Department of Geography, University of Zurich, Center for Remote Sensing, Survey Department of Sri Lanka (1988) Sri Lanka/Swiss remote sensing project. Final report. Survey Department of Sri Lanka, Zurich/Colombo
Di Gregorio A, Jansen LJM (1998) A new concept for a land cover classification system. Land 2 (1):55–65
FAO (1976) A framework for land evaluation, vol 32, Soils bulletin. Food and Agriculture Organization of the United Nations (FAO), Rome
FAO (1984) Land evaluation for forestry, vol 48, Soils bulletin. Food and Agriculture Organization of the United Nations (FAO), Rome
FAO (1985) Assistance to land-use planning, Ethiopia. Guidelines for land-use planning, vol 10, AG:DP/ETH/78/003 technical report. Food and Agriculture Organization of the United Nations (FAO), Rome
FAO (1993) Guidelines for land-use planning, vol 1, FAO development series. Food and Agriculture Organization of the United Nations (FAO), Rome
FAO (1995) Planning for sustainable use of land resources – towards a new approach, vol 2, FAO land and water bulletin. Food and Agriculture Organization of the United Nations (FAO), Rome
FAO (1998) Land and water resources information systems. FAO, Rome, land and water Bulletin 7. Food and Agriculture Organization of the United Nations (FAO), Rome
FAO (1999) The future of our land – facing the challenge. Guidelines for integrated planning for sustainable management of land resources. FAO Natural Resources Management and Environment Department, http://www.fao.org/DOCREP/004/X3810E/x3810e04.htm#e. Checked 2 Sept 2013
FAO (2007) The future of our land: facing the challenge. FAO, Natural Resources Management and Environment Department, Rome
FAO/UNEP (1997) Negotiating a sustainable future for land. FAO Land and Water Development Division in collaboration with the United Nations Environment Programme, Rome
Fürstenau C, Badeck FW, Lasch P, Lexer MJ, Lindner M, Mohr P, Suckow F (2006) Multiple-use forest management in consideration of climate change and the interests of stakeholder groups. Eur J Forest Res 126(2):225–239
Gregory R, Ohlson D, Arvai J (2006) Deconstructing adaptive management: criteria for applications to environmental management. Ecol Appl 16(6):2411–2425
Grove M, Schweik C, Evans T, Green G (2002) Modeling human-environmental systems. In: Clarke KC, Parks BO, Crane MP (eds) Geographic information systems and environmental modeling. Prentice-Hall, Upper Saddle River, pp 160–187
Hammer T (1999) Nachhaltige Entwicklung im Lebensraum Sahel: Ein Beitrag zur Strategietheorie nachhaltiger ländlicher Entwicklung. LIT Verlag, Münster
Healey P (1997) The revival of strategic spatial planning in Europe. In: Healey P, Khakee A, Motte A, Needham B (eds) Making strategic spatial plans. Innovation in Europe. UCL Press, London, pp 3–19
Heller NE, Zavaleta ES (2009) Biodiversity management in the face of climate change: a review of 22 years of recommendations. Biol Conserv 142:14–32
Hill J, Aspinall R (eds) (2000) Spatial information for land use management. Gordon and Breach Science Publishers, London

Hockensmith RD, Steele JG (1943) Classifying land for conservation farming. US Department of Agriculture, Washington, DC

Holling CS (ed) (1978) Adaptive environmental assessment and management. Wiley, London

Hutter G (2006) Strategische Planung. Raumplan 128:210–214

Hutter G, Wiechmann T (2010) Strategische Planung. Zur Rolle der Planung in der Strategieentwicklung für Städte und Regionen. Altrock, Kassel

Ibisch PL, Kunze B, Kreft S (2009) Biodiversitätserhaltung in Zeiten des Klimawandels: Risikomanagement als Grundlage eines systemischen, nichtwissenbasierten Naturschutzes. Wald im Klimawandel – Risiken und Anpassungsstrategien. Eberswalder Forstl. Schriftreihe, Band 42

IPCC (2007) Climate change 2007 – synthesis report. Contribution of working groups I, II and III to the fourth assessment report of the intergovernmental panel on climate. Intergovernmental Panel on Climate Change (IPCC), Geneva

Jacks GV (1946) Land classification for land use planning, vol 43, Technical communication. Imperial Bureau of Soil Science, Harpenden

Jansen LJM (2005) Harmonisation of land-use class sets to facilitate compatibility and comparability of data across space and time. In: 11th CEReS international symposium, Chiba, 13–14 Dec

Jones G (2005) Is the management plan achieving its objectives? In: Worboys GL, Lockwood M, De Lacy T (eds) Protected area management. Principles and practice, 2nd edn. Oxford University Press, Oxford, pp 555–567. http://www.parks.tas.gov.au/index.aspx?base=7679. Checked 11 May 2011

Jones G (2009) The adaptive management system for the Tasmanian Wilderness World Heritage Area – linking management planning with effectiveness evaluation. In: Allan C, Stankey G (eds) Adaptive environmental management. A practitioner's guide. Springer, Dordrecht. http://www.parks.tas.gov.au/index.aspx?base=15123. Checked 9 Aug 2011

Kang BT, Tripathi BR (1992) Technical paper 1: soil classification and characterization. The AFNETA alley farming training manual. International Institute of Tropical Agriculture, Ibadan

Kellomäki S, Peltola H, Nuutinen T, Korhonen KT, Strandman H (2008) Sensitivity of managed boreal forests in Finland to climate change, with implications for adaptive management. Philos Trans R Soc Lond B Biol Sci 363(1501):2341–2351

King KFS (1966) Land capacity classification and land-use planning with special reference to tropical regions. In: Proceedings of the VI world forestry Congress, vol 3. Madrid, pp 3113–3121

Kirchhoff J-F, Nguyen Khac Ninh (2006) Technical report No. 1, KfW Bankengruppe, GfA Consulting Group, Hanoi

Kirchhoff J-F, Nguyen Khac Ninh (2006) Nursery techniques, site mapping and afforestation on degraded forest land in Vietnam. Experiences from the KfW4 afforestation project in Thanh Hoa and Nghe An provinces, Hanoi

Klijn EH (2008) Policy and implementation networks: managing complex interactions. In: Cropper S, Ebers M, Huxham C, Smith Ring P (eds) The Oxford handbook of inter-organizational relations. Oxford University Press, Oxford, pp 118–146

Klingebiel AA, Montgomery PH (1961) Land capability classification, vol 210, Agriculture handbook. Department of Agriculture, Washington, DC

Kraienhorst H et al. (1996) Guidelines for the implementation of micro land-use planning for forest land, Ministry of Agriculture and Rural Development, Hanoi

Lasch P, Lindner M, Erhard M, Suckow F, Wenzel A (2002) Regional impact assessment on forest structure and functions under climate change – the Brandenburg case study. Forest Ecol Manage 162(1):73–86

Lee KN (1993) Compass and gyroscope – integrating science and politics for the environment. Island Press, Washington, DC

Lee KN (1999) Appraising adaptive management. Conserv Ecol 3(2):3. http://www.consecol.org/vol3/iss2/art3/. Checked 30 Jul 2012

Loikkanen T, Simojoki T, Wallenius P (1999) Participatory approach to natural resources management – a guide book. Metsähallitus Forest and Park Service, Vantaa
Machlis GE, Force JE, Burch WR Jr (1997) The human ecosystem part I: the human ecosystem as an organizing concept in ecosystem management. Soc Nat Res 10:347–368
MacIver DC, Dallmeier F (2000) Adaptation to climate change and variability: adaptive management. Environ Monit Assess 61:1–8
MacIver DC, Wheaton E (2005) Tomorrow's forests: adapting to a changing climate. Climatic Change 70(1–2):273–282
Mahaxay M (1996) Application of remote sensing and GIS in forest land use planning: the case of Sangthong forest training site, Laos. M.Sc. thesis, Asian Institute of Technology, Bangkok
Margoluis R, Stem C, Salafsky N, Brown M (2009) Design alternatives for evaluating the impact of conservation projects. In: Birnbaum M, Mickwitz P (eds) Environmental program and policy evaluation: addressing methodological challenges, vol 122, New directions for evaluation. Jossey-Bass, San Francisco, pp 85–96
Marmorek DR, Robinson DCE, Murray C, Greig L (2006) Enabling adaptive forest management – final report. National Commission on Science for Sustainable Forestry. ESSA Technologies, Vancouver
McAllister D (1973) Environment: a new focus for land-use planning. National Science Foundation, Washington, DC
Meliczek H (1993) FAO's approaches and concepts for rural development. Entwickl Ländlicher Raum 27(2):21–23
Millar CI, Stephenson NL, Stephens SL (2007) Climate change and forests of the future: managing in the face of uncertainty. Ecol Appl 17(8):2145–2151
Nair PKR (ed) (1989) Agroforestry systems in the tropics. Kluwer Academic Publishers in cooperation with ICRAF, Dordrecht/Boston/London
Neef E (1967) Die theoretischen Grundlagen der Landschaftslehre. VEB Hermann Haack, Leipzig
Nguyen Tien Hai (2009) Human ecological analysis and forest use by the Hmong people for harmonizing with the governmental reforestation program in Vietnam. Schriftenreihe des Institues für Internationale Forst- und Holzwirtschaft, Tharandt, p 18
Nguyen Tien Hai, Uibrig H (2011) Human ecological analysis of land and forest use by Hmong people for harmonizing with the governmental reforestation program in Lao Cai. In: Proceedings workshop application of the human ecology framework in natural resources management in Vietnam. Water Resources University, Hanoi pp 94–107
Norton EA (1939) Soil conservation survey handbook, vol 352, Miscellaneous publications. US Department of Agriculture, Washington, DC
Obeng HB, Smith JK (1963) Land capability classification of the soils of Ghana. Ghana J Sci 3:52–65
Olschowy G (ed) (1978) Natur- und Umweltschutz in der Bundesrepublik Deutschland. Paul Parey, Hamburg
Peterson CE, Maguire DA (eds) (2005) Balancing ecosystem values: innovative experiments for sustainable forestry. In: Proceedings of a conference August 15–20, 2004. General technical report PNW-GTR-635, Portland
Provan KG, Kenis P (2007) Modes of network governance: structure, management, and effectiveness. J Public Adm Res Theory 18:229–252
Randolph J (2004) Environmental land use planning and management. Island Press, Washington, DC
Rindfuss RR, Walsh SJ, Mishra V, Dolcemascolo GP (2003) Linking household and remotely linked data. In: Fox J, Rindfuss R, Walsh SJ, Mishra V (eds) People and the environment – approaches for linking household and community surveys to remote sensing and GIS. Kluwer, Boston/Dordrecht/London, pp 1–29
Ritchie MW (2005) Effects of thinning and prescribed fire on understory vegetation in interior pine forest of California. In: Peterson CE, Maguire DA (eds) Balancing ecosystem values: innovative experiments for sustainable forestry. Proceedings of a conference. General technical report PNW-GTR-635, Portland

Salet W, Faludi A (eds) (2000) The revival of strategic spatial planning. Royal Netherlands Academy of Arts and Sciences, Amsterdam
Saylath T (1998) Prinzipien der Klassifikation von Wald für Fernerkundung und vergleichende Analyse von Waldtypisierungen in zwei Testgebieten von Laos. Dipl.-Arbeit, TU Dresden
Seymour RS (2005) Integrating natural disturbance parameters into conventional silvicultural systems: experience from the Acadian forest of northeastern North America. In: Peterson CE, Maguire DA (eds) Balancing ecosystem values: innovative experiments for sustainable forestry. Proceedings of a conference general technical report PNW-GTR-635, Portland
Simon D (ed) (1990) Third World regional development. A reappraisal. Paul Chapman Publishing, London
Smith AC (2009) Lessons learned from adaptive management practitioners in British Columbia, Canada. In: Allen C, Stankey GH (eds) Adaptive environmental management: a practitioner's guide. Springer/CSIRO, Collingwood, pp 39–55
Smoke P, Romeo L (1997) Designing intergovernmental fiscal relations and international finance institutions allocations for rural development. Paper prepared for the Technical Consultation on Decentralisation, FAO, Rome, 16–18 Dec
Stamp LD (1950) The land of Britain, its use and misuse. Longmans, London
Stibig H-J (1996) Interpretation and delineation of satellite imagery. Forest monitoring project. Technical notes 2. MRC/GTZ. Vientiane
Sutcliffe K, Vogus TJ (2003) Organizing for resilience. In: Cameron K, Dutton JE, Quinn RE (eds) Positive organizational scholarship foundations of a new discipline. Berrett-Koehler, San Francisco, pp 94–110
Sysavath V (1997) Untersuchungen zur Erfassung von Waldveränderungen an ausgewählten Standorten in der Region Vientiane, Laos. Dipl.-Arbeit, TU Dresden
Tengberg A, Stocking M (2001) Land degradation, food security, and biodiversity – examining an old problem in a new way. In: Bridges EM, Hannan ID, Oldeman LR, de Vries FWTP, Scherr SJ, Sombatpanit S (eds) Response to land degradation. Science Publishers, Enfield/Plymouth, pp 171–185
Torquebiau EF (2000) A renewed perspective on agroforestry concepts and classification. CR Acad Sci Paris Life Sci 323:1009–1017
Uibrig H (1991) Integration of forest with the agricultural land use system in the eastern Ethiopian highlands applying low cost approaches. In: Proceedings IUFRO Centennial Meeting, Subject Group S 6. 12–03, Berlin
UNCED (1992) Agenda 21. UN Conference on Environment and Development. Rio de Janeiro
Van Lier HN, Jaarsma CF, Jurgens CR, de Buck AJ (eds) (1994) Sustainable land use planning: proceedings of an international workshop, 2–4 September 1992, Wageningen. Elsevier, Amsterdam
Van Valkenburg S (1950) The world land use survey. Econ Geogr 26(1):1–5
Vergara NT (1996) Land use conflicts in the Tropics. In: Proceedings of the FORTROP '96: tropical forestry in the 21st century. Kasetsart University, Bangkok, 25–28 Nov 1996, vol 1, pp 86–98
Volberda HW (1998) Building the flexible firm. How to remain competitive. Oxford University Press, New York
Von Gadow K, Kleinn C (2005) Forest management: science-based and understandable. In: Peterson CE, Maguire DA (eds) Balancing ecosystem values: innovative experiments for sustainable forestry. Proceedings of a conference general technical report PNW-GTR-635, Portland
Von Thünen JH (1826) Der isolierte Staat in Beziehung auf Landwirtschaft und Nationalökonomie. Friedrich Perthes, Hamburg
Walters C (1986) Adaptive management of renewable resources. Macmillan, New York
Walters CJ, Holling CS (1990) Large-scale management experiments and learning by doing. Ecology 71(6):2060–2068. Published by: Ecological Society of America. http://www.jstor.org/stable/1938620. Accessed 05 Jan 2010 04:31

WCED (1987) Our common future. World Commission on Environment and Development, Oxford University Press, Oxford

Weber A (1909) Über den Standort der Industrie [1929 Theory of the location of industries]. University of Chicago Press, Chicago

Wehrmann B (2011) Land use planning: concept, tools and application. Deutsche Gesellschaft für Internationale Zusammenarbeit (GIZ), Eschborn

Weigel G (1986) An agro-ecological development plan for the Maybar area, Wello research unit, Ethiopia, vol 9, Soil conservation research report. University of Bern, Bern

Wiechmann T (2008) Strategische Planung. In: Fürst D, Scholles F (eds) Handbuch Theorien und Methoden der Raum- und Umweltplanung. Dortmunder Vertrieb für Bau- und Planungsliteratur, Dortmund

Wild A (2003) Soils, land and food: managing the land during the twenty-first century. Cambridge University Press, Cambridge

Winckler H (ed) (1999) Regional rural development. A milestone on the way to structural change and transformation. Working Group Regional Rural Development and Natural Resources Management, Cologne

Young A (1993) Chapter 13: Land evaluation and forestry management. In: Pancel L (ed) Tropical forestry handbook, vol 2. Springer, Berlin/Heidelberg/New York/London/Paris/Tokyo/Hong Kong/Barcelona/Budapest, pp 812–845

Young A, Goldsmith PF (1977) Soil survey and land evaluation in developing countries: a case study in Malawi. Geophys J Roy Astron Soc 143(3):407–438

Chapter 12
Forest Governance and Sustainable Rural Development

Daniela Göhler, Benjamin Cashore, and Benjamin Blom

Abstract How might forest institutions be designed to encourage long-term and collective natural resource management, while also addressing the needs of local people? This chapter sheds light on this question by reviewing insights of 'good forest governance' scholarship, with a focus on ongoing developments in Southeast Asia. It is argued that building enduring, effective forest management responsive to the needs of local communities requires greater focus on the role of two key governance concepts: institutional intersection and policy learning. From this review, the chapter extrapolates key findings for practitioners seeking to promote 'good forest governance'.

Keywords Forest governance • Forest policy • Networks • International regimes • Non-state governance • Decentralisation • Institutional intersection • Policy learning • Local communities

D. Göhler (✉)
Deutsche Gesellschaft für Internationale Zusammenarbeit (GIZ) GmbH, Dag-Hammarskjöld-Weg 1-5, 65760 Eschborn, Germany
e-mail: daniela.goehler@gmail.com

B. Cashore
Environmental Governance and Political Science, Yale University's School of Forestry and Environmental Studies, 195 Prospect Street, New Haven, CT 06511, USA
e-mail: benjamin.cashore@yale.edu

B. Blom
Alumnus of Yale University's School of Forestry and Environmental Studies, International Labour Organization, 1562 J Street, Arcata, CA 95521, USA
e-mail: blom.benjamin@gmail.com

J. Pretzsch et al. (eds.), *Forests and Rural Development*, Tropical Forestry 9,
DOI 10.1007/978-3-642-41404-6_12, © Springer-Verlag Berlin Heidelberg 2014

12.1 Introduction

One of the most important challenges facing those promoting sustainable development within, and for, rural and local communities is to better understand how to nurture enduring, problem-focused and authoritative institutions. What are the ways in which international, non-state and other governance mechanisms reinforce, rather than detract from, efforts to address local needs and local participation in natural resource management processes? How might institutions be developed that encourage long-term collective efforts instead of short-term calculations that often divide these same communities?

This chapter sheds light on the above questions by focusing on two important concepts: (1) the role of institutional intersection across global, regional, national and local as well as public and private scales, and (2) the role of policy learning networks in shaping these trends.

The 2.4 billion hectares of tropical and subtropical forests in the world, which together cover nearly one-fifth of the earth's land area (FAO 2010), play a crucial role in the economic development of tropical rural economies. They supply a wide variety of forest products for consumption and trade, support food security and energy needs, and play a vital role in ecosystem functions. These ecosystem roles include the sequestration and storage of carbon, soil protection, maintenance of the water cycle and biodiversity. One billion extremely poor[1] people in developing countries directly depend upon forests for their livelihoods, while 350 million people living in and around forests heavily depend on them (World Bank 2006a). Forest management and rural development goals are, thus, highly interrelated in many tropical and subtropical countries around the world.

Yet, ongoing deforestation and forest degradation continue to diminish the availability and quality of forest resources. Major causes of deforestation include the conversion of tropical forests to agricultural land, and increases in demand for forest commodities. Efforts to control or limit these impacts have been impeded by land tenure insecurity and undervalued ecosystem services (Gibbs et al. 2010; Pfaff et al. 2010). Illegal logging and unsustainable timber harvesting practices predominantly cause forest degradation. Efforts to ameliorate both deforestation and forest degradation as put forward in the international REDD+ mechanism (i.e. reducing emissions from deforestation and forest degradation in developing countries; and the role of conservation, sustainable management of forests and enhancement of forest carbon stocks in developing countries, UNFCCC 2008) under the climate change convention are hampered by inappropriate forest institutions (Cashore et al. 2010).

Another key concern is related to the role of local people in forest management (Peluso 1992; Dove 2006, 2011). Communities throughout tropical forest regions rely heavily on forest resources for their livelihoods and for the maintenance of cultural and environmental integrity. The history shows that decision-making in

[1] The World Bank valued the extreme poverty line at US$ 1.25 a day in 2005 prices (World Bank 2010).

forestry has disadvantaged and marginalized the estimated 1.6 billion people living in local communities within forest areas (Sikor et al. 2010) – a governance problem causing major social and economic inequalities.

Adding to the complexity of the situation, the nature of problems is changing. Expanding global consumption patterns, the growth of transnational firms, new financial architectures and the development of low-carbon economies increasingly pose new challenges for rural governance. In addition, the impacts of global climate change and resultant mitigation and adaptation efforts are felt strongly in rural regions (Galloway et al. 2010).

Both practitioners and scholars recognize that efforts to address these thorny yet important challenges must focus on the role that institutions play, or should play, to secure sustainable livelihoods in rural environments (Ostrom 1990; World Bank 2009a; IFAD 2010). Such a focus must also examine the types of learning processes that promote 'good governance'.

12.2 Definition of Terms

This section defines key terms critical for a common understanding of the analytical framework developed in the chapter.

In accordance with Mayntz's (2004) definition, **governance** as opposed to government, is understood as "the entirety of all co-existing modes of collectively regulating social matters". By emphasizing the co-existence of different modes of governance, the literature incorporates the role of the state while also conceptually expanding from sovereign national authority executed by governments to self-regulated competitive systems (markets) and voluntary coordination in networks (Risse 2007).

Next is to adopt the definition of **institutions** as "a cluster of rights, rules, and decision-making procedures that gives rise to a social practice, assigns roles to participants in the practice, and guides interactions among occupants of these roles" (Young et al. 2008). Although characterized by different sets of rules and procedures, international regimes, national and local governments, public-private partnerships as well as policy networks are all considered as institutions. For the purpose of this chapter, it is necessary to describe some of the different types of institutions in more detail.

Referring to Levy et al. (1995), **international regimes** are understood as "social institutions consisting of agreed upon principles, norms, rules, procedures and programs that govern the interaction of actors in specific issue areas." Usually, regimes center on an international agreement (e.g. United Nations Framework Convention on Climate Change, Convention on Biological Diversity) and respective compliance mechanisms (considered as 'hard law'). However, a regime can also be framed around a non-binding agreement (e.g. codes of conduct, standards) when it establishes norms that shape state behavior (Chasek et al. 2010). In accordance with this logic, Dimitrov et al. (2007) have enriched regime theory through the concept of non-regimes defined as "transnational policy arenas characterized by the absence of multilateral agreements for policy coordination among states." The notion of non-regimes stresses voluntary cooperation and 'soft law' approaches, related to changing global power structures as described by Nye (2004).

Regimes differ from international organizations, including regional intergovernmental organizations such as Association of Southeast Asian Nations (ASEAN), in two ways. The first distinction is a regime's focus on a specific issue area such as climate change or biodiversity, whereas international organizations cover a broader scope (Rittberger et al. 2012). Second, regimes do not function as collective actors, unlike international organizations such as the United Nations or European Union (ibid).

The soft law approach also incorporates **policy networks** which Börzel (1998) characterizes as "predominantly informal interactions between public and private actors with distinctive, but interdependent interests, who strive to solve problems of collective action on a central, non-hierarchical level". Compared to the more formalized international regimes and compliance-based outcomes, networks emphasize benefits resulting from informal, non-hierarchical coordination.

Recognition of these fluid processes focuses and justifies our attention to **policy-oriented learning**, which Sabatier and Jenkins-Smith (1999) define as "relatively enduring alterations of thought or behavioral intentions that result from experience and/or new information and that are concerned with the attainment or revision of policy objectives".

12.3 Analytical Framework: Connecting Five Conceptual Pathways

12.3.1 Overview

Scholarly interest in promoting 'good forest governance' is aided by an understanding of five often disconnected sets of literature: (1) governance and institutions, (2) legitimacy/political authority (3) policy networks, (4) institutional intersection, and (5) policy-oriented learning. This section discusses how these conceptual pathways are interconnected and develops the analytical foundation for understanding the subsequent review of forest institutions and learning processes.

12.3.2 From Government to (Good Forest) Governance

In the 1990s, political scientists shifted their focus from governmental efforts to the broader concept of governance. Cashore et al. (2010) present the emergence of governance as a consequence of globalization and increasingly complex problems such as climate change and land-use conflicts.[2] Dissatisfaction with traditional command-and-control approaches of governments (Kooiman 1993; Rhodes 1997) paved the way for new patterns of interaction (governance) that link public and private authority through various forms of policy networks, transnational coalitions and public-private partnerships. Much of this scholarship asserts that compared to state-centered, 'top-down' styles of governing, these new patterns emphasize more 'bottom-up' approaches that include the involvement of non-state actors and non-hierarchical, voluntary processes of coordination (Börzel 2005; Zürn 2008). The interaction of these various modes of governance has been found to be crucial for effective governance (Cashore 2002; Mayntz 2008) and problem solving (Blumenthal 2005).

The emergence of the governance concept is attended by widespread efforts to promote 'good governance', brought forward mainly by institutions such as the Organization for Economic Cooperation and Development (OECD) and the World Bank.[3] Focusing on forests, Contreras-Hermosilla et al. (2008) describe 'good forest governance' as "governance that best meets, in a transparent, equitable and sustainable way, the forest related needs and goals of the population of the country and its constituent parts. Forest governance is about who holds power, who is responsible and how decision makers are held accountable to citizens and to each other." While distinctive interpretations of 'good (forest) governance' exist, Cashore (2009a) reveals that there is a consensus on the broad principles of inclusiveness, transparency and accountability.

Key to understanding how institutions may be designed to promote good (forest) governance are the dual concepts of legitimacy and political authority. These concepts can be understood through the lens of March and Olsen's (1995) distinction between a 'logic of appropriateness' and a 'logic of consequences'. While the latter (consequences) leads stakeholders to maximize the benefits of policy decisions based on individual or organizational strategic self-interest, the former (appropriateness) occurs when cultural norms and values explain support for policy choices (Bernstein and Cashore 2007). This means that any effort to build enduring institutions and good forest governance must pay attention to the process through which stakeholders and communities support specific policy initiatives, as well as the mechanisms through which stakeholders come to accept the political authority of an institution. Political authority is aided by the concept of legitimacy, in which "general support for a regime or governance institution" is denoted by "subjects willing to substitute the regime's decisions for their own evaluation of a situation" (Bodansky 1999).

[2] See also Rosenau (2006) and Benz et al. (2007).

[3] The 'Worldwide Governance Indicators' (WGI) project, supported by the World Bank, measures the following six dimensions of governance: voice and accountability, political stability and absence of violence, government effectiveness, regulatory quality, rule of law and control of corruption (Kaufmann et al. 2010).

Institutions that gain political authority and legitimacy through March and Olsen's logic of appropriateness are suggested to be more enduring than institutions that gain political authority and legitimacy through a logic of consequences (March and Olsen 1995; Bernstein and Cashore 2007). Appropriateness will play an important role in the review of forest institutions in Sect. 12.4 of this chapter.

12.3.3 Policy Networks as a Feature of Governance

In the context of the emergence of governance, policy networks have become increasingly common. In contrast to top-down governmental approaches, a typical network "combines the voluntary energy and legitimacy of the civil society sector with the financial muscle and interest of businesses and the enforcement and rule-making power, and coordination and capacity-building skills of states and international organizations" (Reinicke et al. 2000).

Although the effectiveness of networks in general is contested (Howlett 2002; Provan and Kenis 2007), there is evidence of what appropriately designed networks can achieve. They contribute to developing responses to external factors, address the implementation gap of policies, promote the principle of subsidiarity, organize consultation and learning processes, and support the development of guidelines and standards (Streck 2005; Adam and Kriesi 2007).

One variety of networks are 'knowledge networks' designed specifically to create a culture of policy learning by generating shared knowledge, assessing policy options and transferring lessons learned (Stone 2000). Compared to other network varieties, knowledge networks focus on linking science with policy by promoting intellectual exchange, supplying expertise, coordinating research and communicating knowledge (Stone 2005). They also seek to enhance accountability by encouraging dialogue between state and non-state actors (Maxwell and Stone 2005).

The term 'network governance' (Rhodes 1997) emphasizes the soft character of interaction, as occurring mainly in networks, and hints at expected behavioral change through mutual learning (Börzel 2005; Risse and Lehmkuhl 2006). Although some scholars suggest that soft mechanisms work only (Scharpf 1993; Héritier 2003) or better (Hogl et al. 2008) in the 'shadow of hierarchy', others argue that sanctions are not required if the governance system is based on incentives for learning and provides for flexibility (Radaelli 2003). How the latter can look in reality will be illustrated through the analysis of knowledge networks in Sect. 12.4 of this chapter.

12.3.4 Institutional Intersection

The concepts described in the preceding sections may be applied to institutions at scales that range from global to local and promote greater understanding of

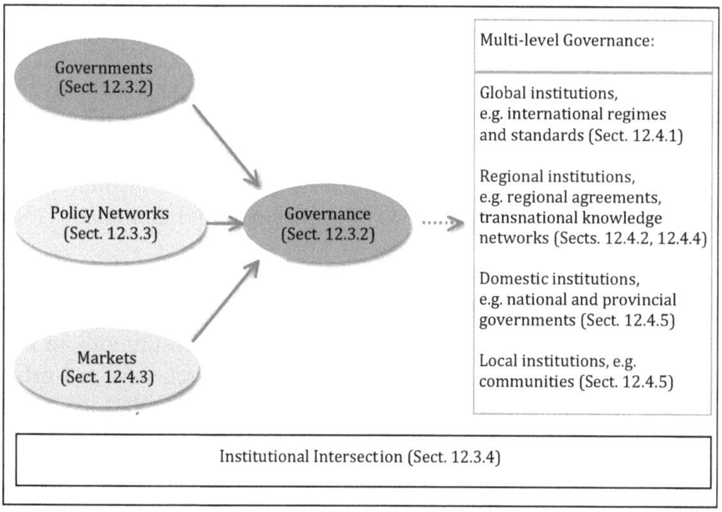

Fig. 12.1 Connecting conceptual pathways – governance and institutions, policy networks, and institutional intersection

'multi-level governance' (Benz 2004; Dunoff 2007).[4] The multi-level perspective offers a framework for governance beyond national boundaries that maintains respect for the principle of subsidiarity and links international deliberations with domestic policy implementation (Zürn 2008). The term shifts the attention away from an exclusive focus on nation-states or an international forest regime to an integrated governance approach. The emergence of multi-level institutional settings has led public policy scholars to place emphasis on 'polycentric' approaches to governance (Ostrom 2009) and what broader policy mix or 'baskets' of policy instruments (Gunningham and Grabosky 1998) can bring about meaningful intersecting effects (del Río 2010; Oikonomou et al. 2010). Figure 12.1 provides an overview of how the conceptual pathways of governance and institutions, policy networks, and institutional intersection relate to each other.

The complex fabric of intersecting institutions and hybrid governance arrangements (Zürn and Koenig-Archibugi 2006; Schuppert 2008) has become increasingly apparent in forest policy worldwide (Glück et al. 2005; Visseren-Hamakers and Glasbergen 2007; Agrawal et al. 2008), and in Southeast Asia in particular (Yasmi et al. 2010). Examples of such an approach include hybrid forms of forest governance (Glück and Rayner 2009) and the 'nested approach' to REDD+ that links global and local stakeholders in national policy-making (Pedroni et al. 2009; Streck 2010). This development in forest policy marks an important paradigm shift, which is built upon in this chapter.

[4] Multi-level governance focuses on the interplay or intersection of institutions throughout all administrative scales (Young 2002).

12.3.5 Policy Learning Across Coalitions

The learning-oriented approach of network governance is key for analyzing forest institutions in general and Southeast Asia forest institutions in particular. In this region, informal institutional environments have emerged and asserted themselves as more effective because of "a cultural aversion to formal institutional arrangements and a reflection of an Asian style of governance and diplomacy" (Nesadurai and Stone 2000).

Learning processes are an important feature of governance institutions that are dynamic and/or adaptive to a changing environment. In their 'dynamic governance' concept, Neo and Chen (2007) emphasize the need for institutions to adapt past policy choices to current realities in order to remain relevant and effective in achieving long-term objectives. In a similar vein, Ostrom (2007) identifies adaptability as indispensible for effective institutional outcomes. McDermott et al. (2011) emphasize the role of policy learning for institutions to contribute to 'problem amelioration' in the context of REDD+. McDermott et al. argue that forest stakeholders deem institutions that are able to adapt to new knowledge (through processes of learning) to be more appropriate, following March and Olsen's logic of appropriateness. Figure 12.2 illustrates how policy learning and the logic of appropriateness are embedded into the chapter's analytical framework.

Several theories exist to explain different notions of learning, often in relation to policy change (Bennett and Howlett 1992). Among the more prominent is the Advocacy Coalition Framework (Sabatier 1988; Bennett and Howlett 1992). The framework explains how policy-oriented learning results in changing belief systems based on experiences and new information about policy objectives (Sabatier and Jenkins-Smith 1999). Drawing on the Advocacy Coalition Framework, Albright (2011) emphasizes professional meetings as a forum for scientists and technical experts (i.e. advocacy coalitions) to discuss (opposing) belief systems and for policy learning to occur, respectively. Regional learning networks, reviewed in Sect. 12.4, reflect this idea of professional forums.

Learning is expected to generate new information, facilitate the interpretation of past policies anew, incorporate new ideas, and contribute to the adaptation of policies over time as prerequisites for policy change (Harguindéguy 2007). Research shows that networks provide greater learning opportunities than traditional governance mechanisms (Streck 2005). The learning capacity of networks is related to their ability to identify and diffuse policy innovations "driven by information flows rather than hierarchical or collective decision making within international institutions" (Busch et al. 2005). In multi-level governance systems, transnational information flows often become more informal through an emphasis on non-imposed mechanisms of coordination (Fogarty 2007).

This approach leads to the question of how transnational networks create venues to nurture learning and behavioral change in the long term. The interconnectedness of cross-sectoral problems in rural development suggests that there is a need to transfer knowledge and enrich learning across institutions to solve problems inside

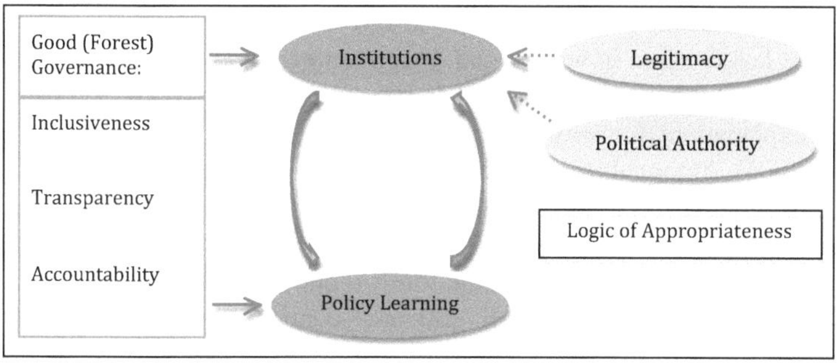

Fig. 12.2 Connecting conceptual pathways – good (forest) governance, policy learning, legitimacy and political authority

and outside the field of forestry. In this context, Theisohn and Land (2009) distinguish vertical communities of practice organized around sectoral, thematic areas such as forestry, agriculture or rural development, and horizontal communities of practice concerned with matters of public management, governance or legitimacy. The authors argue that for communities of practice to improve their practices they must interact more effectively with other communities, and reinforce horizontal and vertical connections where policies intersect. Similarly, Holzinger et al. (2008) find that policy change in the environmental sector often derives from learning processes among overlapping transnational networks. This can be explained with the Advocacy Coalition Framework, where professionalized forums across coalitions provide the potential for minority coalitions to convince other stakeholders of their views (Albright 2011).

12.4 Forest Governance for Rural Development: Review of Intersecting Institutions

12.4.1 Global Intergovernmental Institutions: Efforts to Build a Forest Regime

12.4.1.1 The Forest Non-regime

International forest policy deliberations can be described as a non-regime because systematic efforts to build a regime since the United Nations Conference on Environment and Development (UNCED) in 1992 in Rio de Janeiro have not led to successful formal treaties or compliance mechanisms. Since then, innumerable negotiations and consultations have been held under three consecutive forums – the Intergovernmental Panel on Forests (IPF) from 1995 to 1997, the Intergovernmental Forum on Forests (IFF) from 1998 to 2000, and the United Nations Forum on Forests (UNFF) from 2000 to present. All three have failed to lead to an agreement on a formal forest convention. This failure has resulted in widespread disappointment and a general consensus that existing global forest institutions are ineffective (Humphreys 2006). Scholars provide different explanations for why this has occurred (Rosendal 2001; Davenport 2005; Dimitrov 2005). Among the reasons are discrepancies between developed and tropical developing countries regarding the appropriateness of hard law instruments, particularly in terms of their financial and regulatory burdens (Cashore et al. 2010).

Despite the absence of hard law, the forest regime-making process has impacted rural development policies through 'soft law' contributions to institutional intersection. The following analysis builds on regime outcomes as suggested by Haas (2002) and Vogler (2005).

First, the negotiations resulted in a number of principles and norms, such as the Forest Principles and Chap. 11 of Agenda 21 (adopted in 1992 by the UNCED), the IPF/IFF Proposals for Action (adopted in 1997), the 'non-legally binding instrument on all types of forests' (adopted in 2007 by UNFF), and criteria and indicators for sustainable forest management as benchmarks for countries to design their own domestic policies (Lejano 2006). Although non-binding, these jointly agreed forestry norms respect national sovereignty while also calling for the participation of indigenous peoples and local forest communities in policy development. They also connect forestry objectives with broader development efforts, such as the Millennium Development Goals (MDG).

Second, the idea of 'national forest programs', adopted under the IPF, have stimulated national reforms in some countries, including Indonesia, Vietnam and Cambodia (BMZ 2004; FAO 2006a). National forest programs are designed as a domestic framework for integrated policy processes, which puts forestry in the context of land-use management, poverty alleviation and rural development. This approach shares important features with (new) public management, namely the holistic view of policy-making with interrelated actors (Nelson 1996). The development of national forest programs reflects the move from traditional to network governance based on guiding principles such as adaptive learning and open dialogue with a wide range of actors. National forest programs put national sovereignty at center stage (McDermott et al. 2007), are flexible to accommodate new policy instruments (Rayner and Howlett 2004) and have the potential to improve domestic governance (Pülzl and Rametsteiner 2002; Albrecht and Obser 2003). Although national forest programs have not yet been used by governments to their full potential, they represent a type of institution that may become increasingly important for rural development planning, given their ability to build inter-institutional relationships (Galloway et al. 2010).

A third, related outcome is the involvement of civil actors who have contributed to redefining problems from a bottom-up perspective. Themes put forward by non-state actors include a neo-liberal agenda for economic growth and forest trade, attention on domestic efforts for forest law enforcement and governance, and an emphasis on rights and access to resources for local communities (Cashore et al. 2010).

Fourth, the global debates on forestry promoted the generation and dissemination of scientific knowledge. Examples include the Global Forest Expert Panels initiated by the Collaborative Partnership on Forests (CPF) and the Task Force on International Forest Governance led by the International Union of Forest Research Organizations (IUFRO). The CPF, which groups 14 international forest institutions, was established in 2001 to complement the UNFF as a platform for learning and coordination. Similar to the Intergovernmental Panel on Climate Change (IPCC), the forest expert panels provide independent scientific assessments of key forestry issues such as forest adaptation, an emerging theme for rural development policies, and the international forest regime (GFEP 2009). The CPF uses an interdisciplinary perspective and examines forestry in a wider multi-sectoral context – the so-called 'forests+' approach (Glück et al. 2010). One justification for these collaborations is that it is expected that the CPF will lead to learning about the appropriateness of the existing institutional architecture not only in forestry, but also more widely in rural and sustainable development. This, in turn, may infuse new ideas that will help in shaping more effective global forest institutions, which benefit forest users at local levels (Bozzi et al. 2012).

Hence, although conventional efforts were unable to win broad coalitions of support, they have contributed pieces to the overall forest governance puzzle through their intersection with regional and domestic institutions. As summarized by Cashore et al. (2010) soft law policy tools exerted a 'normative pull' on states and generated new ideas about goals and instrument choices, although these new ideas are focused more on broad principles and objectives and less on practical details and on the ground calibrations. The process also opened new channels for intergovernmental and transnational networks and learning by leading stakeholders to reflect on the appropriateness of forest governance arrangements in the context of broader development objectives. An assessment of the impacts of the international forest regime in Cameroon, Indonesia and Brazil by Singer (2008) found that its informal aspects, such as principles and policy networks, have contributed to shaping forest policy. This is supported by Conca's (2006) analysis of transnational networks in water governance. He criticizes the formality of the regime approach and calls for institutions with "more pluralistic understandings of authority, more flexible conceptions of territorial sovereignty, and more heterogeneous ways of knowing about problems and solutions".

12.4.1.2 Interplay of Environmental Regimes

In parallel to deliberations on a forest treaty, trade or environmental agreements have governed specific forestry-related sub-themes, namely timber trade under the International Tropical Timber Agreement (adopted in 1983, renewed in 1994 and 2006), forest biodiversity under the United Nations Convention on Biological Diversity (adopted in 1992) and sustainable land management under the United

Nations Convention to Combat Desertification (adopted in 1994). The problem of deforestation regained momentum when the Fourth Assessment Report of the IPCC underscored the significant role of forests in a climate change mitigation and adaptation portfolio (IPCC 2007). The inclusion of forests under the United Nations Framework Convention on Climate Change (adopted in 1992) is being debated in the form of an international REDD+ mechanism (UNFCCC 2008). Levin et al. (2008) argue that for REDD+ to be effective it has to be implanted into longer-term strategic thinking regarding forest institutions and overcome persisting short-term perspectives, which hinder sustainable forest governance.

Many scholars have criticized the issue-by-issue patchwork of global forest institutions as a major obstacle to effective governance (Auer et al. 2005; Humphreys 2006; Young 2010). However, some scholars suggest that institutional fragmentation bears potential strengths as well (Young et al. 2008). Advantages may occur from the provision of complementary governance choices and testing of innovative policy instruments with subsequent diffusion to other countries (Kern et al. 2000; Biermann 2004). The International Tropical Timber Organization (ITTO), for instance, was instrumental in developing criteria and indicators for sustainable forest management which, in turn, contributed to more authoritative policy choices through governments and encouraged learning (Cashore et al. 2010). REDD+ adds a new scheme to provide financial incentives for the protection and sustainable management of forests for developing countries (Dutschke et al. 2008). Yet, solutions for appropriate governance of REDD+ that allows for the effective participation of local communities remain critical to gain the support of those stakeholders most affected by the impacts of climate change (Phelps et al. 2010; Angelsen et al. 2011).

A review of the pros and cons of institutional interplay among multilateral environmental agreements affecting forestry is beyond the scope of this chapter. The point to be made here is that, although it occurs as a patchwork of disconnected regime efforts that may support sustainable forest management in one place but hinder it in another, the perspective of institutional intersection offers a framework to explore more carefully where connections to other institutions may be beneficial to improving forest governance.

12.4.2 Regional Institutions: In the Midst of Global and Domestic Forest Institutions

The above review suggests that global-scale institutions are important but insufficient to change behavior and create adequate support for sustainable forest management (SFM). In 2005, the UNFF led a debate to regionalize the global forest policy dialogue (Obser et al. 2005) and regional approaches have been gaining importance since then (RLI 2008; OTCA et al. 2009). Growing regionalism and regional initiatives, both formal and informal, have become important trends in contemporary international relations. This trend is particularly observable in Asia (OECD 2005; ADB 2008), where ASEAN has become a core integration movement (Öjendal 2001; Frost 2008).

Compared to global or domestic efforts, regional initiatives are effective in building coalitions of support for issues of common concern, reducing transaction costs and increasing the likelihood of reaching consensus since fewer countries are involved in negotiations (Martin 2004). In the case of controlling trade in illegal timber products, regional cooperation offers a forum for debate among trading partners, an exchange of tested practices, a platform for common use of data, as well as a framework for enforcement cooperation and an international tracking and/or licensing system to guarantee legality (Brack 2006). Against this background, timber consumers like the European Union (EU) engage regionally to complement their bilateral activities with timber producer countries. Among the limitations of regional approaches are potential conflicts due to interrelations and power dynamics of neighboring countries, as well as weak motivation to coordinate beyond regional boundaries (McDermott et al. 2007).

12.4.2.1 Regional Intergovernmental Organizations

Regional intergovernmental organizations differ greatly in their institutional setups, mandates given by their citizens and business cultures (Lawson 2009). For instance, unlike the forestry-focused Central African Forest Commission (COMIFAC), the Amazon Cooperation Treaty Organization (OTCA) and ASEAN cover all policy fields of common interest to their region. Compared to informal regional processes (next section), regional organizations represent governments politically and provide formal institutional settings.

As the ASEAN case study shows (Box 12.1), intersecting institutional effects are often related to (1) the reinforcement of common interests and power of a common voice of sovereign states in global deliberations; (2) regional soft agreements which adapt global agreements to better meet regional specifications; (3) policy frameworks which support domestic efforts through regional coordination; (4) regional benchmarking and guidelines; (5) monitoring functions; and (6) the participation of, and learning among, various stakeholders through transnational networks that may foster the formation of support coalitions.

Box 12.1 Lessons from ASEAN in the Midst of Global and Domestic Institutions

Southeast Asia represents the second largest tropical forest area in the world. In international deliberations at the global level, ASEAN has underscored the kinds of institutional arrangements the coalition of its ten member states

(continued)

Box 12.1 (continued)

considers appropriate to solve forestry problems. Lessons can be drawn from the establishment of the 'ASEAN Caucus on Forestry at the UNFF' as a cost-effective mechanism to facilitate regional coordination, and from the 'ASEAN Regional Knowledge Network on Forests and Climate Change', which significantly facilitated the deliberations of a common regional position paper on REDD.[5] The network's research agenda also includes the development of minimum standards and methodologies, which signals the member states' interest in addressing REDD+ at the regional scale.

At the regional level, ASEAN achieved the following soft agreements, which in most cases originate from global institutions but are adapted to regional specifications.

First, triggered by regional criteria and indicator (C&I) processes to define sustainable forest management, ASEAN adopted its own C&I distinguishing the appropriateness of each indicator for regional, national and local levels. In addition, it developed a monitoring, assessment and reporting format, overseen by the ASEAN Secretariat, which provides a joint framework for member states to evaluate performance regarding SFM. In order to gain support from forest managers on the ground, ASEAN facilitated capacity building for decentralized forest management that addresses more practical questions such as data collection. Besides monitoring purposes, the C&I were used as benchmarks for a peer review mechanism through which member states share good forestry practices and learn about the innovative policy approaches of their peers. As an example, a review of forest policy in the Philippines reveals how ASEAN motivated one of its members to use the C&I to foster policy coherence in the context of 'forest law enforcement and governance' (FLEG).

Second, ASEAN reached agreement on regional guidelines for a phased approach to forest certification and on criteria and indicators for legality of timber. While such regional standards say little about on the ground results, regional consultations provided a forum for forestry stakeholders to share their experiences about the appropriateness of different tools to verify timber legality. The regional standards also generated learning results that, in turn, may affect the design of domestic verification systems.

Third, a regional tool to assess FLEG at the domestic level was developed, as called for in ASEAN's work plan on FLEG. The regional action program

(continued)

[5] REDD ('reducing emissions from deforestation and forest degradation in developing countries') was later expanded to REDD+.

Box 12.1 (continued)

was prioritized in conjunction with efforts promoted under the East Asia and Pacific FLEG process.

Fourth, ASEAN created transnational expert networks for better informed policy-making on forest-related issues. In the context of ASEAN's move towards a more people-oriented (or legitimized) community, the networks are expected to increase the acceptance of regionally promoted forest institutions by involving non-state actors and by channeling local people's perspectives to influence the choices of their politicians.

Finally, the 'ASEAN Multi-Sectoral Framework on Climate Change: Agriculture and Forestry towards Food Security (AFCC)' gives an idea of how regional organizations can assume a role in supporting domestic efforts through regional coordination. While an overall regional climate change policy in ASEAN is still in the making, the forest policy sector through the AFCC has become a forerunner in integrating sectoral policies into broader strategic frameworks that support greater policy coherence and cross-sectoral coordination. This way, ASEAN attempts to address regional particularities of climate change related to food security and most vulnerable countries and sub-regions in Southeast Asia.

The AFCC bridges the three pillars of the Roadmap for an ASEAN Community 2009–2015.[6] For instance, trade-related issues for forestry products, including forest certification, fall under the ASEAN Economic Community. Sustainable management of forests, other natural resources and biodiversity, as well as climate change, fall under the ASEAN Socio-Cultural Community. Existing regional soft agreements in the forest sector such as the ASEAN C&I, FLEG efforts and the timber legality standard are now to be embedded into the cross-sectoral approach.

Source: ADB (2009), ASEAN (2007), ASEAN (2008a, b, c, d), ASEAN (2009a, b), Eucker and Hein (2010), Fawzia (2010), Göhler et al. (2009b), Hinrichs (2008, 2009), Thang (2009, 2010), Yusuf and Francisco (2009)

12.4.2.2 Regional Criteria and Indicator Processes

This section moves the focus away from both international regimes and intergovernmental organizations and builds the bridge to informal regional processes. The international efforts to establish a forest convention led to the emergence of nine

[6] The three pillars are: Political-Security Community, Economic Community and Socio-Cultural Community.

regional criteria and indicator (C&I) processes,[7] specifically under the auspices of the Intergovernmental Panel on Forests and the Intergovernmental Forum on Forests. There is little doubt that the C&I processes resulted in significant learning about the science of sustainable forest management that permeated a range of stakeholders and more authoritative policy processes (Cashore et al. 2010). They led to broad consensus around seven thematic elements regarding SFM and respective criteria (McDermott et al. 2007).[8] This standard was accepted by FAO member countries as well as the UNFF (FAO 2006b) and strengthens the comparability of domestic forest policies.

Although the C&I provide an important evaluative instrument, comprehensive monitoring, reporting and verification systems remain weak regarding their impact on forest management policies and practices on the ground (Siry et al. 2005). On the other hand, the C&I processes seen as intergovernmental learning networks reinforce national sovereignty by supporting countries in developing their own policies (Cashore et al. 2010). These policies may reach down to local levels through corresponding C&I for forest management units. Partly, processes at the level of forest management units are driven by local communities who want to manage their forests sustainably (Thang 2008a). Although direct causal linkages are difficult to establish, such observations give an indication of how norm development on higher administrative scales, if organized in a participatory manner, may contribute to behavioral change and win the support of local stakeholders.

12.4.2.3 Regional Good Forest Governance Networks

In parallel to the normative-oriented C&I processes, more action-oriented attempts to strengthen forest institutions were made under the auspices of 'forest law enforcement and governance' (FLEG). FLEG was triggered by the G-8 Summit in 1998, when governments called for fighting against illegal logging (G8 2008). Four regional FLEG processes have been initiated, co-hosted by both producer and consumer governments and the World Bank, in East Asia and the Pacific, Africa, Europe and North Asia, and in Central America.

Outcomes of these regional FLEG efforts include political commitments through voluntary ministerial declarations, which prioritize regional/international and national themes (Bali Declaration adopted in 2001, Yaounde Declaration adopted in 2003, St- Petersburg Declaration adopted in 2005), the creation and

[7] The nine processes are: C&I of the International Tropical Timber Organization (ITTO), the Pan-European process through the Ministerial Conference for the Protection of Forests in Europe (MCPFE), the Montreal Process for temperate and boreal forests, the African Timber Organization's C&I, the Dry-Zone Africa Process, the Tarapoto Process for the Amazon region, the Lepaterique Process for Central America, as well as C&I processes in the Near East and on Dry Forests in Asia (FAO 2006b). In the case of ITTO, it is an international process, not regional.

[8] The seven elements are: extent of forest resources, biological diversity, forest health and vitality, productive functions of forest resources, protective functions of forest resources, socio-economic functions, legal, policy and institutional framework.

dissemination of FLEG-specific knowledge (Contreras-Hermosilla 2007) and capacity building (Brown et al. 2008a). In addition, FLEG processes influenced the placement of illegal logging on the agenda of regional organizations in Southeast Asia, Central Africa and Central America (Contreras-Hermosilla 2007) and motivated follow-up initiatives such as the recent EU program on 'Improving Forest Law Enforcement and Governance in the European Neighborhood Policy East Countries and Russia' and the EU Forest Law Enforcement, Governance and Trade (FLEGT) initiative.

On the one hand, follow-up action at country levels and on the ground results have proven elusive (Contreras-Hermosilla 2007; FAO 2009). Despite broad coalitions of support, FLEG processes have not yet achieved the expected results (Cashore and Stone 2012). On the other hand, there is evidence that in contrast to forest convention efforts, FLEG processes enhanced national sovereignty (Cashore et al. 2010) and encouraged governments to develop their own frameworks and standards (McDermott et al. 2010). For instance, in the Europe and North Asia region attention was focused on the development of national action plans (World Bank 2006b, c; Salmi and Samyn 2008).

Importantly, the soft, informal environments under the auspices of FLEG created policy learning networks among countries that involved civil society and private actors (IUCN 2005; TFD 2005; Thang 2008b) and fostered exchange about countries' experiments with various policy instruments (FAO and ITTO 2005; Pescott et al. 2010). The learning process, nurtured by efforts on domestic good forest governance (Glück et al. 2005; Cashore 2009a; World Bank 2009b; WRI 2009), contributed to greater recognition that FLEG needs to be put into a wider public policy perspective. For instance, it must be linked to domestic reforms targeting equitable tenure rights and benefit-sharing systems, societal processes, long-term sustainable development strategies and the modernization of the public administration. Learning revealed how a narrow focus primarily on law enforcement measures is insufficient to address problems such as illegal logging, wildlife poaching, encroachment and corruption (World Bank 2006a; Salmi and Samyn 2008).

The current debate on REDD+ confirms that many countries continue to have an urgent need to clarify forest tenure and use rights for local forest-dependent stakeholders, as well as to set appropriate governance standards for implementation, monitoring and evaluation (Robledo et al. 2008). In this context, ongoing processes on national forest programs are earning greater attention as policy frameworks that are flexible enough to accommodate both national FLEG action plans as well as REDD + approaches (Göhler et al. 2009a).

Some scholars suggest that efforts summarized under FLEG have contributed to substantial progress in the fight against illegal logging. For instance, Lawson and MacFaul (2010) estimate that in the last decade in Indonesia illegal logging has fallen by 75 %, up to 17 million hectares of forest have been protected from degradation and at least 1.2 billion tons of carbon dioxide emissions have been avoided. China and Vietnam have both signed non-binding Memoranda of Understanding on timber with a number of trade partners as a result of their participation in the East Asia and Pacific FLEG process (ibid).

In 2003, the EU adopted its FLEGT Action Plan. The centerpiece of this action plan are 'Voluntary Partnership Agreements' (VPA) between the EU and timber producing countries to avoid the import of illegal timber into the EU market (next section).

In addition to bilateral VPA processes that are underway in Indonesia (VPA signed in September 2013), Malaysia and Vietnam, the EU complements its efforts through a region-specific approach in Southeast Asia, which serves as a venue for learning from the experiences of peers (EFI 2009). While the more technocratic FLEGT approach seems to fall short of addressing the broader challenge of forest governance, it may intersect with other policy instruments by providing useful contributions to baseline problem solving (Cashore et al. 2010).

12.4.3 Non-state Governance: New Ways of Shaping Good Forest Governance

A decade and a half ago, one of the most innovative institutions in global forest management was created through the global supply chain-focused institution known as forest certification or 'non-state market-driven' global governance (Cashore et al. 2007). The idea of product labeling was initially met with resistance from tropical producer countries. Following general frustration of many of the world's leading environmental groups over the failure of intergovernmental efforts to achieve a binding global forest convention, the World Wide Fund for Nature (WWF) spearheaded a coalition of environmental, social and business activists to establish the Forest Stewardship Council (FSC) certification program in 1993. The approach of the FSC was designed to address many of the failures noted above. Importantly, it promoted a governance approach in which business interests could not dominate the policy-making process – a direct rebuke to their concerns that many domestic and intergovernmental efforts appeared captured by the very business interests they sought to regulate (ibid). However, instead of dismissing neo-liberal ideas that many argue were at the heart of the failures noted above, FSC strategists sought instead to embrace global markets by embedding in them socially and environmentally responsible business practices.

The FSC has a top-down approach in that it requires all standards adhere to fundamental principles and criteria, which include requiring that managers address indigenous peoples' and local communities' rights to resources; but also a bottom-up approach in that the specific standards are developed through multi-stakeholder national or subnational working groups.

This means that, if successful, these efforts could create a win-win solution by simultaneously championing the goals of neo-liberal markets; amelioration of deteriorating environmental functions of the world's forests; and the promotion of poverty alleviation, indigenous rights, and community participation. This may explain why the World Bank has been so instrumental in supporting FSC-style certification (Elliott and Schlaepfer 2001; Gullison 2003), as it represents an opportunity to support their ongoing efforts to promote socially and environmentally responsible practices in ways that are consistent with their broader neo-liberal goals.

While many firms in the forest sector initially balked at the idea of outside scrutiny of their forest practices, by the mid-2000s, two discernible trends had emerged. First, most industrialized countries in North America and Europe came to embrace third-party certification, though many supported 'FSC competitors' that emerged in the 1990s as an alternative choice to the FSC. These alternative programs were generally much more flexible than FSC standards, leaving specific decisions about what to do to meet objectives up to the firm, rather than the certification program. Second, uptake in developing countries has been much more controversial and limited. Existing studies have found that lack of capacity, the complicated nature of forest management, lack of clear rights to resources, and ongoing deforestation, explain the more limited uptake (Cashore et al. 2006).

Partly as a result, many strategists are now focusing on addressing good forest governance and illegal logging through promoting efforts such as the EU FLEGT initiative, or 'legality verification' through EU and US import laws. These efforts, are seen as distinct from certification in that they reinforce domestic sovereignty by verifying that forest managers and companies follow the laws that governments develop, but have difficulty enforcing (Bernstein et al. 2010).

The EU began focusing on illegal logging a decade ago. Though most early attempts did not result in discernible changes on the ground (Currey 2001; Tacconi 2007), the EU and the US have been redoubling efforts to promote good forest governance in the tropics through demand-side controls. The FLEGT approach is complemented by public procurement policies and the EU Timber Regulation, which became effective in March 2013. This regulation is similar to the US Lacey Act, which was amended in 2008 to prohibit the sale of illegal timber products into the US market. The EU Timber Regulation obliges operators to exercise due diligence in identifying the legality and traceability of timber products, and to establish, timber legality assurance' systems to prove compliance. The process of legality verification includes third-party auditing of forest practices to assess whether companies and governments meet their commitments.

12.4.4 Transnational Learning Networks: Innovative Bottom-Up Approaches to Forest Governance

While the significance of understanding when and where stakeholders support institutions cannot be overstated, it is equally important to assess how policy learning is reinforcing, or detracting from, efforts towards shaping appropriate institutional settings for SFM and towards integrating rural communities into forest management decision-making. This section focuses on the role of transnational policy networks in providing opportunities for sharing knowledge and encouraging learning about policy solutions, and on how this may help address persistent obstacles to rural forest development.

Global and regional forest institutions are informed by countries' own experiences regarding which policy instruments and institutional arrangements have worked or not worked domestically in the past. Not surprisingly, these experiences vary widely and are highly context-specific (World Bank 2006a). Promising developments in forest governance have been found, for instance, regarding community forest management (Agrawal and Angelsen 2009), forest certification (Cashore et al. 2006) and efforts to combat illegal logging (Tacconi 2007). More critical findings about forest institutions include the disempowerment of local communities in the context of REDD+ (Phelps et al. 2010).

Within Southeast Asia, there are many examples of domestic efforts for good forest governance that were well-intended but poorly implemented. These examples include independent forest monitoring in Cambodia (Brown et al. 2008a), community-based forest management (Yasmi et al. 2010), 'Multi-Sectoral Forest Protection Committees' in the Philippines (Cruz and Tapia 2006), land allocation to households in Vietnam (Yasmi et al. 2010), and decentralization efforts in Indonesia (Dauvergne 1994; Colfer et al. 2008; Nomura 2008).

The question that emerges is: In which ways can institutions learn from each other's experiences in order to support appropriate policy choices in changing environments? Transnational 'knowledge networks' provide an invaluable forum for this to occur. An empirical case analyzed in Box 12.2, namely the 'ASEAN Regional Knowledge Network on Forest Law Enforcement and Governance', provides some insight in answering the question.

The assumption is that research about on the ground realities and evidence about the impacts of forest policies on rural local communities would help to increase the effectiveness of policy formulation and local implementation. This may be done through regional benchmarking and the establishment of common standards, guidelines, and comparative assessments regarding the appropriateness of institutions for forest and rural development policies. For instance, comparison can be geared towards finding explanations for why coalitions of support emerged for a specific policy instrument in one context but not in another. Alternatively, comparison can be geared towards evaluating the effectiveness of existing governance institutions. Learning is expected to result in increased knowledge, revised understanding and modified beliefs as prerequisites for policy change. A study by McDermott et al. (2010) on global forest policies provides evidence regarding the usefulness of comparative analysis.

The case study in Box 12.2 provides empirical observations of how regional policy networks contribute to (1) establishing common standards and guidelines; (2) conducting comparative assessments; (3) bridging the implementation gap of regional policies; (4) guiding consensus-building on benchmarks and respective evaluation tools; and (5) promoting normative discourse on good forest governance to enhance the legitimacy of institutions. Table 12.1 provides an overview on principles and criteria of good forest governance as pushed forward through the ASEAN knowledge network (Cashore 2009a). These functions demonstrate how transnational networks help to better understand the appropriateness of policy instruments and create effective forms of institutional intersection.

Box 12.2 How a Regional Knowledge Network in ASEAN Helped to Improve Forest Law Enforcement and Governance

The 'ASEAN Regional Knowledge Network on Forest Law Enforcement and Governance' was established in 2008 with the primary motivation to better inform decision-makers through organized processes of policy learning. The following contributions by the network to improving forest governance through a regional approach can be observed:

First, the network guided consensus-building on a regional evaluative tool to assess the outcomes of domestic efforts to implement ASEAN's work plan on FLEG. The standard accommodates, among others, principles such as transparency, public disclosure policies, equitable participation, fair tenure rights and the recognition of customary rights. The network extended the learning process beyond abstract policy goals to more practical, on the ground specifications by organizing a structured exchange of countries' experiences about successes and failures of FLEG policies. As an example of how policy innovations may spread through transnational networks, the Philippines made use of an ASEAN peer review mechanism to assess FLEG.

Second, the network initiated a normative discourse on delineating principles of good forest governance that takes into consideration the specific realities of Southeast Asia, as well as international benchmarks and good practices. It encouraged consensus about criteria for inclusiveness, transparency and accountability beyond issues of enforcement and compliance. It also developed recommendations for the ASEAN Senior Officials on Forestry regarding concrete practical interventions and highlighted the clarification of forest property rights and stakeholder support.

Third, the network analyzed various features of forest governance in the region in order to draw decision-makers' attention to intersecting institutions and to influence the political agenda regarding FLEG. Research topics included potential improvements to policy implementation through ongoing processes on national forest programs, implications of new market instruments such as the EU FLEGT Action Plan and the US Lacey Act, the principle of subsidiarity with a focus on local institutions, as well as impacts on local livelihoods. In this capacity, the network supplied information about local conditions, which is indispensible for evidence-based policy making, a public policy movement to enhance the efficiency and effectiveness of institutions.

Finally, an additional mechanism for increased transparency has been created with the internet-based ASEAN Forest Clearing House Mechanism where stakeholders make data and knowledge publicly available.

(continued)

Box 12.2 (continued)

Source: ASEAN (2008d), ASEAN (2009b), Cashore (2009b), Göhler and Schwaab (2009), Göhler et al. (2009), Howlett et al. (2009), Koeng (2009), Koeng and Malessa (2009), Pescott et al. (2010), Soriaga (2010), Soriaga and Cashore (2009), Thang (2009, 2010)

12.4.4.1 Translating Learning into Policy Development in Network Governance

Regional knowledge networks such as the ASEAN knowledge network on FLEG reflect what the essence of policy learning is, namely to inform decision-makers based on evidence (Sanderson 2002). A key management challenge in this regard is to make the knowledge generated usable for policy development. Researchers have found positive effects of transnational communication on domestic environmental public policy (Holzinger et al. 2008). Learning networks are a common form of transnational communication that diffuse ideas that have proven successful and help translate those ideas into country-specific policy responses. In turn, this may foster durable and adaptive forest governance institutions across the region.

In order to explain how favorable conditions can be created for policy learning to occur, it is necessary to examine the relationship between policy-makers and scientists/experts in two types of networks: private and public-private. In private networks such as the Global Forest Expert Panel (GFEP), learning occurs first of all among independent experts. In public-private networks such as the ASEAN knowledge network, the learning process includes a transparent policy process that intersects more directly with other governing institutions. In concrete terms, a public-private network such as the one described in Box 12.2 may be composed of government officials from forestry administrations of a number of countries as well as experts from non-governmental organizations, academia, think tanks and the private sector. On the one hand, such networks represent an important forum of bottom-up policy networks. On the other hand, they build strong ties with top-down intergovernmental institutions. Interaction between scientific knowledge and policy-making is, thus, more frequent and more common in public-private networks than in private networks.

In order to explain further how the functions of transnational networks actually result in policy learning, Howlett and Joshi-Koop (2011) suggest shifting attention to policy analytical capacity. In the ASEAN network, analytical capacity is 'outsourced' to the policy network. At the same time, strong ties between those who generate knowledge and those who will use it are maintained. One means to ensure this is to restrict the actual membership to experts from within the transnational geographical area, whereas external experts (from outside that region) participate in a looser form (i.e. not as members but, for instance, as partners).

Table 12.1 Principles and criteria of good forest governance and some examples

Principles	Criteria and examples
Inclusiveness	(a) Formalized venue for participation, e.g. as members of knowledge networks, formalized consultations for national forest programs (b) Widespread invitations (c) Notification of policy deliberations (d) Meaningful participation rather than 'lip-service' • Permit stakeholders to deliberate on substantive questions • Allow input on different policy options • Include knowledge raised by stakeholders in addition to knowledge generated within the policy subsystem
Transparency	(a) Notification • Of intent to create a policy • Of draft proposals • Of final policy decision (b) Conflict of interest rules • Eliminate personal gain of policy maker for a particular decisions (c) Make (legal) influence transparent • Clear identification of groups involved • Identify their interests • Document their efforts (d) Eliminate secret influence • Undocumented lobbying • Side payments (corruption) (e) Information/knowledge sharing, e.g. through ASEAN Forest Clearing House Mechanism • Quantitative metrics • Measurement
Accountability	(a) Explain rationale • For initial policy deliberations, e.g. scientific analysis by Global Forest Expert Panels • Through draft comments, e.g. organized through ASEAN knowledge networks • For final decision (b) Institutionalize public comment • Formally recognize • Formally respond (c) Make it clear who is making the decisions (d) Power sharing

Source: Compiled and adapted from Cashore (2009a) and Esty (2006)

The emphasis on the mobilization of analytical capacity within the institutional boundaries of a region like ASEAN is in line with recent findings that decision-makers are rather unlikely to use knowledge from external sources (Howlett and Joshi-Kopp 2011).

12.4.4.2 Intersecting Policy Networks for Sustainable Rural Development

The previous section uncovered how regional knowledge networks accomplish their mission of promoting policy learning in a multi-level governance landscape in order to reorient forest institutions to be more appropriate and more legitimate. The learning process in a transnational public-private context is informed by experiences beyond domestic jurisdictions and a variety of interest groups. While recent findings suggests that such networks and coalitions serve as bridging institutions for effective policy learning, more polycentric approaches, i.e. engaging various levels of governance, are necessary to address the complex, interrelated problems in rural areas in times of change (Cheng et al. 2011).

Beyond the contributions of issue-specific policy networks, cross-sectoral problems like climate change and rural development require learning across coalitions themselves. As an example, scholars point out that weak overall governance poses high risks for REDD+, in particular from the perspective of forest communities (Saunders and Nussbaum 2007; Angelsen 2008; Peskett et al. 2008; Cotula and Mayers 2009; Springate-Baginski and Wollenberg 2010).

Comprehensive strategies like the ASEAN Multi-Sectoral Framework on Climate Change and Food Security (Box 12.1) provide one opportunity to adapt learning structures such as transnational knowledge and other forms of policy networks in order to support appropriate, long-lasting institutions for sustainable rural development. The task for individual institutions – be it those which focus on forest governance, REDD+ or public policy – is to invite perspectives from other governance actors in order to reflect on and adapt previous policy choices. As earlier sections revealed, problems like deforestation and forest degradation demonstrate the need to connect learning networks in order to increase the impact of forest institutions.

12.4.5 Domestic Governance and Local Institutions: The Backbone of Policy Implementation

While in many cases forest-dependent communities have historically acted as effective stewards of their forests, these communities have frequently been excluded from forest management decision-making, as well as from the benefits gained from forest resources, under the pretext that they are irresponsible forest stewards (Dove 1983; Scott 1998). As a result, communities have often managed forests in a non-sanctioned manner that precludes them from engaging in broader institutional networks (Wollenberg et al. 2009).

This section first describes forest management trends at the community level, and then outlines some opportunities for improving forest management outcomes at this level. It concludes with a case study taken from Indonesia that applies, in a local context, the concepts of institutional intersection, authority and legitimacy, as well as the logic of appropriateness.

12.4.5.1 Decentralization

Over the last 20 years, trends towards decentralization have redefined the relationship between local communities and forest management institutions. Decentralization was initially encouraged as a means to bring decision-making closer to the communities that depend on forests for their livelihoods (Larson 2005). Because decentralization has been applied using a number of different processes, it is helpful to break the term down into multiple sub-definitions. Wollenberg et al. (2009) distinguish between state-sanctioned community-based natural resource management (CBNRM), whereby state institutions grant formal rights for management to community-level institutions (a process defined as devolution), and local governance, whereby formal control is transferred from centralized state institutions to local state institutions such as provinces or districts (a process defined as deconcentration).

While communities have benefited in some instances from increased accessibility to decision-making institutions, decentralization processes have been hindered by the retention of control and resources by central governments (Ribot et al. 2006). Additionally, decentralization has often been instituted without adequate preconditions, such as the presence of local institutions that are both downwardly accountable and responsive to the interests of communities (Ribot 2005). In Indonesia, decentralization led to an explosion of resource conflicts, forest extraction and environmental degradation, as previously excluded local elites vied for the benefits of newly accessible and abundant forest resources (Peluso and Harwell 2001; Curran et al. 2004; McCarthy 2004). In Malinau District, East Kalimantan, in particular, decentralization led to newfound opportunities for community revenue generation from forest resources, but resulted in widespread ethnic conflict, corruption and the rapid consumption of generated revenue (Limberg et al. 2007; Sudana 2009).

12.4.5.2 Community Land Tenure

One of the most significant obstacles to successful integration of local communities into the broader landscape of forest institutions has been the lack of clarity for community land tenure rights. Without adequate protections, local communities have often been excluded from the benefits of forest resource extraction (Dove 1983; Scott 1998; Sikor et al. 2010). In addition, without adequate land tenure rights, traditional forest uses and forest management practices are often ignored or undervalued (Colfer 2005). Efforts to improve the environmental and social outcomes of forest management have also been consistently impeded by unclear community land tenure (Peluso 1993; Colfer 2005; Boyd et al. 2007). While the forest-related needs of local communities have gained more recognition as a result of decentralization and some of the global and regional initiatives described in previous sections, accompanying land tenure clarity has remained elusive.

Ongoing developments regarding REDD+ have rekindled the debate regarding the role of forest-dependent communities in a global institutional landscape. While REDD+ has the potential to provide significant benefits to forest-dependent communities, a large number of scholars and advocates have expressed alarm over the potential for REDD+ to lead to restrictions on communities' continued use of, and access to, forests (Brown et al. 2008b; Seymour 2008; Blom et al. 2010; Okereke and Dooley 2010). While Chhatre and Agrawal (2009) found that increased local autonomy over forests resulted in greater carbon storage benefits, it remains an open question as to whether REDD+ will result in significant clarification over the land tenure and property rights of forest-dependent communities. This has major implications for the effectiveness of REDD+, as a lack of clarity regarding land tenure rights for communities has prevented CBNRM and Payment for Environmental Services (PES) initiatives from meeting rural development objectives in the past (Nanang and Inoue 2000; Boyd et al. 2007). More details on PES are given in Chap. 9.

12.4.5.3 Opportunities for Institutional Change at the Community Scale

A common critique from practitioners and scholars is that governance institutions are often designed in ways that truncate their ability to be seen as appropriate, or to gain legitimacy, at the local and/or community level. Greater care, some argue, needs to be placed in linking SFM institutional networks with the economic development goals of local communities. Because of the complexity associated with incorporating such goals, the thorny tradeoffs between SFM and rural development are often unacknowledged and unaddressed (Blom et al. 2010). In addition, the uncertainties surrounding land tenure and property rights are often major barriers to the successful incorporation of communities into larger institutional networks. Equally important is the fact that local-level forest management requires site-specific approaches that are context-appropriate and provide adequate incentives under a broad range of cultural and environmental conditions (Ostrom 1990; Ostrom 2000). Without the ability of institutions to remain adaptive and appropriate within these diverse contexts, studies have shown they will face challenges in producing marked benefits for rural communities (Salafsky and Margoluis 2004; Colfer et al. 2008).

Box 12.3 provides an on the ground example taken from Malinau District, East Kalimantan, Indonesia. This case is used because the challenges facing the rural communities of Malinau, such as conflicting demands for environmental protection and economic development, insecure land tenure, poor institutional legitimacy, and the struggle to retain traditional ways of life, are representative of many forest-dependent regions (Belcher et al. 2004; Peskett et al. 2008; Seymour 2008). The conclusions drawn from this case may be applicable to other forest-dependent regions and serve to highlight the relevance of the conceptual pathways used throughout the chapter.

Box 12.3 highlights the difficulties inherent to the incorporation of local and/or community institutions into broader forest governance institutions. Based on this real case, how might REDD+ intersect with existing forest management institutions in order to reduce deforestation and forest degradation, while also ensuring that forest-dependent communities receive an equitable share of the benefits from REDD+ ? How might REDD+ come to be viewed as an appropriate and legitimate institution by forest-dependent communities?

First, given pervasive distrust and lack of transparency at the community level, REDD+ would need to establish significant preconditions on existing forest management institutions to ensure that REDD+ comes to be viewed by forest-dependent communities as both appropriate and legitimate. In the absence of such preconditions, REDD+ may come to be seen by local communities as another institution designed to expropriate the forest resources on which they depend.

Second, unless the role of customary institutions and state law is clarified, existing forest management institutions will not likely gain legitimacy from the perspective of forest-dependent communities. To gain such legitimacy would require a radical transformation within existing local institutions to a system that clearly incorporates elements of customary forest management. If successful, such a transformation would allow for a shift from forest governance predicated on the logic of consequences to one based on a logic of appropriateness, in which forest-dependent communities support policy choices based on cultural norms and values (March and Olsen 1995; Bernstein and Cashore 2007). Such a shift is believed to yield more enduring forest institutions (Bernstein and Cashore 2007).

Box 12.3 On the Ground Forest Management in Malinau District, East Kalimantan, Indonesia

Despite widespread deforestation throughout Indonesia over the past 20 or more years, 90 % of Malinau's area (~3.62 million ha) remains forested. While international non-governmental organizations (NGOs) seek to protect Malinau's remaining forests, there is significant pressure for deforestation in Malinau, particularly for the establishment of oil palm and *Acacia mangium* plantations. In recent years, the international community has identified the potential for conserving Malinau's remaining forestlands through the REDD+ mechanism in which Malinau's forest management stakeholders would be given financial incentives for forest conservation.

Forest management in Malinau is currently highly fluid and chaotic, which has resulted in an atmosphere that encourages short-term, unsustainable resource extraction and short-term, unsustainable resource consumption. In

(continued)

Box 12.3 (continued)

instances where forest resources have been commercially extracted by timber concession holders, communities have experienced high frequencies of conflict and rapid environmental and social change. Many residents of Malinau claim to feel conflicted between desiring to protect remaining forests in order to maintain traditional cultures and traditional forest uses, while also desiring to benefit from forest resource extraction and deforestation before someone else does.

Corruption and a lack of transparency are perceived to characterize the decision-making processes of most, if not all, forest management institutions, including local NGOs. Despite some decentralized decision-making processes, forest management decisions in Malinau's District Government are still made with limited transparent consultation with local communities. As a result, many residents of rural Malinau's forest-dependent communities believe that district forest management institutions (such as administrative bodies of the Ministry of Forestry) make decisions for personal gain. One resident of Paking, a rural Malinau village, stated the following when asked about the potential for his village to receive financial incentives under the REDD+ mechanism:

> Money should go directly to the villages. People here are pessimistic about the district government giving compensation to villages. Compensation from the government is always too small. There is no transparency. (Translated from Indonesian).

Insufficient transparency and corruption also often characterize decisions made by forest management institutions within communities themselves. Suspicion regarding the mishandling of logging revenues received by village leaders in the late 1990s/early 2000s led to rampant short-term consumption that could have otherwise been used for community-wide improvements in health, education and infrastructure. Often, these suspicions pit different ethnic groups in the region against each other.

In addition, unclear and inconsistent recognition of the joint roles of *adat* (customary) and state institutions have also contributed to forest management inequity and environmental degradation in Malinau. While decentralization reforms that were passed in Indonesia in the late 1990s recognized *adat*, or customary law, this recognition continues to be subject to the application of district governments themselves and decentralization has not allowed for a much greater recognition of *adat* than previous governance regimes. The uncertainty surrounding the role of *adat* in forest management decisions manifests itself most clearly in ongoing land tenure disputes in Malinau. While land tenure claims have traditionally been based on *adat* rules, state institutions do not often recognize these claims, or do so inconsistently.

(continued)

Box 12.3 (continued)

One outcome of this uncertainty is increased environmental degradation, as some village leaders explained that they supported forest conversion and exploitation as a means to affirm their village's tenuous land claims.

Source: Kamelarczyk and Strandby (2009), Limberg (2009), Malinau District Government (2007), Moeliono and Limberg (2009), Palmer (2009), Rhee (2009), Sandker et al. (2007), Sudana (2009), Thorburn (2002), Interviews with village residents as well as local government and NGO officials (July-August, 2009)

12.5 Outlook

This chapter reviewed forest institutions at multiple scales of governance, ranging from global intergovernmental institutions to domestic and local institutions. It emphasized how these institutions may overlap or interact with each other to result in favorable or unfavorable outcomes for sustainable forest management. The review required an interdisciplinary analytical perspective, which was realized through the lenses and connections of five conceptual pathways. In addition, selected case studies of regional learning networks in Southeast Asia and district level forest governance in Indonesia were presented to exemplify how different forest institutions recognize the needs of local communities and concerns of the global community alike.

The main question guiding the review was how enduring, problem-focused and authoritative forest institutions can be encouraged in order to promote good forest governance. This was mainly assessed through March and Olsen's logic of appropriateness and the related concepts of political authority and legitimacy. Three key findings of this chapter are summarized below and provide recommendations for practitioners:

First, individual forest institutions are unable by themselves to address the interrelated economic, environmental and social challenges that global change poses for rural economies. As an example, efforts to establish a global forest convention failed because they fell short of sufficient support from key stakeholders. However, they contributed, among other efforts, to establish international principles and norms. At the regional level, FLEG processes were able to reinforce domestic sovereignty and, thus, win the support of governments and other interest groups in spearheading domestic good forest governance initiatives.

Because individual institutions have proven insufficient to address complex forest governance problems, the task for policy-makers remains to foster synergistic and avoid perverse effects among various institutions. A promising approach to overcome institutional fragmentation and related 'governance gaps' is to reorient the focus from isolated institutions and either/or solutions towards intersecting institutional arrangements across administrative scales of governance. It is argued in this chapter that the concept of institutional intersection is well suited to build upon the strengths, and mitigate the weaknesses, of both traditional and new forms of governance. It offers a framework that favors the co-existence of hierarchical, market-oriented and voluntary institutions and connects authoritative policy responses by sovereign governments with the energy of bottom-up policy networks and local institutions.

Second, more attention should be paid to the processes through which institutions gain authority and come to be seen as legitimate. Political authority and legitimacy are prerequisites for good forest management and require the involvement of local communities and transparent policy processes. To gain legitimacy, stakeholders must deem their institutions to be appropriate to their respective environmental, political and social contexts. The case study from Indonesia illustrated that it would be difficult in that particular setting for the REDD+ mechanism to achieve political authority and legitimacy at the community scale. This, in turn, could impair the legitimacy and political authority of REDD+ at national, regional and global scales. On the other hand, if REDD+ were to be incorporated into an institutional landscape that met preconditions of participation and transparency, it could intersect with existing institutions in a way that would foster good forest governance.

Third, scholars and practitioners must place greater emphasis on policy learning processes. A key aspect of policy learning is its ability to promote behavioral change of key governance actors. The review of transnational policy networks with the case study from Southeast Asia reveals that flexible and informal institutional arrangements, in contrast to formal regimes, can play an important role in policy learning. Learning about the promises and pitfalls of policy choices and innovations, as well as the mechanisms through which coalitions of policy support emerge, are indispensable to capitalize on synergies between existing forest institutions.

Reinforcing this chapter's focus on intersection, the authors argue that more effective horizontal and vertical interaction among stakeholders, both within and beyond forestry, is paramount to encouraging sustainable rural livelihoods in a globalized world.

References

Adam S, Kriesi H (2007) The network approach. In: Sabatier PA (ed) Theories of the policy process. Westview Press, Boulder

ADB (2008) Emerging Asian regionalism. A partnership for shared prosperity. Highlights. Asian Development Bank (ADB), Manila

ADB (2009) The economics of climate change in Southeast Asia: a regional review. Asian Development Bank (ADB), Manila

Agrawal A, Angelsen A (2009) Using community forest management to achieve REDD+ goals. In: Angelsen A (ed) Realising REDD+: national strategy and policy options. Center for International Forestry Research (CIFOR), Bogor, pp 201–211

Agrawal A, Chhatre A, Hardin R (2008) Changing governance of the world's forests. Science 320 (5882):1460–1462

Albrecht J, Obser A (2003) Institutional dimensions of international public forest policies. Open meeting of the Human Dimensions of Global Environmental Change research community, panel on "Institutional Dimensions of Global Environmental Change: Fit, Interplay and Scale". Montréal, 16–18 Oct 2003

Albright EA (2011) Policy change and learning in response to extreme flood events in Hungary: an advocacy coalition approach. Policy Stud J 39(3):485–511

Angelsen A (ed) (2008) Moving ahead with REDD. Issues, options and implications. Center for International Forestry Research (CIFOR), Bogor

Angelsen A, Boucher D, Brown S, Merckx V, Streck C, Zarin D (2011) Modalities for REDD+ reference levels: technical and procedural issues. Meridian Institute, Washington, DC

ASEAN (2007) ASEAN criteria and indicators for sustainable management of tropical forests. Association of Southeast Asian Nations (ASEAN), Jakarta

ASEAN (2008a) ASEAN common position paper on reducing emission from deforestation and forest degradation (REDD) in developing countries. Association of Southeast Asian Nations (ASEAN), Jakarta

ASEAN (2008b) ASEAN inputs to the country-led initiative (CLI) in support to the United Nations Forum on Forests (UNFF). Financing for sustainable forest management: an international dialogue. Association of Southeast Asian Nations (ASEAN), Jakarta

ASEAN (2008c) ASEAN inputs to the eights session of the United Nations Forum on Forests (UNFF). Forests in a changing environment and means of implementation for sustainable forest management. Association of Southeast Asian Nations (ASEAN), Jakarta

ASEAN (2008d) Work plan for strengthening forest law enforcement and governance (FLEG) in ASEAN 2008–2015. Association of Southeast Asian Nations (ASEAN), Jakarta

ASEAN (2009a) ASEAN joint statement on climate change to the 15th session of the Conference of the Parties to the United Nations Framework Convention on Climate Change and the 5th session of the Conference of the Parties serving as the Meeting of the Parties to the Kyoto Protocol. Association of Southeast Asian Nations (ASEAN), Jakarta

ASEAN (2009b) A proposed outline "Format for assessing forest law enforcement and governance (FLEG) implementation in ASEAN Member States". ASEAN technical workshop on the thematic elements for assessing FLEG implementation in ASEAN member states in Kuala Lumpur. Association of Southeast Asian Nations (ASEAN), Jakarta

Auer MR, Karr-Colque C, Samman S (2005) Taking stock of the international arrangements on forests. J Forest 103(3):126–133

Belcher B, Rujehan NI, Achdiawan R (2004) Rattan, rubber, or oil palm: cultural and financial considerations for farmers in Kalimantan. Econ Bot 58(Supplement):S77–S87

Bennett CJ, Howlett M (1992) The lessons of learning: reconciling theories of policy learning and policy change. Policy Sci 25(3):275–294

Benz A (2004) Multilevel Governance – Governance in Mehrebenensystemen. In: Benz A (ed) Governance. Regieren in komplexen Regelsystemen. VS Verlag für Sozialwissenschaften, Wiesbaden

Benz A, Lütz S, Schimank U, Simonis G (2007) Einleitung. In: Benz A, Lütz S, Schimank U, Simonis G (eds) Handbuch Governance. Theoretische Grundlagen und empirische Anwendungsfelder. VS Verlag für Sozialwissenschaften, Wiesbaden, pp 9–25

Bernstein S, Cashore B (2007) Can non-state global governance be legitimate? An analytical framework. Regul Gov 1:347–371

Bernstein S, Cashore B, Atyi RE, Maryudi A, McGinley K (2010) Examination of the influences of global forest governance arrangements at the domestic level. In: Rayner J, Buck A, Katila P (eds) Embracing complexity: meeting the challenges of international forest governance. International Union of Forest Research Organizations (IUFRO), Vienna, pp 111–135

Biermann F (2004) Global environmental governance. Conceptualization and examples. Global governance working paper no. 12. The Global Governance Project, Amsterdam/Berlin/Oldenburg/Potsdam

Blom B, Sunderland T, Murdiyarso D (2010) Getting REDD to work locally: lessons learned from integrated conservation and development projects. Environ Sci Policy 13(2):164–172

Blumenthal J (2005) Governance – eine kritische Zwischenbilanz. Z Polit 4(15):1149–1180

BMZ (2004) National forest programmes: instruments for improving sector governance? Experiences of German development cooperation. Bundesministerium für wirtschaftliche Zusammenarbeit und Entwicklung (BMZ), Bonn

Bodansky DM (1999) The legitimacy of international governance: a coming challenge for international environmental law? Am J Int Law 93:596–624

Börzel TA (1998) Organizing Babylon – on the different conceptions of policy networks. Public Adm 76(2):253–273

Börzel TA (2005) European Governance – nicht neu, aber anders. In: Schuppert GF (ed) - Governance-Forschung: Vergewisserung über Stand und Entwicklungslinien. Nomos, Baden-Baden, pp 72–94

Boyd E, Gutierrez M, Chang M (2007) Small-scale forest carbon projects: adapting CDM to low-income communities. Glob Environ Chang 17:250–259

Bozzi L, Cashore B, Levin K, McDermott C (2012) The role of private voluntary climate programs affecting forests: assessing their direct and intersecting effects. In: Ronit K (ed) Business and climate policy: the potentials and pitfalls of private voluntary programs. United Nations University Press, Tokyo/New York/Paris

Brack D (2006) Regional cooperation to combat illegal logging. Options for ASEAN and the East Asia region. Chatham House, London

Brown D, Schreckenberg K, Bird N, Cerutti P, Del Gatto F, Diaw C, Fomete T, Luttrell C, Navarro G, Oberndorf R, Hans T, Wells A (2008a) Legal timber. Verification and governance in the forest sector. Overseas Development Institute (ODI), London

Brown D, Seymour F, Peskett L (2008b) How do we achieve REDD co-benefits and avoid doing harm? In: Angelsen A (ed) Moving ahead with REDD. Center for International Forestry Research (CIFOR), Bogor, pp 107–118

Busch P-O, Jörgens H, Tews K (2005) The global diffusion of regulatory instruments: the making of a new international environmental regime. Ann Am Acad Polit Soc Sci 598(1):146–167

Cashore B (2002) Legitimacy and the privatization of environmental governance: how non-state market-driven (NSMD) governance systems gain rule-making authority. Gov Int J Policy Adm Inst 15(4):503–529

Cashore B (2009a) Key components of good forest governance in ASEAN, vol 6, Exlibris series. ASEAN-German Regional Forest Programme (ReFOP), Jakarta

Cashore B (2009b) Structure for 'FLEG' at a glance. ASEAN Regional Knowledge Network on FLEG. Association of Southeast Asian Nations (ASEAN), Jakarta

Cashore B, Stone M (2012) Can legality verification rescue global forest governance? Analyzing the potential of public and private policy intersection to ameliorate forest challenges in Southeast Asia. Forest Policy Econ 18:13–22

Cashore B, Gale F, Meidinger E, Newsom D (2006) Forest certification in developing and transitioning countries. Environ Sci Policy Sust Dev 48(9):6–25

Cashore B, Auld G, Bernstein S, Mc Dermott C (2007) Can non-state governance 'ratchet up' global environmental standards? Lessons from the forest sector. Rev Eur Commun Int Environ Law 16(2):158–172

Cashore B, Galloway G, Cubbage F, Humphreys D, Katila P, Levin K, Maryudi A, McDermott C, McGinley K (2010) Ability of institutions to address new challenges. In: Mery G, Katila P, Galloway G, Alfaro RI, Kanninen M, Lobovikov M, Varjo J (eds) Forests and society – responding to global drivers of change. International Union of Forest Research Organizations (IUFRO), Vienna, pp 441–486

Chasek PS, Downie DL, Welsh Brown J (2010) Global environmental politics, 5th edn. Westview Press, Boulder

Cheng AS, Danks C, Allred SR (2011) The role of social and policy learning in changing forest governance: an examination of community-based forestry initiatives in the U.S. Forest Policy Econ 13:89–96

Chhatre A, Agrawal A (2009) Trade-offs and synergies between carbon storage and livelihood benefits from forest commons. Proc Natl Acad Sci 106(42):17667–17670

Colfer CJP (2005) The complex forest. Resources for the Future, Washington, DC

Colfer CJP, Capistrano D, Dabal GR, Moeliono M (eds) (2008) Lessons from forest decentralization. Money justice and the quest for good governance in Asia-Pacific. Earthscan, London

Conca K (2006) Governing water. Contentious transnational politics and global institutions building. MIT Press, Cambridge/London

Contreras-Hermosilla A (2007) Forest law enforcement and governance program. Review of implementation. World Bank, Washington, DC

Contreras-Hermosilla A, Gregersen HM, White A (2008) Forest governance in countries with federal systems of government. Lessons and implications for decentralization. Center for International Forestry Research (CIFOR), Bogor

Cotula L, Mayers J (2009) Tenure in REDD. Start-point or afterthought? vol 15, Natural Resource Issues. International Institute for Environment and Development (IIED), London

Cruz RVO, Tapia MA (2006) A review of the multi-sectoral forest protection committees in the Philippines, vol 6, ODI Forestry Briefing. Overseas Development Institute (ODI), London

Curran LM, Trigg SN, McDonald AK, Astiani D, Hardiono YM, Siregar P, Caniago I, Kasischke E (2004) Lowland forest loss in protected areas of Indonesian Borneo. Science 303:1000–1003

Currey D (2001) Timber trafficking: illegal logging in Indonesia, South East Asia and international consumption of illegally sourced timber. Retrieved 20 Feb 2002, from http://www.eia-international.org/campaigns/forests/reports/timber/index.html

Dauvergne P (1994) The politics of deforestation in Indonesia. Pac Aff 66(4):497–518

Davenport DS (2005) An alternative explanation for the failure of the UNCED forest negotiations. Global Environ Polit 5(1):105–130

del Río P (2010) Analysing the interactions between renewable energy promotion and energy efficiency support schemes: the impact of different instruments and design elements. Energy Policy 38(9):4978–4989

Dimitrov RS (2005) Hostage to norms: states, institutions and global forest politics. Global Environ Polit 5(4)

Dimitrov RS, Sprinz DF, DiGiusto GM, Kelle A (2007) International nonregimes: a research agenda. Int Stud Rev 9:230–258

Dove M (1983) Theories of swidden agriculture and the political economy of ignorance. Agroforest Syst 1:85–99

Dove MR (2006) Indigenous people and environmental politics. Annu Rev Anthropol 35:191–208

Dove MR (2011) The banana tree at the gate: a history of marginal peoples and global markets in Borneo. Yale University Press, New Haven

Dunoff JL (2007) Levels of environmental governance. In: Bodansky D, Brunnée J, Hey E (eds) The Oxford handbook of international environmental law. Oxford University Press, Oxford, pp 85–106

Dutschke M, Wertz-Kanounnikoff S, Peskett L, Luttrell C, Streck C, Brown J (2008) Financing REDD. Linking country needs and financing sources, vol 17, Info Brief. Center for International Forestry Research (CIFOR), Bogor

EFI (2009) FLEGT Asia regional programme. Supporting responsible trade for Asia's forest. European Forest Institute (EFI), Kuala Lumpur

Elliott C, Schlaepfer R (2001) Understanding forest certification using the advocacy coalition framework. Forest Pol Econ 2:257–266

Esty DC (2006) Toward good global governance: the role of administrative law. Yale Law J 115

Eucker D, Hein J (2010) Klimawandel in Südostasien. Die ASEAN als Wegbereiter einer regionalen Klimapolitik? vol 4, GIGA Fokus Asien. German Institute of Global and Area Studies (GIGA), Hamburg
FAO (2006a) Understanding national forest programmes. Guidance for practitioners. Food and Agriculture Organization of the United Nations (FAO), Rome
FAO (2006b) Global forest resources assessment 2005 – progress towards sustainable forest management, vol 147, FAO Forestry Paper. Food and Agriculture Organization of the United Nations (FAO), Rome
FAO (2009) State of the world's forests. Food and Agriculture Organization of the United Nations (FAO), Rome
FAO (2010) Global forest resources assessment 2010. Key findings. Food and Agriculture Organization of the United Nations (FAO), Rome
FAO, ITTO (2005) Best practices for improving law compliance in the forest sector, vol 145, FAO Forestry Paper. Food and Agriculture Organization of the United Nations (FAO), Rome
Fawzia F (2010) What's next after Copenhagen? A brief analysis on the outcomes of the Copenhagen climate conference and REDD-plus for ASEAN. ARKN-FCC Briefing Paper No. 5. Association of Southeast Asian Nations (ASEAN), Jakarta
Fogarty EA (2007) Transnational governance. In: Bevir M (ed) Encyclopedia of governance. Sage, London
Frost EL (2008) Asia's new regionalism. Lynne Rienner Publishers, Boulder
G8 (2008) Chair's summary G8 environment ministers meeting. Kobe, 24–26 May 2008
Galloway G, Katila P, Krug J (2010) The need for new strategies and approaches. In: Mery G, Katila P, Galloway G, Alfaro RI, Kanninen M, Lobovikov M, Varjo J (eds) Forests and society – responding to global drivers of change. International Union of Forest Research Organizations (IUFRO), Vienna, pp 441–486
GFEP (2009) CPF Global Forest Expert Panel (GFEP). Expert panel on the international forest regime. Terms of reference. Collaborative Partnership on Forests (CPF), Rome
Gibbs HK, Ruesch AS, Achard F, Clayton MK, Holmgren P, Ramankutty N, Foley JA (2010) Tropical forests were the primary sources of new agricultural land in the 1980s and 1990s. Proc Natl Acad Sci (PNAS) 107(38):16732–16737
Glück P, Rayner J (2009) Governance and policies for adaptation. In: Seppälä R, Buck A, Katila P (eds) Adaptation of forests and people to climate change – a global assessment report, vol 22, IUFRO World Series. International Union of Forest Research Organizations (IUFRO), Vienna
Glück P, Rayner J, Cashore B (2005) Changes in the governance of forest resources. In: Mery G, Alfaro R, Kanninen M, Lobovikov M (eds) Forests in the global balance – changing paradigms, vol 17, IUFRO World Series. International Union of Forest Research Organizations (IUFRO), Vienna, pp 39–74
Glück P, Angelsen A, Appelstrand M, Assembe-Mvondo S, Auld G, Hogl K, Humphreys D, Wildburger C (2010) Core components of the international forest regime complex. In: Rayner J, Buck A, Katila P (eds) Embracing complexity: meeting the challenges of international forest governance. A global assessment report prepared by the Global Forest Expert Panel on the international forest regime. International Union of Forest Research Organizations (IUFRO), Vienna, pp 37–55
Göhler D, Schwaab J (2009) Managing knowledge and regional policy advice: ASEAN forest clearing house mechanism. ASEAN high-level seminar on climate adaptation and mitigation. Association of Southeast Asian Nations (ASEAN), Jakarta, 23–25 Mar 2009
Göhler D, Liss B-M, Thunberg J (2009a) Implementation of forest law enforcement and governance (FLEG) through national forest programs (nfp). ARKN Policy Brief. Association of Southeast Asian Nations (ASEAN), Jakarta
Göhler D, McGill H, Obser A (2009b) Development interfaces: peer review and the development potential of regional institutions, policies and programs, vol 4, Exlibris series. ASEAN-German Regional Forest Programme (ReFOP), Jakarta

Gullison RE (2003) Does forest certification conserve biodiversity? Oryx 37(2):153–165
Gunningham N, Grabosky PN (eds) (1998) Smart regulation. Designing environmental policy. Oxford socio-legal studies. Oxford University Press, New York
Haas PM (2002) UN conferences and constructivist governance of the environment. Glob Gov 8:73–91
Harguindéguy J-B (2007) Policy learning. In: Bevir M (ed) Encyclopedia of governance. Sage, London
Héritier A (2003) New modes of governance in Europe: increasing political capacity and policy effectiveness? In: Börzel TA, Cichowski R (eds) The state of the European Union, vol 6, Law, politics, and society. Oxford University Press, Oxford, pp 105–126
Hinrichs A (2008) Training workshop on timber verification of legality systems, Brunei Darussalam. Briefing Paper No. 4. Association of Southeast Asian Nations (ASEAN), Jakarta, 21–23 July 2008
Hinrichs A (2009) Briefing note on ASEAN criteria and indicators for legality of timber, vol 5, Briefing Paper. Association of Southeast Asian Nations (ASEAN), Jakarta
Hogl K, Nordbeck R, Pregernig M (2008) GoFOR – new modes of governance for sustainable forestry in Europe. Project under the sixth EU framework programme for research and technology development. Publishable final activity report as of 13 Aug 2008. University of Natural Resources and Applied Life Science, Vienna
Holzinger K, Knill C, Sommerer T (2008) Environmental policy convergence: the impact of international harmonization, transnational communication, and regulatory competition. Int Organ 62(3):553–587
Howlett M (2002) Do networks matter? Linking policy network structure to policy outcomes: evidence from four Canadian policy sectors 1990–2000. Can J Polit Sci 35(2):235–267
Howlett M, Joshi-Koop S (2011) Transnational learning, policy analytical capacity, and environmental policy convergence: survey results from Canada. Glob Environ Chang 21(1):85–92
Howlett M, Ramesh M, Perl A (2009) Studying public policy. Policy cycles & policy subsystems. University Press, Oxford
Humphreys D (ed) (2006) Logjam: deforestation and the crisis of global governance. Earthscan, London
IFAD (2010) Rural poverty report 2011. International Fund for Agricultural Development (IFAD), Rome
IPCC (2007) Climate change 2007: synthesis report. Intergovernmental Panel on Climate Change (IPCC), Geneva
IUCN (2005) The beginning of the ENA-FLEG process in Russia: civil society insights. International Union for Conservation of Nature (IUCN), Moscow
Kamelarczyk K, Strandby U (2009) The effect of IPPKs on household livelihoods in Malinau district, East Kalimantan. In: Moeliono M, Wollenberg E, Limberg G (eds) The decentralization of forest governance; politics, economics and the fight for control of forests in Indonesian Borneo. Earthscan, London
Kaufmann D, Kraay A, Mastruzzi M (2010) The worldwide governance indicators: methodology and analytical issues. Global Economy and Development at Brookings Institutions, Washington, DC
Kern K, Jörgens H, Jänicke M (2000) Die Diffusion umweltpolitischer Innovationen. Ein Beitrag zur Globalisierung von Umweltpolitik. Forschungsstelle für Umweltpolitik (FFU). Freie Universität Berlin, Berlin
Koeng CH (2009) Policy brief on the impact of EU FLEGT action plan on ASEAN Member States intra- and international timber trade flow to EU. ARKN Policy Brief. Association of Southeast Asian Nations (ASEAN), Jakarta
Koeng CH, Malessa U (2009) Policy brief on the impact of the Lacy Act (amended 2008) on ASEAN Member States intra- and international timber trade flow to the USA. ARKN Policy Brief. Association of Southeast Asian Nations (ASEAN), Jakarta
Kooiman J (ed) (1993) Modern governance: new government – society interactions. Sage, London

Larson AM (2005) Democratic decentralization in the forestry sector. In: Colfer CJP, Capistrano D (eds) The politics of decentralization. Earthscan, London, pp 32–62
Lawson FH (2009) Comparative regionalism. Ashgate, Aldershot
Lawson S, MacFaul L (2010) Illegal logging and related trade. Indicators of the global response. Chatham House, London
Lejano R (2006) The design of environmental regimes: social construction, contextuality, and improvisation. Int Environ Agreem 6(2):187–207
Levin K, McDermott C, Cashore B (2008) The climate regime as global forest governance: can reduced emissions from deforestation and forest degradation (REDD) initiatives pass a 'dual effectiveness' test? Int For Rev 10(3):538–549
Levy MA, Young OR, Zürn M (1995) The study of international regimes. Eur J Int Relat 1 (3):267–330
Limberg G (2009) Impact of small timber harvest permits on villages in the Malinau watershed. In: Moeliono M, Wollenberg E, Limberg G (eds) The decentralization of forest governance; politics, economics and the fight for control of forests in Indonesian Borneo. Earthscan, London
Limberg G, Iwan R, Moeliono M, Sudana M, Wollenberg E (2007) Community-based forestry and management planning. In: Gunarso P, Setyawati T, Sunderland T, Shackleton C (eds) Managing forest resources in a decentralized environment. Center for International Forestry Research (CIFOR) & International Tropical Timber Organization (ITTO), Bogor
Malinau District Government (2007) Baseline data on carbon stocks; identification of potential forest use for carbon sequestration in Malinau, East Kalimantan. District Agency of Environmental Impact Mitigation (BAPEDALDA), Malinau District Government, Malinau
March JG, Olsen JP (1995) Democratic governance. Free Press, New York
Martin RM (2004) Regional approaches: bridging national and global efforts. Unasylva 55(3):3–8
Maxwell S, Stone D (2005) Global knowledge networks and international development. Bridges across boundaries. In: Stone D, Maxwell S (eds) Global knowledge networks and international development: bridges across boundaries. Routledge, London, pp 1–17
Mayntz R (2004) Governance im Modernen Staat. In: Benz A (ed) Governance – Regieren in komplexen Regelsystemen. Eine Einführung. Verlag für Sozialwissenschaften, Wiesbaden, pp 63–76
Mayntz R (2008) Von der Steuerungstheorie zu Global Governance. In: Schuppert GF, Zürn M (eds) Governance in einer sich wandelnden Welt. VS Verlag für Sozialwissenschaften, Wiesbaden, pp 43–60
McCarthy JF (2004) Changing to grey: decentralization and the emergence of volatile socio-legal configurations in Central Kalimantan, Indonesia. World Dev 32(7):1199–1223
McDermott CL, O'Carroll A, Wood P (2007) International forest policy – the instruments, agreements and processes that shape it. United Nations Forum on Forests (UNFF), New York
McDermott C, Cashore B, Kanowski P (2010) Global environmental forest policies. An international comparison. Earthscan, London
McDermott CL, Levin K, Cashore B (2011) Building the forest-climate bandwagon: REDD+ and the logic of problem amelioration. Glob Environ Polit 11(3):85–103
Moeliono M, Limberg G (2009) The geography of Malinau. In: Moeliono M, Wollenberg E, Limberg G (eds) The decentralization of forest governance; politics, economics and the fight for control of forests in Indonesian Borneo. Earthscan, London
Nanang M, Inoue M (2000) Local forest management in Indonesia: a contradiction between national forest policy and reality. Int Rev Environ Strateg 1(1):175–191
Nelson BJ (1996) Public policy and administration: an overview. In: Goodin RE, Klingemann H-D (eds) A new handbook of political science. Oxford University Press, New York, pp 551–592
Neo BS, Chen G (2007) Dynamic governance. Embedding culture, capabilities and change in Singapore. World Scientific Publishing, Singapore

Nesadurai HES, Stone D (2000) Southeast Asian research institutes and regional cooperation. In: Stone D (ed) Banking on knowledge. The genesis of the global development network. Routledge, London/New York, pp 183–201

Nomura K (2008) The politics of participation in forest management: a case from democratizing Indonesia. J Environ Dev 17(2):166–191

Nye JS (2004) Soft power. The means to success in world politics. Public Affairs, New York

Obser A, van Dijk K, Liss B-M, Wibowo PM, von Pfeil E (2005) Regionalization of the international forest policy dialogue: options for functions and structures (non-paper). Country led initiative (CLI) in support of the United Nations Forum on Forests (UNFF): scoping for a future agreement on forests. Federal Ministry of Consumer Protection, Food and Agriculture (BMVEL) and Federal Ministry for Economic Cooperation and Development (BMZ), Berlin, 16–18 Nov 2005

OECD (2005) Regional integration in the Asia Pacific. Organisation for Economic Cooperation and Development (OECD), Paris

Oikonomou V, Flamos A, Grafakos S (2010) Is blending of energy and climate policy instruments always desirable? Energy Policy 38(8):4186–4195

Öjendal J (2001) Southeast Asia at a constant crossroads: an ambiguous 'new region'. In: Schulz M, Söderbaum F, Öjendal J (eds) Regionalization in a globalizing world. A comparative perspective on forms, actors and processes. Zed Books, London/New York, pp 147–172

Okereke C, Dooley K (2010) Principles of justice in proposals and policy approaches to avoided deforestation: towards a post-Kyoto climate agreement. Glob Environ Chang 20:82–95

Ostrom E (1990) Governing the commons. The evolution of institutions for collective action. Cambridge University Press, Cambridge

Ostrom E (2000) Collective action and the evolution of social norms. J Econ Perspect 14 (3):137–158

Ostrom E (2007) Institutional rational choice. An assessment of the institutional analysis and development framework. In: Sabatier PA (ed) Theories of the policy process. Westview Press, Boulder

Ostrom E (2009) A polycentric approach for coping with climate change. Policy research working paper 5095. Background paper to the 2010 World Development Report. World Bank, Washington, DC

OTCA, CCAD, FAO/COFLAC (2009) Regionalization of the forest dialogue. The Puembo initiative. Lessons learnt and outlook. XIII World Forestry Congress, side event. Buenos Aires, 19 Oct 2009

Palmer C (2009) Who, how and for what? Negotiations and outcomes from IPPK agreements between communities and brokers in Malinau. In: Moeliono M, Wollenberg E, Limberg G (eds) Decentralization of forest governance; politics, economics and the fight for control of forests in Indonesian Borneo. Earthscan, London

Pedroni L, Dutschke M, Streck C, Porrura ME (2009) Creating incentives for avoiding further deforestation: the nested approach. Clim Policy 9:207–220

Peluso NL (1992) Rich forests, poor people. Resource control and resistance in Java. University of California Press, Berkeley/Los Angeles

Peluso NL (1993) Coercing conservation? The politics of state resource control. Glob Environ Chang 3(2):199–217

Peluso NL, Harwell E (2001) Territory, custom, and the cultural politics of ethnic war in West Kalimantan, Indonesia. In: Peluso NL, Watts M (eds) Violent environments. Cornell University Press, Ithaca, pp 83–116

Pescott MJ, Durst PB, Leslie RN (eds) (2010) Forest law enforcement and governance: progress in Asia and the Pacific. Asia-Pacific Forestry Commission (APFC), Bangkok

Peskett L, Huberman D, Bowen-Jones E, Edwards G, Brown J (2008) Making REDD work for the poor. A Poverty Environment Partnership (PEP) report. https://cmsdata.iucn.org/downloads/redd_paper___web.pdf

Pfaff A, Sills EO, Amacher GS, Coren MJ, Lawlor K, Streck C (2010) Policy impacts on deforestation. Lessons learned from past experiences to inform new initiatives. Nicholas Institute for Environmental Policy Solutions, Duke University, Durham
Phelps J, Webb EL, Agrawal A (2010) Does REDD+ threaten to recentralize forest governance? Science 328:312–313
Provan KG, Kenis P (2007) Modes of network governance: structure, management, and effectiveness. J Public Admin Res Theory 18:229–252
Pülzl H, Rametsteiner E (2002) Grounding international modes of governance into national forest programmes. Forest Pol Econ 4(4):259–268
Radaelli CM (2003) The open method of coordination: a new governance architecture for the European Union? Swedish Institute for European Policy Studies (SIEPS), Stockholm
Rayner J, Howlett M (2004) National forest programmes as vehicles for next generation regulation. In: Humphreys D (ed) Forests for the future. National forest programmes in Europe. Country and regional reports from COST Action E19. European Cooperation in the Field of Scientific and Technical Research (COST), Luxembourg
Reinicke WH, Deng F, Witte JM, Benner T, Whitaker B, Gershman J (2000) Critical choices. The United Nations, networks, and the future of global governance. International Development Research Centre (IDRC), Ottawa
Rhee S (2009) The cultural politics of collaboration to control and access forest resources in Malinau, East Kalimantan. In: Moeliono M, Wollenberg E, Limberg G (eds) The decentralization of forest governance; politics, economics and the fight for control of forests in Indonesian Borneo. Earthscan, London
Rhodes RAW (1997) Understanding governance. Policy networks, governance, reflexivity and accountability. Open University Press, Buckingham/Philadelphia
Ribot JC (2005) Choosing representation: institutions and powers for decentralized natural resource management. In: Colfer CJP, Capistrano D (eds) The politics of decentralization. Forests, power and people. Earthscan, London
Ribot JC, Agrawal A, Larson AM (2006) Recentralizing while decentralizing: how national governments reappropriate forest resources. World Dev 34(11):1864
Risse T (2007) Zur “Reisefähigkeit” des Governance-Konzeptes. SFB-Governance Working Paper Series Nr. 5. Deutsche Forschungsgemeinschaft (DFG) Sonderforschungsbereich (SFB) 700, Berlin
Risse T, Lehmkuhl U (2006) Governance in areas of limited statehood – new modes of governance? SFB-Governance Working Paper Series No. 1. Deutsche Forschungsgemeinschaft (DFG) Research Center (SFB) 700, Berlin
Rittberger V, Zangl B, Kruck A (2012) International organization, 2nd edn. Palgrave Macmillan, Basingstoke
RLI (2008) Report of the Australian – Swiss region led initiative on regional input in support of the United Nations Forum on Forests. Region led initiative (RLI) co-hosted by the governments of Australia and Switzerland. Geneva, 28–30 Jan 2008
Robledo C, Blaser J, Byrne S, Schmidt K (2008) Climate change and governance in the forest sector: an overview of the issues on forests and climate change with specific consideration of sector governance, tenure and access for local stakeholders. Intercooperation, Berne
Rosenau JN (2006) The study of world politics, vol 2, Globalization and governance. Routledge, London
Rosendal KG (2001) Overlapping international regimes. The case of the International Forum on Forests (IFF) between climate change and biodiversity. Int Environ Agreem Polit Law Econ 1 (4):447–468
Sabatier PA (1988) An advocacy coalition framework of policy change and the role of policy-oriented learning therein. Policy Sci 21:129–168
Sabatier PA, Jenkins-Smith HC (1999) The advocacy coalition framework. An assessment. In: Sabatier PA (ed.) Theories of the policy process. Westview Press, Boulder, pp. 117–166

Salafsky N, Margoluis R (2004) Using adaptive management to improve ICDPs. In: McShane TO, Wells MP (eds) Getting biodiversity projects to work. Columbia University Press, New York, pp 372–394
Salmi J, Samyn J-M (2008) National action plans to combat illegal logging and other forest crimes – progress and lessons learnt. Expert meeting on forest law enforcement and governance – progress and lessons learned. Bucharest, 13–15 Oct 2008
Sanderson I (2002) Making sense of 'what works': evidence based policy making as instrumental rationality? Public Policy Admin 17(3):61–75
Sandker M, Suwarno A, Campbell B (2007) Will forests remain in the face of oil palm expansion? Simulating change in Malinau, Indonesia. Ecol Soc 12(2):37
Saunders J, Nussbaum R (2007) Forest governance and reduced emissions from deforestation and degradation (REDD). Chatham House, London
Scharpf F (1993) Positive und negative Koordination in Verhandlungssystemen. In: Héritier A (ed) Policy-Analyse. Kritik und Neuorientierung (PVS-Sonderheft 24/1993). Westdeutscher Verlag, Opladen, pp 57–83
Schuppert GF (2008) Governance –auf der Suche nach Konturen eines "anerkannt uneindeutigen Begriffs". In: Schuppert GF, Zürn M (eds) Governance in einer sich wandelnden Welt. VS Verlag für Sozialwissenschaften, Wiesbaden
Scott JC (1998) Seeing like a state; how certain schemes to improve the human condition have failed. Yale University Press, New Haven
Seymour F (2008) Forests, climate change, and human rights: managing risk and trade-offs. Human rights and climate change. Center for International Forestry Research (CIFOR), Bogor
Sikor T, Stahl J, Enters T, Ribot JC, Singh N, Sunderlin WD, Wollenberg L (2010) REDD-plus, forest people's rights and nested climate governance. Glob Environ Chang 20(3):423–425
Singer B (2008) Putting the national back into forest-related policies: the international forests regime and national policies in Brazil and Indonesia. Int For Rev 10(3):523–537
Siry J, Cubbage F, Ahmed M (2005) Sustainable forest management: global challenges and opportunities. Forest Policy Econ 7(4):551–561
Soriaga R (2010) A review and analysis of the impacts of forest law enforcement and governance (FLEG) on sustainable livelihoods. ASEAN Regional Knowledge Network on FLEG Policy Brief. Association of Southeast Asian Nations (ASEAN), Jakarta
Soriaga R, Cashore B (2009) Contribution and effectiveness of local governments in forest law enforcement and governance in ASEAN. ASEAN Regional Knowledge Network on FLEG Policy Paper. Association of Southeast Asian Nations (ASEAN), Jakarta
Springate-Baginski O, Wollenberg E (eds) (2010) REDD, forest governance and rural livelihoods. The emerging agenda. Center for International Forestry Research (CIFOR), Bogor
Stone D (2000) Non-governmental policy transfer: the strategies of independent policy institutes. Gov Int J Policy Adm 13(1):45–70
Stone D (2005) Knowledge networks and global policy. In: Stone D, Maxwell S (eds) Global knowledge networks and international development: bridges across boundaries. Routledge, London, pp 89–105
Streck C (2005) Governments and policy networks: chances, risks, and a missing strategy. In: Wijnen F, Zoeteman K, Pieters J (eds) A handbook of globalization and environmental policy. Egward Elgar, Cheltenham
Streck C (2010) Reducing emissions from deforestation and forest degradation: national implementation of REDD schemes. Clim Change 100(3):389–394
Sudana M (2009) Winners take all: understanding forest conflict in the era of decentralization in Indonesia. In: Moeliono M, Wollenberg E, Limberg G (eds) The decentralization of forest governance; politics, economics and the fight for control of forests in Indonesian Borneo. Earthscan, London
Tacconi L (ed) (2007) Illegal logging: law enforcement, livelihoods and the timber trade. Earthscan, London

TFD (2005) Practical actions to combat illegal logging. The Forests Dialogue, Yale University, New Haven

Thang HC (2008a) Linkages between the ASEAN criteria and indicators for sustainable forest management of tropical forests and the Intergovernmental Panel on Forests (IPF)/ Intergovernmnetal Forum on Forests (IFF) proposals for action. Association of Southeast Asian Nations (ASEAN), Jakarta

Thang HC (2008b) Forest law enforcement, governance and trade in Asia. An update. VERIFOR project/FAO international workshop on legality of traded timber: the development challenges. Rome, 24–26 Nov 2008

Thang HC (2009) Managing regional expert pools through regional knowledge networks in ASEAN. In: Van Wart M, Göhler D, Fawzia F (eds) Seminar proceedings ASEAN high-level seminar. Climate change adaptation and mitigation: towards a cross-sectoral programme approach in ASEAN. Association of Southeast Asian Nations (ASEAN), Jakarta

Thang HC (2010) The use of the ASEAN peer consultation framework (PCF) in forest law enforcement and governance (FLEG) assessment in the Philippines. Association of Southeast Asian Nations (ASEAN), Jakarta

Theisohn T, Land T (2009) Connecting communities of practice. Practice reports capacity development knowledge architecture, issue 36. http://www.capacity.org/en/layout/set/print/content/view/full/5533. Accessed on 18 Dec 2009

Thorburn C (2002) Regime change–prospects for community-based resource management in post-new order Indonesia. Soc Nat Resour 15:617–628

UNFCCC (2008) Report of the conference of the parties on its thirteenth session, held in Bali, from 3 to 15 Dec 2007. Addendum part two: action taken by the conference of the parties at its thirteenth session. United Nations Framework Convention on Climate Change (UNFCC), Bonn

Visseren-Hamakers IJ, Glasbergen P (2007) Partnerships in forest governance. Glob Environ Chang 17:408–419

Vogler J (2005) In defense of international environmental cooperation. In: Barry J, Eckersley R (eds) The state and the global ecological crisis. The MIT Press, Cambridge/London, pp 229–253

Wollenberg E, Moeliono M, Limberg G (2009) Between state and society: decentralization in Indonesia. In: Moeliono M, Wollenberg E, Limberg G (eds) The decentralization of forest governance; politics, economics and the fight for control of forests in Indonesian Borneo. Earthscan, London

World Bank (2006a) Strengthening forest law enforcement and governance. Addressing a systematic constraint to sustainable development. World Bank, Washington, DC

World Bank (2006b) The Antalya workshop – from declaration to action. Implementing the St. Petersburg declaration on forest law enforcement and governance (FLEG) in Europe and North Asia (ENA). Antalya, 16–18 May 2006

World Bank (2006c) Guidelines. Formulation and implementation of national action plans to combat illegal logging and other forest crime. Draft for discussion, revised June 2006. World Bank technical papers, sustainable development series. World Bank, Washington, DC

World Bank (2009a) World development report 2010. Development and climate change. World Bank, Washington, DC

World Bank (2009b) Roots for good forest outcomes: an analytical framework for governance reforms. World Bank, Washington, DC

World Bank (2010) World development indicators 2010. World Bank, Washington, DC

WRI (2009) The governance of forests initiative. Second meeting of the ASEAN Regional Knowledge Network on FLEG. Association of Southeast Asian Nations (ASEAN), Manila, 10–12 Feb 2009

Yasmi Y, Broadhead J, Enters T, Genge C (2010) Forest policies, legislation and institutions in Asia and the Pacific. Trends and emerging needs for 2020. Asia-Pacific forestry sector outlook

study II. Working Paper Series No. APFSOS II/WP/2010/34. Bangkok: Food and Agriculture Organization of the United Nations (FAO)
Young OR (2002) The institutional dimensions of environmental change. Fit, interplay, and scale. MIT Press, Cambridge
Young W (2010) Rio conventions redux: an argument for merging the trio into a single convention on environmental management. J Sustain Dev 4(1):134–154
Young OR, Leslie AK, Schroeder H (eds) (2008) Institutions and environmental change. Principal findings, applications, and research frontiers. MIT Press, Cambridge
Yusuf AA, Francisco H (2009) Climate change vulnerability mapping for Southeast Asia. Economy and Environment Program for Southeast Asia (EEPSEA), Singapore
Zürn M (2008) Governance in einer sich wandelnden Welt – eine Zwischenbilanz. In: Schuppert GF, Zürn M (eds) Governance in einer sich wandelnden Welt. VS Verlag für Sozialwissenschaften, Wiesbaden, pp 553–580
Zürn M, Koenig-Archibugi M (2006) Conclusion II: the modes and dynamics of global governance. In: Koenig-Archibugi M, Zürn M (eds) New modes of governance in the global system. Exploring publicness, delegation and inclusiveness. Palgrave Macmillan, New York, pp 236–254

Chapter 13
Prospects for Forest-Based Rural Development

Jürgen Pretzsch, Dietrich Darr, André Lindner, Holm Uibrig, and Eckhard Auch

Keywords Co-evolution • Institutions • Action orientation • Innovation • Scenarios • Value chains • Green economy • Forestry development theory

This synopsis provides an outlook for the complex issue of tropical forestry and rural development. It is based on the previous chapters, and incorporates historical facts, rural development politics and policies, and management systems and instruments. The mix between factual and instrumental aspects rules out a simple conceptual framework. In this synopsis the co-evolution model presented at the outset is taken up, to represent the complex reality of tropical forestry and rural development (Fig. 13.1). The highly dynamic and action-oriented character of this outlook is rooted in the discussion of paradigm changes, leading to design elements for a prospective tropical forestry and rural development theory.

The classical concept of sustained yields from forests and the comprehensive sustainable forestry concept (Monserud 2003, p. 36) do not wholly satisfy the unique philosophy required to design forest management and forest-based development strategies for the tropics. The numerous external influencing factors and trade-offs between distinctive objectives such as biodiversity conservation and poverty alleviation are only barely integrated. Present day long term planning is extremely vague and measures adopted to optimize planning are often substituted in favor of the principle of keeping open reasonable and realistic options for the future. The co-evolution model accounts for the necessarily adaptive character of forest

J. Pretzsch (✉) • A. Lindner • H. Uibrig • E. Auch
Institute of International Forestry and Forest Products, Technische Universität Dresden, Postfach 1117, Tharandt 01735, Germany
e-mail: juergen.pretzsch@forst.tu-dresden.de; andre.lindner1@mailbox.tu-dresden.de; holm.uibrig@forst.tu-dresden.de; eckhard.auch@tu-dresden.de

D. Darr
Faculty of Life Sciences, Hochschule Rhein-Waal, Marie-Curie-Str. 1, Kleve 47533, Germany
e-mail: dietrich.darr@hochschule-rhein-waal.de

J. Pretzsch et al. (eds.), *Forests and Rural Development*, Tropical Forestry 9, DOI 10.1007/978-3-642-41404-6_13, © Springer-Verlag Berlin Heidelberg 2014

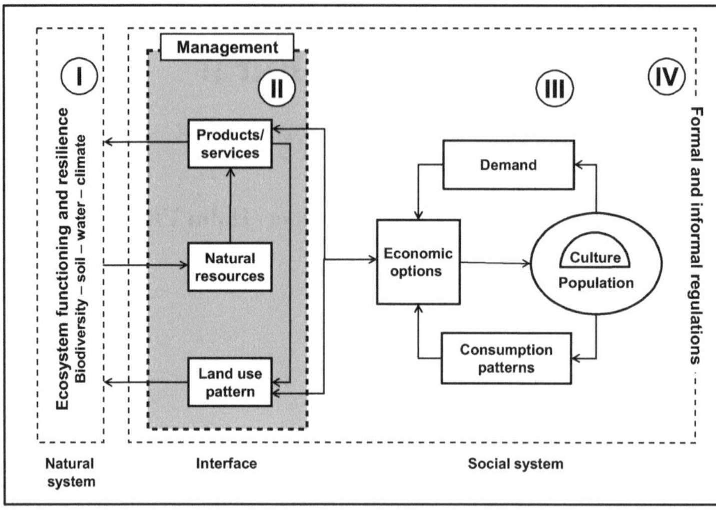

Fig. 13.1 The co-evolution model as a national model for developing countries. Roman numerals indicate the spheres of action for future tropical forest management and rural development (Source: Derived from Hurni and Messerli (1981), Norgaard (1994), Berkes and Folke (1998))

management. It also coincides well with the perspective that ecological and human systems are resilient (Chap. 1; Norgaard 1994). In accordance with the spheres of action presented in Fig. 13.1, the outlook is structured into four complexes.

Complex I deals with ecosystem functioning and the corresponding interactions. As a consequence of the numerous anthropogenic and natural influences, forest ecosystems are permanently under threat. Although one third of the global land area is still covered by forest, deforestation and land degradation are ongoing and a large part of the forests in tropical countries are not integrated in sustainable management regimes. Influences that are difficult to predict, such as weather conditions related to climate change, fire and direct disturbances by humans, are increasing (Chap. 3). As a direct consequence, there is a great risk that ecosystem services are not provided in a sustainable way. A continuous supply of forest products and drinking water, biodiversity conservation and climate regulation can no longer be assured. The conventional understanding is that the functioning of an ecosystem depends on the combined, related activities of its subsystems. Our understanding of the disruption caused by species loss and the resultant forfeiture of the corresponding goods and services provided by the ecosystem is still limited, however (Rinawati et al. 2013). Ecosystems respond to cumulative alterations. Despite all of the advantages provided by modeling, up to now the effects brought about by gradual changes remain unpredictable. These changes influence the carrying capacity and overall resilience of the system. Where natural systems such as forests are exposed to fast paced, high impact events, they may degrade to a lower state or even collapse. The shift to a lower state and towards a less resilient ecosystem results in a decreased supply of services and resources (Scheffer et al. 2001). The overall coherency between ecosystems and social systems remains complex; the causes and effects are difficult to forecast. Predictions made for ecosystem mechanisms are affected by individual human actions. The difficulty predicting precisely these impacts in space and time, and their extent, leads to increased uncertainty. Where prediction is not really possible, it may be an option to use scenarios as a framework for the identification and development of applicable measures (Carpenter 2002). Balanced future development requires that fundamental adjustments occur at all levels in the management of ecological and social systems.

Complex II focuses on the need to develop mechanisms for the adaptation of forest and tree-based management systems to the changing ecological and socio-economic environment. The growing global demand for forest and tree-based products and services coincides with the expanding complexity of land use patterns and results in the involvement of an increasing number of stakeholders. As a consequence, innovative forest management systems have evolved dynamically in recent decades (Chap. 5). Enterprises of all kinds have to build capacity in terms of organization, controlling and change management. Examples are systems based on state/community relations such as joint forest management and collaborative forest management schemes, and different types of community or association-based forestry (Ostrom 1990). The increasing competition for land has led to the establishment of different forms of private-private partnerships like out-grower schemes. Big enterprises contract small farmers to enlarge their resource base following the principle of economies of scale. State forestry bodies employ private accounting systems to become more competitive. Particular attention must be focused on the development paths of small farms as these are responsible for the management of a major proportion of the forests and trees in developing countries. In the past, these farms were frequently marginalized under rural development strategies, in spite of the fact that they support rural livelihoods, ensure ecological services and provide food for urban populations (Ashley and Maxwell 2001). Diversification strategies, as a means of adaptation to the effects of external influences like climate change and the globalization of agricultural markets, need to be advanced (Ellis 1998). Home-gardens and other agroforestry practices meet this requirement, but under these complex systems labor productivity is comparatively low. Where production involves a huge number of heterogeneous products, the options for mechanization and market access are limited (Chap. 4). For this reason, the further development of management systems must follow a holistic approach. The creation of social capital and the strengthening of trade chains are indispensable complementary activities. Landscape-oriented forest management models must be developed, and integrated within a theory framework that follows the requirements for continuous adaptation to a changing framework.

With the rising world population and the gradual substitution of energy derived from non-renewable sources by renewable sources of energy, forest and tree-based products and services are growing in importance. In the meantime, however, the coordination of the respective markets and value chains is lagging behind. The wide spectrum of products and services from trees and forests requires tailor-made commercialization and distribution channels (Chap. 6). Value chain development is more likely to be successful if directed towards existing markets, or market niches, that can be developed easily. This is more difficult in the case of products with an unclear or unknown market potential. The creation of functioning markets, encouragement of entrepreneurial activities, and higher rates of market participation by smallholders induce overall socio-economic development in forest dominated areas. Value chains must be integrated in rural development concepts as components of a holistic model of a 'green economy' opening up new development options. Using this model it is unrealistic to drive development by means of subsistence economy alone, embedded only in local traditional networks. The supply of goods and services, the market demand, financial incomes and the interaction with urban centers already play a key role in rural economies. This leads to increasing expectations with respect to livelihood, especially on the part of young people, and a corresponding rapid change to consumption patterns. The volatile nature of international markets must be buffered by a certain diversification of value added chains and the creation of a regional green economy.

Technical and organizational innovation is the most important prerequisite for rural development in the rapidly globalizing world. Initiatives are doomed to failure if rural populations are not convinced of the benefits and long term profitability of the related activities. Innovative approaches are more likely to be adopted if they have a clear market potential and offer commercial benefits (Ellis and Biggs 2001). This in turn can only be realized where there are inventive, skilled and motivated laborers. In rural areas the creation of job opportunities on one's own farm, in small scale industry and business, or within cooperatives and associations is a pre-condition for maintaining a minimum level of infrastructure and options for future development. Only by significantly increasing livelihood development can the exodus of rural young people be controlled and reduced. Outward migration may lead to sparsely populated regions, which can result in fundamental shifts in land use management, towards reduced-labor technologies. Inherent in the green economy concept is a profound trade-off in the choice of technologies. Labor intensive technologies create jobs and may be environmentally adapted, but often they are perceived as being backward. Advanced industrial technologies are capital intensive. They lead to high technology standards and the generation of sophisticated products. However, labor demand is low and often the rural labor force is not sufficiently skilled. The trade-off between both paths has been an important point of discussion in recent times. The holistic model of a green economy employed in tropical countries may overcome this contradiction, because both types of technology may be applied at the different levels of product and service chains (UNEP 2011).

Complex III deals with social systems, which are determined by location-specific cultural traditions and practices. Commercial global forest management thinking often clashes with these specific local practices (Chap. 7). This may be related to the conservation of sacred forests, subsistence use, traditional local institutions and reciprocity mechanisms (Agrawal 1995; Berkes 2012). Increase in the complexity of the use of forests and trees, as well as new challenges to future-oriented rural development, require innovative strategies linked to specific state interventions and instruments. At this instrumental level, future development paths can be built on the rich experience gleaned over decades. The increasing complexity of land use combining small scale land utilization and advanced technologies in large, market-oriented enterprises will pose a challenge to future rural development and land use planning (see Chap. 11). In addition, the stronger linkage between goods and services provided by rural regions and consumption in urban centers serves to increase the number of relevant stakeholders; and stakeholders will often possess distinct and often competing interests. In striving for sustainable land use, conventional technocratic and top-down planning approaches are insufficient to reconcile the diverse stakeholder interests (Uphoff 1993). Equally, and contrary to initial expectations, the various recent participatory or so-called bottom-up political planning approaches, based on the involvement of technocrats and various stakeholders, including the land users, have not provided the necessary results either. Often these are linked to short term projects and once funding stops the institutional setting collapses. Such projects must build primarily on endogenous dynamics and motivations. Innovative approaches characterized by flexibility and lasting adaptation to changing conditions and risks have been developed and are to be applied in rural regions, involving the combination of ecological, social and economic aspects at various spatial scales (Escobar 1988). This adaptive planning is expected to prove a successful tool in the context of increasing threats of extreme natural events such as drought, flood, storm, variable dry and rainy seasons and higher temperatures, as well as a rapidly evolving human society. An emphasis is placed on common learning along causal event chains and innovative economic branches, which must involve the people affected and requires a new type of planner (Gelder and O'Keefe 1995; Habermas 1988). The necessary skill set includes not only profound technical knowledge but also excellent communication, conflict resolution and leadership capabilities. Furthermore, planning will preferably function as a continued adaptive process rather than adhering strictly to a particular blueprint.

Innovative extension approaches have developed towards a synthesis of local knowledge and external interventions based on advanced, mostly Western technologies (Chap. 8). To cope with changing ecological and human conditions, innovative technologies need to be developed and implemented. In many cases they are still imposed by Westerners, today including modern China (Fals-Borda and Mora-Osejo 2003; Ferguson 2012). Endogenous strategies require more attention. There have been few studies to investigate systematically the technology gap between Western and developing countries. The funding conditions in projects may be insufficient over time to motivate a real appreciation of the impacts of external technology transfer. For this reason, specific local and regional knowledge must be activated and integrated in innovation processes. The creation of social capital within the scientific community plays an essential part in the initiation of successful cooperation in developing countries in the longer term. The creation of knowledge typically constitutes a collaborative process. Farmers, research institutions, entrepreneurs and other private sector players, authorities, consumers and NGOs need to interact in order to develop, improve, disseminate and adopt technological, social and organizational innovations in rural areas. Fostering opportunities for collaboration and knowledge exchange between these stakeholders usually requires the conscious effort of a dedicated innovation broker and should become a core task of rural advisory services. These services can help rural people to compile, consolidate and document the explicit and tacit knowledge accumulated locally over generations, to assess its usefulness in light of technological progress and advancing globalization, and to cross-fertilize modern scientific knowledge with the wisdom and experience of rural populations (Warren et al. 1995). Leveraging the wealth of local knowledge can also help maintain the niches and biotopes that foster socio-cultural diversity and help build a counterpoise to the homogenizing and unifying forces of globalization.

Adaptive planning and extension facilities are basic prerequisites of successful decentralization and devolution processes. The last two decades were characterized by a shift of forest management rights and responsibilities from state institutions to local actors such as communities, user groups and individuals. This shift has created new livelihood and development options at the local level. However, in many cases the land handed over to local communities and individuals was degraded and they first had to invest substantial labor and even capital resources in the regeneration and rehabilitation of the land and forests. Thus, the evaluation of the impacts of decentralization requires a holistic analysis, whereby further investigation is especially necessary in relation to the congruence of costs and benefits and benefit sharing rules.

Financing instruments for forest and tree-related activities in tropical and subtropical countries are still at an initial stage of development. The reasons for this are the long duration of capital fixation, high natural risks and low returns on investment in comparison to agriculture (Chap. 10). Investments in medium and large scale forestry have increased during recent global financial and economic crises, with funds coming mainly from the private sector. Given the low or even negative returns from conventional investments, institutional investors have increasingly reshuffled their portfolios and invested in assets that are more independent of the overall macro-economic situation, such as investments in forest plantations. In light of the global environmental and sustainability debate, small private investors are increasingly taking into consideration the social and environmental impacts of their investment decisions, leading to the emergence of a sustainable investment sector, in which forestry projects play a major role. Despite the fact that these trends have the potential to significantly contribute to social and economic development in tropical and subtropical forest regions, questions arise as to how stable these funding sources will be in the long term. The relatively modest profitability, high risk and long horizon of many forestry projects mean this is not always an attractive option. Due to the mobility and proverbial flightiness of international capital in a globalized world, investors might in time return to other more profitable investments. This could happen once the economic conditions have stabilized, or when new trends and fashions make alternative investments more lucrative. Tightening the regulative framework for global capital flows so as to increase incentives for long term, responsible investments may serve to benefit forestry projects and rural development.

The initial euphoria surrounding the valuation of non-market services from forestry by means of payment for environmental service (PES) instruments has dimmed and assessments have become more realistic (Chap. 9). To date, most of the established PES schemes are still governed by subsidies and not by market demand and respective payments, as was once expected. This is especially true for the instruments related to international conventions.

In addition to adequate technologies, market chains, education and infrastructure, the development of social capital is another important cornerstone in the creation of innovation capacity. Investigations of how different forms of social capital may create nuclei for rural development and integrated forestry, how these institutions may be installed between tradition and modernity, and with a degree of high flexibility, is a challenging and urgent task. For reasons that are historically justified, cooperatives have a largely negative image. As a consequence, the concept requires renovation, based on positive experiences. Ultimately cooperatives may prove to be very effective instruments in coping with climate change and other threats to rural development, and support in the creation of innovative product chains. Integrated within local networks, they can respond to market demands in a flexible way, buffer negative effects of globalization such as price fluctuations and thus contribute to local development by providing better livelihoods.

Complex IV deals with institutions, which are the most important drivers of rural development (North 1990, 1991). The co-evolution model is strongly influenced and driven by formal and informal institutions. These are determined by a mix of policy decisions, politics and development models and paradigms. It has been demonstrated that rural development processes, and the historical development of tropical forestry, can be well explained by political ecology studies, comprising historical facts and the alteration of power structures (Blaikie 1985; Peluso 1992; Peet and Watts 1996). Historical steps and respective underlying models were characterized as paradigms (Chap. 2). The number of actors involved in tropical forest management systems and policy making has increased substantially over time. Studies are necessary to understand in greater depth the respective interactions between state institutions, various types of communities, private actors and civil society. Interaction and common organization are indispensable tools in the formulation and implementation of an innovative rural development policy. Forestry actors, who often keep themselves relatively isolated, must actively and fully integrate in the policy and decision making networks of rural development. The local development patterns discussed, such as culture, economics and social organization, forest management systems and the instruments of rural development like regional and land use planning, extension, financing and the decentralization of decision making, should all play a part in the overall rural land use policy, which is closely linked to rural development policy.

Summing up, the review of rural development and its differentiation in paradigms has permitted a theory-oriented explanation of rural development and tropical forestry. In recent times the underlying development theories and models have been gradually replaced by instruments with limited outreach and preferably short term impacts; often these are merely market mechanisms. This circumstance gives rise to a great need for innovative theory development. Markets tend to under-deliver in terms of social justice, poverty alleviation and/or environmental sustainability. Rural development, therefore, cannot be left to market forces alone

(Eucken 1990). Phenomena such as information imbalances, distorted market structures and external effects diminish the market power of smallholders and typically work to the disadvantage of rural populations.

In most countries of the world, rural regions are undergoing fundamental changes. In tropical and subtropical countries regional and technological dualism has led to an extremely unequal development between rural and urban areas in terms of educational facilities, infrastructure, health care and market access. A primary task of government is to establish framework conditions that allow rural populations to participate in markets on an equal footing and to reap the benefits of increased market involvement. Providing and enforcing the rules and legal (Choussudovsky 1997; McCarthy and Prudham 2003) frameworks required to create a level playing field and to balance these structural disadvantages is, therefore, an indispensable precondition of successful forest-based rural development.

This issue is exemplified by the market euphoria accompanying the implementation of international conventions. While the political and scientific communities dominate the major part of the corresponding negotiations, the local stakeholders from the rural population remain largely unrepresented and must contend with insufficient communication. An accompaniment to these conditions is the enduring invisibility of existing trade-offs and conflicts. The local implementation of convention-based measures is as a result often short term and lacking in sustainability.

This is in accordance with the outcomes of many projects failing to live up to expectations regarding forestry development. An explanation might be found in the negative consequences of mainly Western interventions that have resulted in deforestation, forest degradation and the misuse of resources. Therefore, greater emphasis must be placed on endogenous strategies, largely independent of technology transfer and external support. Positive examples of development succeeding without a predomination of Western ideologies offer potential (Roe 1994). These must accompany the prospective development of new models and theories, and eventually allow for the formulating of a new paradigm for future development of tropical forestry.

Experience gleaned from two decades of market dominance has revealed the need for common rules at the state and the community level. Foresters must enter into this discourse proactively and provide best practice cases to underline the potential that resides in future forest and tree management. Any such approach should be capable of successfully combining forestry with socio-economic and cultural development in rural regions in the tropics and subtropics. Furthermore, the current dominance of rather simple market mechanisms must be replaced by innovative, complex forestry development theories, and allow for consideration of an advanced combination of regulating state interventions, market mechanisms and a more integrative involvement of civil society and local communities.

References

Agrawal A (1995) Dismantling the divide between indigenous and scientific knowledge. Dev Change 26:413–439

Ashley C, Maxwell S (2001) Rethinking rural development. Dev Policy Rev 19(4):395–425

Berkes F (2012) Sacred ecology, 3rd ed. Routledge, New York/London

Berkes F, Colding J, Folke C (eds) (1998) Linking social and ecological systems. Management practices and social mechanisms for building resilience. Cambridge University Press, Cambridge

Blaikie P (1985) The political economy of soil erosion in developing countries. Longman, London

Carpenter SR (2002) Ecological futures: building an ecology of the long now. Ecology 83(8): 2069–2083

Choussudovsky M (1997) The globalisation of poverty: impacts of IMF and World Bank reforms. Third World Network, Penang

Ellis F (1998) Household strategies and rural livelihood diversification. J Dev Stud 35:1–38

Ellis F, Biggs S (2001) Evolving themes in rural development 150s–2000s. Dev Policy Rev 19(4): 437–448

Escobar A (1988) Power and visibility: development and the intervention and management of the Third World. Cultur Anthropol 3(4):428–443

Eucken W (1990) Grundsätze der Wirtschaftspolitik, 6th edn. J. C. B. Mohr (Paul Siebeck), Tübingen

Fals-Borda O, Mora-Osejo LE (2003) Context and diffusion of knowledge. A critique of Eurocentrism. Action Res 1(1):29–37

Ferguson N (2012) Civilization. The six killer apps of western power. Penguin Books, London

Gelder BV, O'Keefe P (1995) The new forester. Intermediate Technology, London

Habermas J (1988) Theorie des kommunikativen Handelns, 1 & 2nd edn. Suhrkamp, Frankfurt

Hurni H, Messerli B (1981) Mountain research for conservation and development in Simen, Ethiopia. Mt Res Dev 1(1):49–54

McCarthy J, Prudham S (2003) Neoliberal nature and the nature of neoliberalism. Geoforum. http://dx.doi.org/10.1016/J.geoforum.2003.07.003

Monserud RA (2003) Evaluating forest models in a sustainable forest management context. Forest Biometry Model Inf Sci 1:35–47

Norgaard RB (1994) Development betrayed the end of progress and a coevolutionary revisioning of the future. Routledge, London/New York

North DC (1990) Institutions, institutional change and economic performance. Cambridge University Press, Cambridge

North DC (1991) Institutions. J Econ Perspect 5(1):97–112

Ostrom E (1990) Governing the commons: the evolution of institutions for collective action. Cambridge University Press, New York

Peet R, Watts M (1996) Liberation ecologies: environment, development, social movements. Routledge, London/New York

Peluso N (1992) Rich forests, poor people. Resource control and resistance in Java. University of California Press, Berkley/Los Angeles/London

Pretzsch (2003) Methodological aspects of tropical forest management research. In: Contribution to the XII World Forestry Congress, Quebec City. http://www.fao.org/DOCREP/Article/WFC/XII/0791-C4.HTM. Last Access 5 Jan 2013

Rinawati F, Stein K, Lindner A (2013) Climate change impacts on biodiversity – the setting of a lingering global crisis. Diversity 5:11–123

Roe E (1994) Narrative policy analysis. Theory and practice. Duke University Press, Durham/London

Scheffer M, Carpenter S, Foley JA, Folke C, Walker B (2001) Catastrophic shifts in ecosystems. Nature 413:591–596

UNEP (2011) Green Economy Report. Towards a green economy: pathways to sustainable development and poverty eradiction. http://www.unep.org/greeneconomy/GreenEconomyReport. Pree release 2011, last access October 2013

Uphoff N (1993) Grassroots organizations and NGOs in rural development: opportunities with diminishing states and expanding markets. World Dev 21(4):607–622

Warren DM, Slikkerveer LJ, Brokensha D (1995) The cultural dimension of development. Indigenous knowledge systems. Intermediate Technology Publications, London

Index

DOI 10.1007/978-3-642-41404-6, © Springer-Verlag Berlin Heidelberg 2014

D

G

H

I

Q

R

S

T

U

V

MIX
Papier aus verantwortungsvollen Quellen
Paper from responsible sources
FSC® C105338
FSC
www.fsc.org

If you have any concerns about our products,
you can contact us on
ProductSafety@springernature.com

In case Publisher is established outside the EU,
the EU authorized representative is:
Springer Nature Customer Service Center GmbH
Europaplatz 3, 69115 Heidelberg, Germany

Printed by Libri Plureos GmbH
in Hamburg, Germany